Narratives, Political Representation, and Extremism in Polish Anti-science Movements

The titles published in this series are listed at *brill.com/vibs* and *brill.com/cevs*

Narratives, Political Representation, and Extremism in Polish Anti-science Movements

"Plandemic," "Depopulation," and "The Free People"

By

Przemysław Witkowski

BRILL

LEIDEN | BOSTON

Cover illustration: Photo by Pramod Tiwari on Unsplash.

Library of Congress Control Number: 2025949037

Typeface for the Latin, Greek, and Cyrillic scripts: “Brill”. See and download: brill.com/brill-typeface.

ISSN 0929-8436
ISBN 978-90-04-74222-2 (hardback)
ISBN 978-90-04-74785-2 (e-book)
DOI 10.1163/9789004747852

This book is printed on acid-free paper and produced in a sustainable manner.

Contents

Introduction

Technological progress was so slow for centuries that its effects on most of the Earth were barely noticeable. As late as the 19th century, one could still find many vast spaces where people lived immersed in a sacred sense of time,[1] working using the simplest, unchanged for centuries tools. With the Industrial Revolution, scientific progress accelerated significantly, and the 20th century brought modernization, often rapidly, to many regions. As a result, at the beginning of the 21st century, only a few people inhabiting isolated islands of the Andaman archipelago, mountainous areas of New Guinea, or deep forests of the Amazon rainforest are beyond the reach of global communication and modern technologies. However, these are the last such groups and spaces. The Internet, telecommunications satellites, drones, and easily accessible phones equipped with cameras and Wi-Fi have made it practically impossible to escape from the panopticon.

However, not only the macro scale has become the domain of supervision, control, and surveillance. Even on the micro scale of our bodies, we have become traceable, transparent, and controllable, down to the smallest cell of our body. Genome maps, nanotechnology, cyborgization, the development of AI, and cloning technology have made knowledge about ourselves and our interior and the capabilities or composition of our bodies more precise and accessible, incomparably to any moment in human kind history.[2] If we add to this the countless amounts of data that we share about ourselves, whether in the form of content published on social media, where we post thousands of photos, opinions, "likes," or through "watches" measuring our blood pressure, heart rate and even type of sleep, it turns out that the amount of information about us available to external actors is already starting to exceed human capabilities of understanding. The reality of science fiction novels by Stanisław Lem, William Gibson, or Philip K. Dick is slowly becoming our reality.

At the same time, the previously mass and dominant monotheistic religions have problems effectively explaining today's world to their believers.[3] They

1 Mircea Eliade, *Sacrum, mit, historia. Wybór esejów* (Warsaw:PIW, 1974), 86–113.

2 Adam Greenfield, *Radical Technologies. The Design of Everyday Life* (London, Verso Books, 2017) Kindle.

3 Susan Petrilli, "Multicaultural Societies, Monotheistic Religions, and Globalization: Semioethic Vistas", *The Journal of Dialogic Ethics: Interfaith and Interhuman Perspectives* 1, no. 1 (2022):46–77.

 | DOI:10.1163/9789004747852_002

respond to this with secularization, which is particularly intensifying in the countries of the wealthy North.[4] Universalistic ideologies of the 20th century, such as socialism, communism or liberalism, have also succumbed to the deconstructive tools of post-structuralism, postmodernism, post-colonial studies, or feminist readings. Formerly dominant meta-narratives, which allowed for the integration of various fields of science and created a framework for the functioning of societies, have become in this light only a temporary emanation of class, race, and gender – one of many stories.[5] As a result, fear, conditioned by changeability and information overload, spreads.

However, people strive to eliminate it as quickly as possible.[6] Today's socio-political reality, with its multitude of stimuli, contradictory signals, and a shared sense of lack of control over the ongoing social processes, increases fear. It is deepened by economic, ecological, or epidemic crises and acts of terrorism or armed conflicts. As a result, many people have a permanent sense of threat.[7] If we add to this the access to an incomparable amount of information compared to previous periods, much of which concerns dangers, disasters, accidents, and human tragedies, we can say, following Ulrich Beck, that we live in a "risk society."[8]

Some political actors take advantage of the growing sense of fear in society. Additionally, they try to strengthen this emotion, trying to benefit from the fear of their voters, readers, or viewers. As a result, they use this emotion to influence the decisions and actions of people and societies, "managing fear."[9] A popular tool here is to provoke "moral panic" to defend the weakest (children, seniors, the sick, and various vulnerable social groups) from a supposed

---

4 Egbert Ribberink; Peter Achterberg; Dick Houtman, "Religious polarization: contesting religion in secularized Western European countries", *Journal of Contemporary Religion* 33, no. 2 (2018):209–227.

5 Jean-François Lyotard, "Kondycja ponowoczesna", *Sztuka i Filozofia* 13 (1997):25–35.

6 Ralph Adolphs, "The biology of fear", *Current Biology* 23, no. 2 (2013):79–93.

7 Matti Näsi et al., "Crime News Consumption and Fear of Violence: The Role of Traditional Media, Social Media, and Alternative Information Sources", *Crime & Delinquency* 67, no. 4 (2021):574 600.

8 See further: Ulrich Beck, *Społeczeństwo ryzyka. W drodze do innej nowoczesności* (Warsaw: Wydawnictwo Scholar, 2002), 15–24.

9 Fear management is a political manipulation strategy involving arousing or directing emotions of fear and anxiety towards a selected phenomenon in recipients while simultaneously showing the person who uses this strategy as the best defender against that phenomenon. See further: Paweł Cywiński et al., *Zarządzanie strachem. Jak prawica wygrywa debatę publiczną w Polsce* (Warsaw: Fundacja im. Stefana Batorego, 2019) Kindle.

threat.[10] Such actions sometimes also have the character of a deliberate attempt to influence the situation in third countries from the outside. Strong negative emotions in the political system often cause instability, polarization,[11] and radicalization[12] of its participants, leading to extremism.[13] Popular tools for such action are disinformation[14] and the distribution of so-called fake news, which is part of this process. The latter may be untrue, a mixture of facts and fiction, or just a modification of trustworthy information. However, they are designed to use patterns of irrational perception of reality (e.g., stereotypes, prejudices, or elements of religious faith) and to evoke strong emotions in recipients.[15] Due to the flooding of the media system with fake news, trust in the media, public institutions, and fellow citizens has also eroded.[16] These are not new techniques. They have been used in political messages for centuries.[17] However, it was only with the development of the Internet and social

---

10 Moral panic is a phenomenon of a sudden spread of social anxiety caused by a threat to the essential values of a community. It is a way of expressing disapproval, condemnation, or criticism of someone perceived as a deviant at a given moment. Condemnation and social disapproval constitute a moral element here, while panic expresses hysteria and overreaction. Media involvement is critical here; without them, moral panic cannot develop. The media can initiate it or inflate it. See further: Stanley Cohen, *Folk Devils and Moral Panics: The Creation of the Mods and Rockers* (London and New York: Routledge, 2002), 1–20.

11 Political polarization is the moving of political attitudes away from the centre towards ideological extremes. See further: Morris P. Fiorina and Samuel J. Abrams, "Political Polarization in the American Public", *Annual Review of Political Science* 11, no. 1 (2008):563–588.

12 Radicalization is the process by which an individual or group adopts increasingly radical views opposing the political, social, or religious status quo. It can lead to extremism or acts of terrorism. See further: Alex P. Schmid, "Radicalisation, De-radicalisation, Counter-Radicalisation: A Conceptual Discussion and Literature Review" (Haga: The International Centre for Counter-Terrorism, 2013) Kindle.

13 Extremism, as understood in this work, is the holding of extreme views and or the use of violent means to achieve a political, religious, economic, or ideological goal, including terror, terrorism, or armed change of the political system.

14 Disinformation is false information intentionally spread to deceive people. Disinformation is an organized, adversarial effort in which actors use strategic deception and media manipulation tactics to achieve political, military, or commercial goals. See further: Richard H. Shultz and Roy Godson, *Dezinformatsia: Active Measures in Soviet Strategy* (Washington: Pergamon-Brassey's, 1984), 37–38.

15 Cywiński et al., *Zarządzanie strachem. Jak prawica wygrywa debatę publiczną.*

16 Katherina Ognyanova et al., "Misinformation in action: Fake news exposure is linked to lower trust in media, higher trust in government when your side is in power", *Harvard Kennedy School Misinformation Review* 4, no. 1 (2020), n. pag.

17 See further: Geoffrey Scarre and John Callow, *Witchcraft and Magic in Sixteenth- and Seventeenth-Century Europe* (London: Palgrave Macmillan, 2001).

media that they could achieve an unprecedented scale of impact. The amount of information we receive during the day significantly exceeds what we experienced earlier.[18] The problem is exacerbated by recipients' generally low competencies in critical thinking and information analysis.[19]

In addition, the crisis of metanarratives[20] has brought a paradoxical democratization and pluralization of the narratives. Especially those that society creates about itself. However, contrary to the intentions of many "deconstructors," this has turned out to be a double-edged sword. Not only has this crisis removed the hegemonic status of privileged stories conditioned by class, gender, and race, but it has also meant that any story, if only it could find a group that wanted to believe in it and profess it, can be considered "true." For if the old, modernist "truth" was only the voice of the elites who shaped it using ideological apparatuses[21] and made it the framework for our vision of the world[22] and was only knowledge/power,[23] what prevents another story from replacing it? If science is a construct and not some truth about reality, why can't "another science" replace it? And if school is another kind of implementation of the prison model, why shouldn't we free ourselves from it?

Cyclical economic crises, epidemics, natural disasters, ongoing climate change, and new wars increased uncertainty at the beginning of the 21st century. Conspiracy narratives increasingly became the "answer."[24] Information overload and the many possible interpretations of events began an increasing widespread of simplified stories that provided an uncomplicated explanation

---

18 Mirosław Usidus, "Jak przetrwać tsunami z 120 bitami w głowie? Ludzki mózg i ogrom informacji", *Młody Technik*, no data, accessed 07.09.2024, https://mlodytechnik.pl/technika/31258-jak-przetrwac-tsunami-z-120-bitami-w-glowie-ludzki-mozg-i-ogrom-informacji.

19 Jonas de Keersmaecker and Arne Roets, "»Fake news«: Incorrect, but hard to correct. The role of cognitive ability on the impact of false information on social impressions", *Intelligence* 65 (2017):107–110.

20 Lyotard, "Kondycja ponowoczesna".

21 Louis Althusser, "Ideologia i aparaty ideologiczne państwa." *Recykling Idei*, 14.06.2006, accessed 07.09.2024, http://web.archive.org/web/20180722174007/http://recyklingidei.pl/althusser-ideologie-aparaty-ideologiczne-panstwa.

22 Martyna Bieniek, "Ramowanie a metafora w ujęciu kognitywistycznym – czy mają ze sobą coś wspólnego?", *Philosophical Discourses Prace Naukowe Uniwersytetu Humanistyczno-Przyrodniczego im. Jana Długosza w Częstochowie* 2 (2020):149–164.

23 Adam Ostolski, "Dyskurs u władzy. Filozofia polityczna u Michaela Foucaulta", *Etyka* 36 (2003):159–172.

24 Matthew R.X. Dentith, *The Philosophy of Conspiracy Theories* (London: Palgrave Macmillan, 2015), 1–6.

of multi-threaded and multi-causal events.[25] Even if more elaborate narratives were created, full of unclear, confusing fragments, their construction effectively eluded falsification.[26] A vicious circle of reasoning was created, which nevertheless gave the believer a sense of discovering a hidden truth about reality, even the most terrible, and consequently a sense of meaning and faith in the possibility of repair, order, and change, allegedly restoring peace and sense.

In addition, the visibility and social influence of conspiracy narratives are amplified by the spread of social media, information warfare and disinformation activities of conflicting states. Jews, Freemasons, Illuminati, Jesuits,[27] and other "sinister" forces have been joined in recent decades in the stories of supporters of conspiracy narratives by Big Pharma, Bill Gates, and George Soros.[28]

In many conspiracy narratives, especially the newer ones, irrationalism and anti-scientism are essential components. The lack of scientific evidence to confirm the conspiracy claims does not bother their supporters. On the contrary, there is often a component in them indicating that the lack of evidence for a given concept only confirms the participation of academics in the conspiracy. Attacked by "alternative researchers" are archaeologists (crypto-archaeology)[29], astronomers and physicists (ufology, anti-5G movement), anthropologists (ariosophy, turbo-Slavism),[30] virologists (anti-vaccination

25 Raphael Heiss et al., "Social media information literacy: Conceptualization and associations with information overload, news avoidance and conspiracy mentality", *Computers in Human Behavior* 148 (2023):107908.

26 Jovan Byford *Conspiracy theories: a critical introduction* (Houndmills, Basingstoke, Hampshire: Palgrave Macmillan, 2011), 32–37.

27 Michael Butter and Peter Knight, ed., *Routledge Handbook of Conspiracy Theories*. London (New York: Routledge, 2020), 555–568.

28 Jonathan Key, *Among the Truthers: A Journey through America's Growing Conspiracist Underground* (New York: Harper Collins, 2011) Kindle.

29 Victor Schnirelmann, "Russian response: archaeology, Russian nationalism and Arctic homeland", in: *Selective remembrance: archaeology in the construction, commemoration, and consecration of national pasts*, ed. Philip L. Kohl, Mara Kozelsky, Nachman Ben-Yehuda (Chicago: Chicago University Press, 2007):31–70.

30 Victor Schnirelmann, "How to become the »Slavic Aryans«: The founders of the Russian Neo-paganism and their ambitions", in: *Religion and Magic in Socialist and Post-Socialist Contexts, I. Historic and Ethnographic Case Studies of Orthodoxy, Heterodoxy, and Alternative Spirituality*, ed. Alexandra Cotofana, James M. Nyce (Stuttgart: Ibidem Press, 2017), 75–98.

movement), biologists (anti-GMO movement),[31] and physicians (alternative medicine).

It would seem that as our scientific knowledge of the world grows, the opposite process should take place. The cause-effect and rational description of the world should increasingly dominate. In the meantime, however, a reaction is emerging on the fringes of science. The common element of these actions is opposition to values associated not only with science in general but also with the Enlightenment. The name of this era refers to light, understood as a force that dispels darkness and superstition. In its spirit, reason, freed from bonds and shackles, would be a light illuminating the path to knowing the truth about the world and man. The latter should be guided in its approach to reality by naturalism, rationalism, humanistic and cognitive optimism, faith in progress, criticism, empiricism, sensualism, utilitarianism, and striving to change the world for the better.[32]

At the same time, the Enlightenment was accused of being the root of totalitarianism. Zygmunt Bauman saw it as the source of its right-wing version,[33] Roger Scruton saw it as the source of left-wing,[34] and Michael Foucault named it as a source of both.[35] A strong anti-Enlightenment component, often playing an axial role in these ideologies and concepts, can be found in integral traditionalism,[36] religious fundamentalisms of various religions, including Islam, Christianity, as well as in new spirituality – new age and new religious movements,[37] in monarchism – legitimism,[38] neo-pagan

31 Marcin Rotkiewicz, *W królestwie monszatana. GMO, gluten i szczepionki* (Czarne: Wołowiec, 2022) Kindle.

32 See further: Pierre Chaunu, *Cywilizacja wieku oświecenia* (Warsaw, PIW, 1993).

33 See further: Zygmunt Bauman, *Nowoczesność i Zagłada* (Krakow: Wydawnictwo Literackie, 2009).

34 See further: Gene Callahan and Kenneth B., McIntyre, *Critics of Enlightenment Rationalism* (London: Palgrave, 2020).

35 See further: Michael Foucault, *Historia szaleństwa w dobie klasycyzmu* (Warsaw: PIW, 1987) and Michael Foucault, *Nadzorować i karać. Narodziny więzienia* (Warsaw: Aletheia, 2021).

36 See further: Mark Sedgwick, *Against the Modern World. Traditionalism and the Secret Intellectual History of the Twentieth Century* (Oxford: Oxford University Press, 2004).

37 Marisa Meltzer, "QAnon's Unexpected Roots in New Age Spirituality", *The Washington Post Magazine*, 29.03.2021, accessed 07.09.2024, http://washingtonpost.com/magazine/2021/03/29/qanon–new–age–spirituality.

38 See further: Jacek Bartyzel, *Umierać, ale powoli. O monarchistycznej i katolickiej kontrrewolucji w krajach romańskich 1815–2000* (Krakow: ARCANA, 2006).

neo-fascism,[39] Völkism,[40] Ariosophy,[41] deep ecology,[42] and several other trends of ecologism, in particular those related to new spirituality and new religious movements, as well as in the trend of conservative revolution.[43]

Most currents of thought critical of the Enlightenment are characterized by a greater or lesser distrust of science, and even technophobia, irrationalism, nativism, emotionalism, strong opposition to the idea of progress and modernity, fantasies of an Arcadian past, anti-democracy, and anti-liberalism. Although all the ideologies mentioned above have a strong anti-Enlightenment component, the issues of approach to science are diverse. Some reject a large part of its achievements (e.g., anarcho-primitivism, deep ecology, Völkism, new age), others accept fragmentary use of its achievements, as long as its achievements do not undermine the "truths" they proclaim (e.g., religious fundamentalism, monarchism-legitimism). There are, however, some that have made opposition to science the central point of their "doctrine."

Movements of the far right are particularly anti-Enlightenment, and so people who oppose some vital aspect of the values of the heritage of this era often end up there. But does this mean that opposition to scientific consensus automatically qualifies a group or individual for the far right? It is worth noting that there are also several anti-scientific theories of a similar type, more popular on the left, such as the vision of the harmfulness of genetically modified food (GMO) to health and nature, the threat of using nuclear energy, or the beneficial effect of veganism on health. There are also such conspiracy narratives that are unanimously adhered to by representatives of various political ideologies, such as the concept of a "plandemic," i.e., an alleged planned and hidden agenda to introduce the COVID-19 epidemic. Additionally, each group of supporters had a different motivation for action. For the right wing, it is the fight against restrictions on freedom or tradition; for the left – it is the fight against medical corporations. Hence, it is worth asking ourselves and considering, using specific examples, what the political representation of movements opposing the scientific consensus looks like. What arguments do representatives of such groups use? Do they fit into a left-wing or right-wing

---

39 See further: Tamir Bar-On *Rethinking the French New Right. Alternatives to Modernity* (London, Routledge, 2013) and Tamir Bar-On *Where Have All the Fascists Gone?* (London: Routledge, 2019).

40 See further: George L. Mosse, *Kryzys ideologii niemieckiej* (Warsaw: Czytelnik, 1972).

41 Schnirelmann, "How to become the »Slavic Aryans«".

42 Andy Scerri, "Deep Ecology, the Holistic Critique of Enlightenment Dualism, and the Irony of History", *Environmental Values* 25, no. 5 (2016):527–551.

43 See further: Roger Woods, *The Conservative Revolution in the Weimar Republic* (London: Palgrave Macmillan, 1996).

ideological pattern, or are they just random fellow travellers who share goals rather than an ideological foundation?

As in many other countries of the North, the current reality, also in Poland, is fertile soil for the growth of anti-scientific conspiracy narratives. The crisis of political and religious meta-narratives, especially those with Enlightenment roots, the overload of individuals and societies with information, and narrative pluralization have been compounded in recent years by additional factors. The American international hegemony is wobbling, and the growing aspirations of regional powers and their initiation of the fight for a multipolar world manifested in the intensification of disinformation activities and wars, both informational and military conflicts,[44] also in Poland's immediate vicinity. All of these issues, and at the same time the growing crisis of trust in elites,[45] have provided the subsoil for the flourishing of conspiracy narratives.

This tendency has been additionally reinforced by the COVID-19 pandemic and the often contradictory information appearing in its context, as well as the inconsistent and incoherent attempts by governments of many countries to deal with its effects. Therefore, the context of the pandemic has notably strengthened the visibility and audibility of anti-scientific narratives related to alleged threats to human health. Hence, the choice of the movements studied here – opposition to fifth-generation telephone networks and mandatory vaccinations. Both related, both at the level of activists and leaders, also sharing the main channels of distribution of the narratives, not only became particularly visible in the context of the pandemic, but their demands also gained a clear impact on mainstream politics, blocking or initiating numerous initiatives and actions of even the largest Polish parties. The public activities of both movements also clearly attracted the most significant number of participants among all social movements, challenging the scientific consensus.[46]

At the same time, both movements have appeared in the mainstream of political discourse recently, and they raise issues where the scientific consensus is often still being formed, leaving many areas in their theory and practice

---

44 As understood by Immanuel Wallerstein, See further: Immanuel Wallerstein, *Decline of American Power: The U.S. in a Chaotic World* (New York, New Press, 2003).

45 Nicolò Conti, "National political elites, the EU, and the populist challenge", *Politics* 38, no. 3 (2018):361–377.

46 Agnieszka Demczuk, "Ruch antyszczepionkowy w Polsce i jego kabiny pogłosowe w alt-internet w latach 2020–2022", *Rocznik Instytutu Europy Środkowo-Wschodniej* 3 (2022):9–35 and Agnieszka Demczuk, "Fenomen ruchu antyszczepionkowego w cyberprzestrzeni, czyli fake news i postprawda na usługach hipotezy Andrew Wakefielda", *Studia de Cultura* 4 (2018):92–113.

worth exploring. Especially in the context of the movements opposing fifth-generation telephony and mandatory vaccinations, there are numerous accusations in the media about the influence of external actors (especially the Russian Federation) on their messages, narratives, and actions on the actions of both movements.[47] Hence, one of this work's goals is to closely examine the connections between these environments in Poland and politically active supporters of Russian narratives in matters other than science and public health.

It is also easy to locate the beginnings, emergence, and creation of one of these two narratives that challenge the scientific consensus concerning fifth-generation telephony networks. Its decline was and is also a process that is easy to trace. The relatively short period of its existence (since 2017) allows it to be treated as a model, as an example, allowing us to see in detail how such narratives arise and spread. As a result, such a model can be compared to the second of the movements studied – the opposition to mandatory vaccinations. This one, which has existed for over two hundred years, turns out to, as the work shows, be based on similar patterns. Additionally, the coherent point of both narratives is the COVID-19 pandemic (2019–2022) period, the time of which became the moment when the circles creating both movements began to interpenetrate largely. Not only were the similar pattern of emergence and similar issues questioned by them, but as it turns out, similar channels served to distribute them, and activists were often active in both movements. However, determining to what extent, who precisely, and where the critical points of contact were located are also questions this work aims to answer.

The increasing activity of the movements in question, their intensifying interpenetration with political groups, as well as numerous manifestations of extremist activities, combined with the potential for strengthening the message that the potential next pandemic predicted by scientists[48] may give, mean that in the matter of anti-scientific movements and the conspiracy narratives associated with them, we can expect further development of the visi-

---

47 For example: Julia Dauksza and Mariusz Sepioło, "Orkiestra Putina: Dyrygenci i chórek", *Frontstory*, 31.03.2022, accessed 07.09.2024, https://frontstory.pl/dezinformacja–antyszczepionkowcy–wojna–rosja–ukraina–konfederacja–braun and William J. Broad, "Your 5G Phone Won't Hurt You. But Russia Wants You to Think Otherwise", *The New York Times*, 12.05.2019, accessed 07.09.2024, http://www.nytimes.com/2019/05/12/science/5g–phone–safety–health–russia.html.

48 For example: University of Edinburgh, "Virus ancestry could aid bid to predict next pandemic, study finds", *ScienceDaily*, 05.02.2024, accessed 07.09.2024, http//www.sciencedaily.com/releases/2024/02/240205165913.htm and Maryam Shafaati, "The next pandemic catastrophe: can we avert the inevitable?", *New microbes and new infections* 52 (2023):1–4.

bility and significance of the political entities representing them, and hence we need to examine them as thoroughly as possible, especially in the context of political activity and connections with the political class.

However, the tasks that this work faces are broader. It aims not only to describe and examine the message of the movements opposing 5G and mandatory vaccinations or to indicate the most important organizations, their leaders, and activists of both movements but also to describe the channels through which the anti-scientific narrative penetrates political parties, to establish the political representation of the movements and describe it, and also to consider the connections between anti-scientific movements and the far right. It also tries to answer the question of whether they are ideological or just a coincidental convergence of goals. To do this, it is also necessary to determine the sources of anti-scientific narratives about vaccines and 5G and how they have penetrated Polish political discourse. Also, it is necessary to examine the primary channels for propagating conspiracy narratives related to 5G and vaccinations. It is essential to answer how these narratives penetrate ranks of the political parties, including the mainstream ones. As a result, it must be determined which political parties are primarily associated with the movements opposing the introduction of the 5G network and mandatory vaccinations. And whether we can talk about the convergence of their narratives with Russian narratives. To complete the picture of the political dimension of the movements discussed, it is also essential to deal with media reports on the extremist activities of representatives of the movements mentioned above. Did radicalization only concern specific individuals, or is this process permeating broader milieus? And are extremist tendencies common in them or relatively isolated cases?

Such broad research goals required the author to gather extensive material for multi-threaded analysis.

In the first stage, it was necessary to gather a vast collection of texts, videos, posts in channels, fan-pages and social media profiles (YouTube, X, Facebook), in open groups on Facebook related to alternative medicine, ecology, esoteric topics and other conspiracy theory groups, which concerned conspiracy narratives on vaccines, COVID-19 and 5G in the Polish-language part of these platforms (primarily posted in the 2019–2024). Accounts of events related to the protest movements against mandatory vaccinations and the introduction of 5G technology, such as protests, meetings, demonstrations, lectures, festivals and rallies, were also collected. The national and local media were also reviewed for the presence of narratives and events related to the protest movements against mandatory vaccinations and the introduction of 5G technology.

After reviewing the most popular conspiracy narratives – appearing in social media, publications and statements of leaders and activists of movements opposing mandatory vaccinations and the introduction of 5G telephony, several key ones were selected. Observing the dynamics and trajectory of the spread of conspiracy theory content and their thematic and semantic categorization was also crucial for understanding.

For this purpose, the leading organizations associated with activists of movements opposing mandatory vaccinations and the introduction of 5G telephony were also selected. The main criterion was the media activity of their leaders, presence at demonstrations related to the topic of the study, connections with politicians of political parties (formal and informal), presence in parliament and discussions organized there, presence on the electoral lists of political parties, activity in their media and media unrelated to the movements. By observing their activity, connections and interactions, working maps were constructed showing connections between various conspiratorial groups and bridge communities, such as alternative medicine, ecological and esoteric groups, and also alternative right media, thanks to which the message of the former could penetrate other milieus, including far-right parties, which in turn transferred them to the mainstream of Polish political life. Discussion cascades modelling was also used for this purpose, i.e., tracking how a given content spreads in the social network – who and when shared a post, who reacted to it and what comments appeared in its milieu. The organization's narrative, organizational connections, movements, and individuals promoting them were also examined. The content published by various influencers was also analysed, paying attention to how they adapt English-language conspiracy theories to Polish cultural and political contexts.

As a result, a collection of materials was created consisting of over 150 YouTube channels, over 200 Facebook pages, around 70 Facebook groups, over 200 Twitter accounts, reports of around 150 demonstrations of the movements and around 20 acts of political extremism committed by Polish participants of the movements opposing mandatory vaccinations and 5G studied.

The main narratives spread by the above entities were compared with the narratives promoted by the Russian public media and known cases of Russian disinformation. From the collected material, entries consistent with Russian narratives were selected to examine the quantitative occurrence of at least a correlation between the dominant narratives of the movements studied and the narratives of the Russian state media and associated politicians and publicists.

To analyse the collected materials, as thoroughly as possible, it was necessary to adopt an interdisciplinary approach in the work, combining political, sociological, cultural and historical analysis. In the context of political science, the key field was an attempt to understand how disinformation and conspiracy theories can serve to destabilize democratic systems and weaken trust in institutions, as well as an analysis of the tools of soft power and information warfare used by some state actors, with particular emphasis on Russia. A sociological approach was necessary to examine the formation of community mechanisms in conspiracy theorist groups and analyse the role of the crisis of trust and authority in elites in the emergence of alternative narratives undermining the scientific consensus. Cultural elements were necessary to analyse the symbolism, myths and cultural codes in conspiracy narratives and to conceptualize alternative media as a "counter-hegemonic culture." The historical approach enabled the contextualization of conspiracy narratives, indicating their continuity with the past (e.g., old anti-Masonic, anti-Semitic narratives) and allowing them to be inscribed into historical patterns, for example, Russian propaganda and disinformation.

A set of scientific tools was selected, which included analyses from the field of qualitative content analysis, qualitative Critical Discourse Analysis, qualitative network analysis, Qualitative Rhetorical Analysis, qualitative Open Source Intelligence analysis, hermeneutics and interpretive analysis, semiology and myth analysis, ethnographic methods and participant observation combining digital ethnography with ethnographic content analysis, and meta-analyses and systematic reviews.

To identify systematically recurring motifs, narratives, figures and rhetorical and linguistic structures ("globalists," "puppets," "dark forces," "covidians," "truth," "awakening") and the emotions (anger, fear) and values (freedom, choice, independence) emerging from them, it was necessary to use Qualitative Content Analysis[49] for the analysed messages – texts, memes, films, social media posts – and then categorize statements and texts and interpret them in terms of context and culture. Qualitative Content Analysis was also helpful in analysing the similarities between the narratives of opponents of mandatory vaccinations and 5G and those disseminated by Russia.

Qualitative Rhetorical Analysis helped examine how persuasive messages are constructed. Searching for rhetorical devices: hyperbole, irony, allusions, anecdotes, appeals to emotions and analysis of statements using emotional

---

49 See further Philipp Mayring, *Qualitative Content Analysis: Theoretical Foundation, Basic Procedures and Software Solution* (Klagenfurt: SSOAR, 2014), accessed 20.04.2025, https://www.ssoar.info/ssoar/bitstream/handle/document/39517/ssoar-2014-mayring-Qualitative_content_analysis_theoretical_foundation.pdf or Margrit Schreier, *Qualitative Content Analysis in Practice* (London: SAGE Publications, 2012).

language,[50] and specific stylistic figures and methods of argumentation helped to identify persuasion strategies,[51] i.e., how creators of conspiracy narratives (e.g., bloggers, alternative "experts," pro-Russian politicians) use metaphors, analogies, repetitions, hyperbole and other rhetorical figures to convince recipients, how they present the "guilty" of the "plandemic" (elites, WHO, the West) and the "wronged" (nation, individual, "woke") and how conspiracy theories convince recipients to believe in alternative facts. In turn, the study of topoi and cultural stereotypes, i.e., socially established "thinking patterns" (e.g., "the government is hiding the truth," "the media is lying"), searched for their presence in Russian propaganda or texts of the movements studied, additionally looking for similarities and analogies.

Qualitative Critical Discourse Analysis provided a deeper, qualitative approach to analysing the thesis's language and social and ideological context.[52] It enabled a more effective search for metaphors, patterns, and linguistic strategies in the materials studied that build a sense of threat, conspiracy, or hidden truth. It is essential to look at texts' structure, headings, repeated motifs, ways of describing opponents, and their contextualization in power relations, history, and ideology. As a result, it was possible to effectively compare the discourses of mainstream media, those of the "alternative" and those associated with the Russian centre of power, as well as effectively recognize how the narratives of "alternative truth" are constructed and better understand how language serves to legitimize disinformation, how it creates the identities of "us" versus "them." The latter was heard, for example, in the analysis of texts by opponents of vaccinations, showing doctors as corrupt and "closed-minded."

Qualitative network analysis – identifying relationships between movements, such as common leaders, citations, transmission of narratives, who shares them, who is the source, how information spreads[53] – was crucial for

50 See further: Ann Gill, *Rhetoric and Human Consciousness: A History* (Long Grove, IL: Waveland Press, 2006).

51 See further: Jonathan Charteris-Black, *Politicians and Rhetoric: The Persuasive Power of Metaphor* (Basingstoke: Palgrave Macmillan, 2011).

52 See further: Norman Fairclough, *Discourse and Social Change* (Cambridge: Polity Press, 1992); Norman Fairclough, *Critical Discourse Analysis: The Critical Study of Language* (London: Longman, 1995); Ruth Wodak and Michael Meyer, eds., *Methods of Critical Discourse Analysis*, 2nd ed. (London: SAGE Publications, 2009) and Ruth Wodak, *The Discourse of Politics in Action: Politics as Usual* (Basingstoke: Palgrave Macmillan, 2009).

53 See further: Mario Diani, "The Concept of Social Movement," *The Sociological Review* 40, no. 1 (1992): 1–25, https://doi.org/10.1111/j.1467-954X.1992.tb02943.x and Nick Crossley, *Towards Relational Sociology* (London: Routledge, 2011).

determining the "nodes" that are key to the spread of narratives of the studied movements.[54] As a result, it was possible to indicate connections between the creators of narratives that undermine the scientific consensus, right-wing politicians sympathetic to Russia and media associated with them, and to determine the central nodes of this network in Poland. Manual analysis of Open Source Intelligence sources[55] was also helpful in this, i.e., analysis of publicly available data (websites, social media accounts, domain registries, metadata, journalistic investigations, research on so-called "false amplification" – i.e., small sources cited by others as "scientific evidence," reconstruction of the origin of the narrative: manual comparison of citations, sources, publication dates), which allowed for the identification of organizations, individuals and sources involved in disinformation campaigns.

An additional asset of this study was ethnographic methods and participant observation, combining digital ethnography with ethnographic content analysis.[56] By joining dozens of open and closed Facebook groups to observe and analyse discussions, it was possible to understand better the motivations of activists of the movements studied, the culture of this environment and the mechanisms of creating and perpetuating alternative narratives and disinformation. Immersion in these communities gave better access to their language, motivations, and emotions, allowed for tracking communication rituals, building the position of "experts" in them and mechanisms, what practices (e.g., sharing links, joint "investigations") support the functioning of disinformation movements, but also to answer what emotions, frustrations, a sense of exclusion or fears drive conspiratorial thinking.

Hermeneutics and interpretive analysis also proved helpful in analysing the material in this publication, allowing for a deep analysis of conspiracy narratives. The hermeneutical approach assumes that texts must be interpreted in a

---

54 Roger Häussling, "A Network Analytical Four-Level Concept for an Interpretation of Social Interaction in Terms of Structure and Agency" In *Mixed Methods Social Networks Research: Design and Applications*, edited by Silvia Domínguez and Betina Hollstein (Cambridge: Cambridge University Press, 2014), 90–118.

55 See further: Daniel Mider, "Open Source Intelligence on the Internet – Categorisation and Evaluation of Search Tools," *Przegląd Bezpieczeństwa Wewnętrznego* 31 (2024), 383–412 and Michael Bazzell, *Open Source Intelligence Techniques: Resources for Searching and Analyzing Online Information* (Charleston: Createspace Independent Publishing Platform, 2018).

56 See further: Sarah Pink, *Doing Visual Ethnography*, 4th ed. (London: SAGE Publications, 2021) and Tom Boellstorff et al., *Ethnography and Virtual Worlds: A Handbook of Method* (Princeton, NJ: Princeton University Press, 2012).

historical, political, and social context.[57] In this case, it helped extract hidden intentions and symbolic meanings in conspiracy narratives. Therefore, the analysis of conspiracy narratives as forms of mythologizing modernity, reconstructing for current controversial issues related to vaccinations and 5G the topos of the fight between good and evil in an apocalyptic spirit or a quasi-religious epiphanic revelation[58] of truth in the "discovery" of the demonic global conspiracy of "big pharma" and "world government," as well as the possibility of setting these narratives in a broader context, in this case, for example, in Russia's disinformation strategies, turned out to be important for this book. Thanks to this, it was possible to distinguish quasi-mythical stories, such as "awakening," "secret governments," "hidden history" or "natural medicine." The study's indication of how mythical elements in conspiracy narratives (e.g., "almighty elites," "global conspiracies") are constructed and how they give the messages the appearance of obviousness was a handy analytical tool for examining the disinformation threads described in the study. At the same time, reaching for ingrained cultural codes ("pure nature," "Christian Europe," "traditional family") allowed us to capture them as constructs helpful in achieving political goals by political entities influencing and associated with the movements under study. Identifying recurring mythical narrative patterns (heroic protagonist vs. oppressive system, nation betrayed by elites, enlightened individual vs. befuddled masses) and noticing their cultural roots allowed us to link popular narratives in the movements under study (activists vs. system, corrupt elites, rebels vs "sheep") with their structural roots. In this context, the narratives of "Putin as a defender of traditional values" and "Russia as a stronghold of normality" in the face of the "Western madness of sanitization" that appear in the context of 5G and vaccinations are examples of the vitality of the messianic myth in the studied movements, where all those opposed to the far-reaching anti-pandemic restrictions represent good, and governments or the WHO represent evil. Similar examples of the analytical use of semiological tools[59] can also be indicated here, noticing and analysing the myth of innocent nature present in the narratives of the studied movements, where positively valorized nature (the human body, food, the environment) is con-

57 See further: Hans-Georg Gadamer, *Truth and Method* (London: Bloomsbury Academic, 2004).

58 See further: Lech M. Nijakowski, *Świat po apokalipsie: Społeczeństwo w świetle postapokaliptycznych tekstów kultury popularnej* (Warszawa: Wydawnictwo Naukowe Scholar, 2019).

59 See further: Roland Barthes, *Mythologies* (New York: Hill and Wang, 1972) and Umberto Eco, *A Theory of Semiotics* (Bloomington: Indiana University Press, 1976).

trasted with dark and dangerous technology (vaccines, 5G, HAARAP). As a result, it was also possible to notice how contemporary conspiracy theories function as a cultural myth, giving people a sense of meaning, identity, and a story to follow.

The above approaches were additionally supplemented by the author's meta-analysis and systematic review of the results of other studies from the researched thematic area. They allowed for the synthesis of the results of many studies on the effectiveness of disinformation, the demonstration of trends and research gaps, and the assessment of which threads and strategies are repeated in movements of opposition to mandatory vaccinations and 5G telephony in other countries similar to Poland.

As a result of intertwining these tools, a multi-level analysis field was created, where the language and narratives of the studied movements were qualitatively analysed by Critical Discourse Analysis, rhetorical analysis, content analysis, semiology and hermeneutics, mutual connections, sources, and structures of narratives and entities carrying them were examined by source analysis and Open Source Intelligence. The community and functioning of groups of opponents of mandatory vaccinations and the introduction of 5G telephony were analysed by ethnographic methods and participant observation, combining digital ethnography with ethnographic content analysis. The whole was given a geopolitical context by political science and historical tools.

In addition, just a few scientific publications are analysing the activities, and even more so the political connections, of Polish anti-scientific movements. Interesting works on the narrative and self-narrative of the anti-vaccination movement were published in Polish by Agnieszka Demczuk (in the form of individual articles[60] and monographs[61]) and Helena Chmielewska-Szlajfer.[62] Similar, narrowly focused, were the otherwise valuable reports "The time of the COVID-19 pandemic in Polish social media," "Perception of the »Coronavirus« on the Polish Internet until the confirmation of the first case

---

60 Demczuk, "*Ruch antyszczepionkowy w Polsce" and Demczuk, "Fenomen ruchu antyszczepionkowego w cyberprzestrzeni*"

61 See further: Agnieszka Demczuk, SARS-CoV-2 i COVID-19. Plotki, dezinformacje i narracje spiskowe w polskim dyskursie publicznym (Lublin, Wydawnictwo UMCS, 2021).

62 Helena Chmielewska-Szlajfer et al., "Badania jakościowe na temat postaw rodziców mających wątpliwości czy szczepić swoje dzieci oraz postaw antyszczepionkowych w Internecie – raport z badań ALK", *Reaserchgate*, 09.12.2022, accessed 07.09.2024, https://www.researchgate.net/publication/366155169_Badania_jakosciowe_na_temat_postaw_rodzicow_majacych_watpliwosci_czy_szczepic_swoje_dzieci_oraz_postaw_antyszczepionkowych_w_Internecie_-_raport_z_badan_ALK.

of the virus in Poland – Initial results"[63] or "The popularity of conspiracy narratives in Poland during the COVID-19 pandemic. A report from representative survey studies."[64] However, in Polish science, there are no attempts to indicate any other type of political and media connections apart from the concordances in the propagated narratives.

Once again, the fascinating research by Dariusz Jemielniak indicates that in the case of communication about the AstraZeneca vaccine, the high presence of manipulation and attempts to arouse fear in social media correlated with the high intensity of alarmist messages in the Russian media was symptomatic.1 However, in Polish science, there are no attempts to indicate any other type of political and media connections apart from the mere congruences in the propagated narratives. Individual articles, such as those by Piotr and Paweł Żuk,[65] are insufficient in this matter due to the broadness of the topic. This situation occurred because the narratives of the anti-vaccination and 5G movements were primarily studied, not their organization or political representation. As a result, there were signals, whether scientific[66] or journalistic,[67] about the compatibility of these narratives, but no in-depth analysis of this matter was created. Again, a more systematic presentation of these possible connections is necessary to obtain a clearer picture of this matter. This work, therefore, aims to supplement knowledge about these aspects.

Outside Poland, the activities of social movements contesting the scientific consensus, and in particular anti-vaccination activism, have been researched to a much greater extent. Before the peak of interest in the anti-vaccination movement caused by the COVID-19 pandemic, a series of popular science works was devoted to describe it, written by doctors such

63 Andrzej Jarynowski; Monika Wójta-Kempa; Vitaly Belik, "Percepcja "koronawirusa" w polskim Internecie do czasu potwierdzenia pierwszego przypadku zakażenia SARS-CoV-2 w Polsce", *Pielęgniarstwo i Zdrowie Publiczne* 10, no. 2 (2020):89–106.

64 Franciszek Czech and Paweł Ścigaj, "Popularność narracji spiskowych w Polsce czasu pandemii COVID-19. Raport z reprezentatywnych badań sondażowych", *Instytut Studiów Międzynarodowych Uniwersytetu Jagiellońskiego*, 2020, accessed 07.09.2024, http://www.ism.uj.edu.pl/wp–content/uploads/2020/07/covid–narracje–spiskowe.pdf.

65 Piotr Żuk and Paweł Żuk, "Right-wing populism in Poland and anti-vaccine myths on YouTube: Political and cultural threats to public health". *Global Public Health* 15, no. 6 (2020):790–804 oraz Piotr Żuk; Paweł Żuk, Justyna Lisiewicz-Jakubaszko, "The anti-vaccine movement in Poland: The socio-cultural conditions of the opposition to vaccination and threats to public health", *Vaccine* 37, no. 11 (2019):1491–1494.

66 See further: Karina Stasiuk-Krajewska and Michał Wenzel, ed., *Dezinformacja w czasach kryzysu* (Toruń, Wydawnictwo Adam Marszałek, 2024)

67 See further: Jacek Hugo Bader, *Strażnicy Wolności* (Warsaw, Agora, 2022).

as Jonathan M. Berman[68] or Paul A. Offit[69] or journalists such as Seth Mnookin[70] or Brian Deer.[71] Over time, academic works devoted to the movement appeared, such as those by Bernice L. Hausman[72] or Maya J. Goldenberg.[73] However, these were presentations of the phenomenon of the anti-vaccination movement itself, or more broadly, anti-scientific movements, not focusing on the political aspect of its activities, and connections with politicians and parties were not discussed there in more detail or broadly, but rather only mentioned.

There are also already valuable individual academic works dealing with the relations of the anti-scientific movement (mainly anti-vaccination movement) with organizations of a strictly political nature. Such cases have been described in context of Brazil,[74] Spain, Italy,[75] and Finland.[76] Notably, in each of these countries, the main political partners of the movements were the parties positioned furthest to the right in the local political spectrum. However, none of these studies went beyond the scale of one or possibly a few

68 See further: Jonathan M. Berman, *Anti-vaxxers. How to Challenge a Misinformed Movement* (Cambridge: The MIT Press, 2000).

69 See further: Paul A. Offit, *The Cutter Incident. How America's First Polio Vaccine Led to the Growing Vaccine Crisis* (New Haven and London: Yale University Press, 2005) and Paul A. Offit, *Deadly Choices. How the Anti-vaccine Movement Threatens Us All* (New York: Basic Books, 2010).

70 See further: Seth Mnookin, *The panic virus. The True Story behind the Vaccine-autism Controversy* (New York: Simon & Schuster, 2011).

71 A series of press articles from 2006–2010, summarized in a book by Brian Deer, *The Doctor Who Fooled the World: Science, Deception, and the War on Vaccines* (Baltimore: Johns Hopkins University Press, 2020).

72 See further: Bernice L. Hausman, *Antivax. Reframing the Vaccine Controversy* (Ithaca, ILR Press, 2019).

73 See further: Maya J. Goldenberg, *Vaccine Hesitancy Public Trust, Expertise, and the War on Science* (Pittsburgh, Pittsburgh University Press, 2021).

74 Régis Ebeling et al., "Analysis of the Influence of Political Polarization in the Vaccination Stance: The Brazilian COVID-19 Scenario", *Proceedings of the International AAAI Conference on Web and Social Media* 16, no. 1 (2022):159–170 oraz Marina Verjovsky et al., "Political quarrel overshadows vaccination advocacy: How the vaccine debate on Brazilian Twitter was framed by anti-vaxxers during Bolsonaro administration", *Vaccine* 41, no. 39 (2023): 5715–5721.

75 Luca Raffini and Clemente Penalva-Verdú, "The Problematic Relationship Between Science, Politics and Public Opinion in Late Modernity: The Case of the Anti-Vax Movement in Spain and Italy", in: *Populism and Science in Europe*, ed. Hande Eslen-Ziya, Alberta Giorgi (Cham: Palgrave Macmillan, 2022), 141–162.

76 Tuija Saresma and Emilia Palonen, "On the Emergence of Alt-Science Counterhegemony: The Case of the Finns Party", in: *Populism and Science in Europe*, ed. Hande Eslen-Ziya, Alberta Giorgi (Cham: Palgrave Macmillan, 2022), 117–140.

scientific articles without undertaking a more detailed analysis in the form of a stand-alone monograph.

The process of radicalization leading to extremism is also a recurring pattern in published works on various regional branches of the anti-vaccination movement. Academic texts show this trend in the USA,[77] Canada,[78] Australia,[79] and Germany.[80] Once again, however, these findings need to be summarized as a stand-alone monograph or at least to be put in the broader analysis of the connections between the movements and political organizations. Also, the issues of Russian inspirations and influences on the antiscience movement's (anti-vaccination and anti-5G) activities, most likely due to the great interest of public opinion, the media, and the political class, have been discussed in at least a dozen or so interesting scientific articles[81] published from "Foreign Policy"[82] to "Vaccine."[83] Yet again, there has been virtually no broader analysis in the form of a monograph examining the narratives, the media involved, and the direct political connections. Additionally, their fragmentary nature allows us to consider this book a pioneering analysis. An interesting context for this radicalization and extremism is provided by the obser-

77 Dana Beth Weinberg and Jessica Dawson, "From Anti-vaxxer Moms to Militia Men: Influence Operations, Narrative Weaponization, and the Fracturing of American Identity", *The Brookings Institution*, 01.11.2020, accessed 07.09.2024, https://www.brookings.edu/articles/from-anti-vaxxer-moms-to-militia-men-influence-operations-narrative-weaponization-and-the-fracturing-of-american-identity.

78 Sibo Chen, "Far-Right Political Extremism and the Radicalisation of the Anti-vaccine Movement in Canada", in: *Communicating COVID-19*, ed. Monique Lewis, Eliza Govender, Kate Holland (Cham: Palgrave Macmillan, 2024), 303–323.

79 Lydia Khalil and Joshua Roose, "Anti-government Extremism in Australia: Understanding the Australian Anti-lockdown Freedom Movement as a Complex Anti-government Social Movement", *Perspectives on Terrorism* 17, no. 1 (2024):144.

80 Daniel Koehler, "Siren calls of anti-government extremism: Far-right influences on the German anti-vax (»Querdenken«) protest milieu through music", *Behavioral Sciences of Terrorism and Political Aggression* 16, no. 2 (2023):1–22.

81 David A. Broniatowski et al., "Weaponized Health Communication: Twitter Bots and Russian Trolls Amplify the Vaccine Debate", *American Journal of Public Health* 108 (2018): 1378–1384; Kun Yan et al., "Russian Troll Vaccine Misinformation Dissemination on Twitter: The Role of Political Partisanship", *Health Communication* 38, no. 8 (2023):1591–1600 and Amir Karami et al., "Identifying and Analyzing Health-Related Themes in Disinformation Shared by Conservative and Liberal Russian Trolls on Twitter", *International Journal of Environmental Research and Public Health* 18, no. 4 (2021):2159.

82 Katherine Kirk, "How Russia Sows Confusion in the U.S. Vaccine Debate", *Foreign Policy*, no data, accessed 07.09.2024, https://foreignpolicy.com/2019/04/09/in-the-united-states-russian-trolls-are-peddling-measles-disinformation-on-twitter.

83 Echo L. Warner et al., "Vaccine Misinformation Types and Properties in Russian Troll Tweets", *Vaccine* 40, no. 6 (2022):953–960.

vations in the text by Sanders and Burnett, which see the anti-vaccination movement as a specific result of applying a radical liberal position emphasizing the importance of issues such as individualism, deregulation, and decentralization to health regulations.[84] When this approach is combined with the conclusions of the text by Russell, who explored the political significance of the interconnected anti-vaccination and anti-lockdown protest movements and described them in terms of the concept of "anti-politics" – the theory of the growing impossibility of political action except in opposition to the political system itself,[85] we obtain an interesting interpretation of the political alliances of these movements.

Conspiracy narratives about 5G have been a common analysis element in numerous texts examining conspiracy narratives about the COVID-19 pandemic. It was primarily because it was one of the most common elements in the stories of advocates trying to discover the "hidden truth" about the mysterious "Wuhan virus."[86] Another reason was the extremist element in the activities of this movement, manifested primarily by the destruction of telecommunications equipment.[87]

However, due to the thematic scope of this book, publications that discuss the narratives or the global dimension of the activities of these movements could be directly useful for this analysis. Others had to serve more as a point of reference or comparison of the situation and the development of connections with the political sphere and considerations of whether the Polish movement

---

84 Chris Sanders and Kristin Burnett, "The Neoliberal Roots of Modern Vaccine Hesitancy", *Journal of Health and Social Sciences* 4, no. 2 (2019):149–156.

85 Francis Russell, "Pox populi: Anti-vaxx, anti-politics", *Journal of Sociology* 59, no. 3 (2023):699–715.

86 Beatriz Buarque, "Is It Conspiracy or »Truth«? Examining the Legitimation of the 5G Conspiracy Theory during the Covid–19 Pandemic", *Social Epistemology* 36, no. 3 (2023):317–328;Johannes Langguth, Petra Filkuková and Stefan Brenner et al, "COVID-19 and 5G conspiracy theories: long term observation of a digital wildfire", *International Journal of Data Science and Analytics* 15 (2023):329–346; Alex Bruns et al., "Covering Conspiracy: Approaches to Reporting the COVID/5G Conspiracy Theory", *Digital Journalism* 10, no. 6 (2022):930–951; Daniel Jolley and Jenny L. Paterson, "Pylons ablaze: Examining the role of 5G COVID-19 conspiracy beliefs and support for violence", *British Journal of Social Psychology* 59, no. 3 (June 2020):628–640; Alex Bruns et al., "»Corona? 5G? or both?«: the dynamics of COVID-19/5G conspiracy theories on Facebook", *Media International Australia* 177, no. 1 (2020):12–29; John Bodner et al., *COVID-19 Conspiracy Theories: Qanon, 5G, the New World Order and Other Viral Ideas.* Jefferson: McFarland & Company, 2020. Kindle

87 Saddiq Basha, "»Death to the Grid«: Ideological Narratives and Online Community Dynamics in Encouraging Far-Right Extremist Attacks on Critical Infrastructure", *Counter Terrorist Trends & Analysis* 15, no. 4 (2023):17–24.

reaches for similar actions, solutions, and above all, whether it creates political alliances with parties of an ideologically analogous nature.

An interesting context for this work are the texts of Alice E. Marwick. In her work, she does not treat disinformation as cultural phenomenon – embedded in specific social structures, user practices and broader political contexts. Such an approach to disinformation removes it from the sphere of influence in hybrid conflict and information warfare and, at the same time does not make it only a manifestation of an individualistic approach, not reducing it to the intellectual or cultural competencies of users to participate in the discourse. Marwick talks about the attention economy that governs social media, which strengthens user engagement by promoting emotionally stronger content and messages' identity-based nature, which dominates their actual truthfulness. Outrage, being part of a larger group fighting for some positive goal or with a supposed threat in its interpretation, is more important to many users than the truthfulness of information.[88] When facts lose dominance in favour of emotional statements, disinformation symbolizes rebellion against the "elites" and institutions – media, science, and politics; it is not only a cognitive problem but a political and identity act. In this context, the goals of political actors meet the "fertile" soil of spontaneous social processes in disinformation. Thanks to her works, we are not dealing, also in the context of this book, with just a "flat" space of flows and influences but with a living and spontaneously reacting mass to which political centres attach "informational electrodes." Social media such as Facebook, Twitter, YouTube, and Reddit enable the creation of counter-publics – alternative public spheres that, operating outside the mainstream of public debate, undermine the authority of traditional political media, which is visible in conservative and far-right circles in the United States, and especially in the so-called alt-right. They operate with different epistemic norms, different sources of authority, and completely different understandings of what "credible information" is. In them, traditional journalistic standards are rejected as forms of elite domination and a means of manipulation by the political, academic, and media establishment. In this narrative, mainstream media are not only wrong but are conscious tools of propaganda, and alternative media are not only not disinformation but a form of resistance to an oppressive system.[89] As a result, the examples of the movement against mandatory vacci-

88 Alice E. Marwick, "Why Do People Share Fake News? A Sociotechnical Model of Media Effects," Georgetown Law Technology Review 2, no. 2 (2018): 474–512.

89 Alice E. Marwick and William C. Partin, "Constructing Alternative Facts: Populist Expertise and the QAnon Conspiracy," *New Media & Society* 26, no. 5 (2022): 2535–2555.

nations and the introduction of fifth-generation telephones show how the Foucault approach of undermining power-knowledge[90] and the Althusserian opposition to the oppressive work of the ideological apparatuses of the dominant way of thinking[91] turns into its caricatural distortion and leads to informational and epistemological chaos.

An interesting addition to the image of the contemporary media system, as the natural space for the spread of disinformation, are the theses put forward by Benkler, Faris, and Roberts. The authors of *Network Propaganda: Manipulation, Disinformation, and Radicalization in American Politics* emphasize in their analyses of the American political media topography founded on the study of millions of stories and social media posts, demonstrating a highly polarized and asymmetric media ecosystem, not external influences, but the natural tendency of the contemporary media system, susceptible by its very nature to the spread of fake news and fake narratives, their rapid spread and legitimization. In the reality they describe, they are not a coincidence but part of the strategies of political actors to create the phenomenon of "network propaganda," i.e., dispersed action, where users, social media, blogs and portals mutually reinforce the message, often without a central coordinator, which creates the impression of grassroots and authenticity.[92] This approach proved helpful for better understanding how anti-scientific conspiracy narratives, especially those aligned with the interests of the Russian Federation, build their credibility in the contemporary media landscape.

Claire Wardle's work focuses on a slightly different aspect of disinformation. She considers it as a part of a broader media and social ecosystem where technology intertwines with emotions, the framework of which amplifies the resonance of false information. However, unlike Benkler and his team, she emphasizes the architecture of platforms in the context of their impact on users. While Benkler speaks directly about the systemic asymmetry in the information ecosystem, where institutions, political divisions, media funding, and platform logic lead to "fertilizing" the soil for politically motivated, in his view, primarily right-wing, disinformation, Wardle focuses on the "micro" scale of the system – connections of platform architecture, that is, the set of algorithmic, interface, and regulatory solutions that determine what we see on social media and lead to the dominance of more clickable emotional and

90 Ostolski, "Dyskurs u władzy", 165.

91 Althusser, "Ideologia i aparaty ideologiczne państwa."

92 Yochai Benkler, Robert Faris, and Hal Roberts, *Network Propaganda: Manipulation, Disinformation, and Radicalization in American Politics* (New York, 2018, Oxford Academic) Kindle.

radicalized content.[93] If we treat both "macro" and "micro" approaches as complementary, a mutually complementary picture emerges, where the system architecture to which information engines are connected begins to fill up with messages fuelled by extreme emotions, the truth of which is no longer examined by anyone. This results in a media landscape in which statements challenging the scientific consensus not only do not encounter gates verifying their claims but are multiplied and amplified, increasingly blurring the picture.

Joan Donovan's work resonates with Benkler and his approach. This researcher, who deals with online extremism and disinformation, points to the political motivation in disinformation and its tools – memes, virals, conspiracy narratives, and fake news – and sees it as the effect of an entire industry connected to politics, business and media, supported by social media platforms, whose very structure influences its spread.[94] Donovan also shows in her work the "life cycle" of media manipulation from planning by small, anonymous groups through dissemination by social media to be sucked in by mainstream media.[95] These reflections were particularly valuable when applied to the "life cycle" of conspiracy narratives related to vaccines and 5G in the Polish media ecosystem and as such. Donovan shows how YouTubers, streamers, and micro-celebrities play a key role in legitimizing disinformation.[96] This process is also the case in the Polish media framework of the conspiracy narratives analysed in this work. By analysing case studies from various political campaigns (e.g., QAnon, anti-vaccination campaigns, deepfakes, political trolling) and showing their tactics, strategies, platforms, and effects of their actions, he shows how QAnon and Stop the Steal memes and narratives shaped the emotional ground for the attack on the Capitol on January 6, 2021,[97] which is particularly valuable for this work, where the connections between online mobilization and conspiracy narratives and political extremism are also indicated.

93 Claire Wardle, "Misunderstanding Misinformation," *Issues in Science and Technology* 39, no. 3 (Spring 2023), accessed 16.04.2025, https://issues.org/misunderstanding-misinformation-wardle.

94 Joan Donovan, Emily Dreyfuss, and Brian Friedberg, *Meme Wars: The Untold Story of the Online Battles Upending Democracy in America* (New York: Bloomsbury Publishing, 2022), 108–155.

95 Ibid, 239–268.

96 Anthony Nadler, Matthew Crain, and Joan Donovan, *Weaponizing the Digital Influence Machine: The Political Perils of Online Ad Tech* (New York: Data & Society Research Institute, 2018), accessed 14.04.2025, https://datasociety.net/wp-content/uploads/2018/10/DS_Digital_Influence_Machine.pdf.

97 Donovan, Dreyfuss, and Friedberg, *Meme Wars: The Untold Story ...*, 303–332.

To better understand Russian influence on the disinformation process and the shaping of the contemporary Western political scene and public discourse, it is essential to read Kathleen Hall Jamieson's texts, both on the issue of conspiracy narratives and on Russia. These two key issues intersect in this book. Jamieson does not claim that Russia created the problems of American democracy but indicates that its interference deepened them, exploiting the existing divisions and weaknesses of the media system. However, she indicates and documents how the Kremlin conducted a multi-layered disinformation campaign in the US and how content through its media and bots was consumed and shared en masse in America and directed at residents of key states and demographic groups for the presidential elections. Of course, she notes that it is impossible to determine precisely how these actions influenced voters' decisions. Still, they strengthened American political polarization, and we can assume that they impacted the electoral situation in swing states.[98]

Her research was also valuable for better understanding the role of the conspiracy narratives they propagated and professed in communities opposing mandatory vaccinations and 5G. Jamieson explains their appeal by reducing the uncertainty created by complex, disturbing events such as the pandemic, terrorist attacks, and political changes. In her opinion, they provide simplified explanations of complex events and organize chaos, giving the impression of regaining control over reality, which is also a close interpretation for the author of this book. A crisis or political decision becomes part of a larger plan, the discovery and announcement of which to others gives the impression of the possibility of change, a reality perceived as hostile or permeated with bad influences. As a result, people prone to conspiratorial thinking feel they have "real" knowledge – often in opposition to official sources.[99] This approach fits perfectly with protest movements contesting the scientific consensus on vaccinations or telephony, cementing the community of beliefs and values among the followers of conspiracy narratives and reinforcing the strong division into "us" and "them."

In understanding how claims from conspiracy theory registers penetrate mainstream media and mainstream public discourse, Johnson's et al. research provided an important insight. It showed the connections between Facebook groups and users by creating special network maps, focusing on the interac-

98 Kathleen Hall Jamieson, *Cyberwar: How Russian Hackers and Trolls Helped Elect a President; What We Don't, Can't, and Do Know* (New York: Oxford University Press, 2018), 255–266.

99 Dolores Albarracin, Julia Albarracin, Man-pui Sally Chan & Kathleen Hall Jamieson, *Creating Conspiracy Beliefs: How Our Thoughts Are Shaped* (Cambridge: Cambridge University Press, 2022), 11–54.

tions between various conspiracy groups (anti-vaccination movements, conspiracy theories around COVID-19, 5G, QAnon movement, and disinformation related to the 2020 US presidential election) and bridge communities, such as wellness, parenting, and spirituality groups. As a result, the growth rate and reach of specific narratives were examined. As it turned out, the groups spreading disinformation were much better organized and coordinated than institutional sources of information, creating dispersed but highly branched structures that effectively reached many independent environments. A key finding of the team's research was the role of groups with a seemingly neutral character (e.g., natural health, spirituality, motherhood) in connecting users with more radical and conspiratorial narratives. These groups were bridging communities that allowed disinformation to reach new, non-political audiences.[100] An analogous process was noticeable in the Polish context. Also, conspiracy themes migrated from conspiracy registers, through intermediaries, to the mainstream. In the Polish case, the pro-Russian nature of some right-wing media turned out to be additionally important, as they served to transfer content that was inconsistent with scientific consensus from the margins to the centre.

An essential study to explain the persistence and speed of the spread of anti-scientific conspiracy narratives was Yafei Zhang et al.'s study of the dynamics of the spread of conspiracy and science content on the Internet. Large-scale quantitative analysis using data from over 100 million cascades, discussion cascade modelling, social network analysis, content analysis, and machine learning showed, confirming the findings of the aforementioned disinformation theorists, that conspiracy content spreads faster and more widely than science content tends to generate more emotional and polarized responses, and also exists in information bubbles rarely penetrated by other types of arguments – the key emotions for increasing reach were anger and fear,[101] which are key to many conspiracy narratives. A team led by Zollo reached similar conclusions regarding the emotional charge of conspiracy messages and the resulting greater engagement of commentators and social media users.[102]

100 Neil F. Johnson et al., "Mainstreaming of Conspiracy Theories and Misinformation," *arXiv* (2021), accessed 14.04.2025, https://arxiv.org/abs/2102.02382.

101 Yafei Zhang et al., "Conspiracy vs. Science: A Large-Scale Analysis of Online Discussion Cascades," *World Wide Web* 24, no. 2 (2021): 585–606, https://doi.org/10.1007/s11280-021-00862-x.

102 Fabio Zollo et al., "Emotional Dynamics in the Age of Misinformation," *PLOS ONE* 10, no. 9 (2015): e0138740.

Many parallels in the transfer of conspiracy narratives from the margins to the centre of public discourse appear in the Polish case in relation to the Czech and Slovak digital spaces examined in this context by Heřmanová. She shows how conspiracy content moves from marginal platforms to the mainstream. Initially, the narratives appeared in English on YouTube channels, alternative medicine, and ecological websites. They are then translated and published on local, marginal platforms (e.g., Reddit, 8chan, Telegram). After this stage, they are shared in semi-closed communities (e.g., Facebook groups, TikTok), often in memes or banners, and then reposted by influencers on mainstream platforms, where the content is adapted to local narratives. The study emphasizes the role of influencers as "cultural translators" who adapt and disseminate conspiracy narratives, giving them local meaning and increasing their reach.[103] A similar thing happened in the Polish case, where these narratives moved from niche groups with the particular participation of alternative right-wing politicians, often with a pro-Russian political turn.

Also worth mentioning as a context for this work are the works that organize knowledge about the intersecting registers of disinformation, conspiracy narratives and extremism, whether in terms of psychological motives for participating in conspiracy theory circles, such as the works of Douglas, Sutton, and Cichocka,[104] or cultural studies, such as the already classic work by Fenster.[105] An interesting context strongly present in conspiracy theorists is apocalyptic culture, which Barkun introduced and developed. He describes how the belief in the coming catastrophe (e.g., significant social changes, cosmic phenomena, civilization wars) shapes American society[106] and, as a pattern, also occurs in other countries, being a motivator, justification or explanation for the involvement of people in conspiracy theorists.

As a result of the theoretical approaches mentioned above, a structure was created which elements include a presentation of the conspiracy narratives themselves and their genesis, the leading Polish entities propagating them, and the channels of their dissemination, as well as an indication and discussion of the political representation of the mentioned movements.

---

103 Marie Hermanova, "From Fringe to Mainstream: QAnon Narratives in Central European Digital Spaces". *AoIR Selected Papers of Internet Research* (March 2022), https://doi.org/10.5210/spir.v2022i0.13020.

104 Karen M. Douglas, Robbie M. Sutton, and Aleksandra Cichocka, *The Psychology of Conspiracy Theories* (London: Routledge, 2024).

105 Mark Fenster, *Conspiracy Theories: Secrecy and Power in American Culture*, 2nd ed. (Minneapolis: University of Minnesota Press, 2008).

106 Michael Barkun, *A Culture of Conspiracy: Apocalyptic Visions in Contemporary America* (Berkeley: University of California Press, 2013).

Chapter 1 (NWO, Big Pharma, and Electromagnetic Weapons: The content and origin of conspiracy narratives that are the theoretical basis for Movements Opposing the Introduction of Fifth Generation Telephony (5G) and Mandatory Vaccinations) indicates related to the anti-science movements conspiracy narratives, such as the so-called "electromagnetic weapons" and "torsion fields" as well as the ufological roots of the narrative about the harmfulness of 5G and the main fake news related to it. This chapter also briefly discusses the history of the anti-vaccination movement with the most important events that shaped its activities, such as media scandals related to the DTP vaccine or the issue of the presence of Thimerosal in vaccines, as well as the case of an article by Andrew Wakefield in the *Lancet*, suggesting connections between vaccinations and autism. This part also indicates the most widespread conspiracy narratives promoted by the movement against mandatory vaccinations, such as the concept of "Big Pharma" or accusations that preparations contain graphene, aluminium, or fragments of foetuses from an abortion procedure. This chapter also discusses the participation of states such as the Russian Federation in publicizing and promoting conspiracy narratives related to 5G and protective vaccinations.

Chapter 2 (The 5G Files: Poland's Anti-5G Movement: Activists, Major Organizations, and "Experts") indicates numerous conspiracy narratives shared by activists of the movement, including those of an anti-Semitic and anti-scientific nature, their connections with the ufological movement, new religious cults, activity in political groups, and businesses related to opposition to 5G. The crucial organizations associating activists of the movement were also discussed in this chapter, such as the STOPZET STOP Organized Electronic Tortures Association, the National Association for Combating Electrocontamination "Right to life," We The Free People, the James Clerk Maxwell Electromagnetic Research Institute Foundation and the 5G Free Poland Coalition.

Chapter 3 (Turn Off the TV – Turn On Your Thinking: The Polish Anti-vaccination Movement: Activists, Significant Organizations, "Experts") presents the Polish part of the movement opposing mandatory vaccinations like "expert" organizations such as: Polish Association of Independent Physicians and Scientists (PSNLiN), the Ordo Medicus Institute, the Society of Integrated Medicine Physicians (TLMZ), and the Association Physicians of Hope. The main motivations important for the leaders involved in the movement like deep religiosity and faith in alternative medicine solutions, are also indicated in this part.

Chapter 4 (Conspiracy Influencers: Alternative Media and Non-governmental Organizations Promoting the Postulates of Movements Fighting against Mandatory Vaccinations and 5G) describes the media and non-governmental organizations promoting the postulates of the movements fighting against mandatory vaccinations and 5G and analyses their significance for the dissemination of the narratives of these groups. This part indicates the main channels for the dissemination of the message of both movements studied, including media associated with the ufological milieu, alternative medicine, new spirituality, and nationalist media, and tries to answer why these entities played a crucial role in reaching the narratives of the studied movements to recipients beyond the activists of the movements discussed in this work. This part also indicates and describes the non-governmental organizations supporting the studied groups. In the above cases, the links between the groups mentioned above and the media with the Russian Federation and the convergence of the narratives they disseminated with the Russian ones were also analysed.

Chapter 5 (Vaccination Macht Frei: The Political Representation of Anti-Vaccination and Anti-Fifth Generation Telephony Movements) presents groups that can be considered a political representation of the movements opposing mandatory vaccination and the fifth-generation telephone network. It lists and describes organizations that cooperated so often and intensively with representatives of the movements studied that their demands permeated not only the public activities of these formations but also became part of their political program. This part primarily highlights the parties of the so-called anti-system right wing, i.e., 1Polska.pl, Kukiz'15 and Confederation. This chapter also indicates and describes in more detail the politicians of these formations, who at the same time act as activists of the movements studied. An additional context discussed in this chapter is the contacts of politicians of the mentioned parties, active in opposition to introducing a fifth-generation network and mandatory vaccinations to Poland, with institutions, politicians, and media of the Russian Federation and its allied countries.

Chapter 6 ("Covidians" and "Corona-Sceptics": Mainstream Parties and the Anti-Vaccination and Anti-5G Telephony Movements) describes the mainstream parties and their ambivalent attitude towards the movement opposing mandatory vaccinations and fifth-generation telephony. These parties, unlike those discussed in the previous chapter, did not directly act in the interests of the movements opposing fifth-generation telephony networks and mandatory vaccinations, but among their politicians one could also find significant representation of people active in issues crucial to the movements studied. The chapter identifies essential actors in this matter and analyses their significance in permeating into the mainstream narratives of the milieu opposed to mandatory vaccinations and the introduction of fifth-generation telephony networks, including those of a conspiratorial and anti-scientific nature.

Chapter 7 (You'll Be Hanged When Poland Returns: Protests, Radicalization, and Extremism in Movements against 5G and Mandatory Vaccinations) discusses the radicalization and extremism in the movements opposing the 5G network and mandatory vaccinations. This part describes and analyses demonstrations against the introduction of the 5G network to Poland and mandatory vaccinations in Poland, as well as slogans that appeared during them, the size of these demonstrations, their leaders, the dynamics of the protests, periods of intensification and calm, their connections with the introduction of government restrictions and protests against the lockdown of other social groups. The argument ends with an analysis of the process of radicalization leading to extremism in the movement opposing the introduction of fifth-generation telephony to Poland and mandatory vaccinations, placing the events from Poland in an international context. It distinguishes the critical cases and groups involved in violent actions that consider themselves part of the movements studied.

It is worth noting that the increasing activity of the movements discussed, their intensifying interpenetration with political groups, as well as numerous manifestations of extremist activities, combined with the potential for strengthening the message that the potential next pandemic predicted by scientists may give,[107] mean that in the matter of anti-scientific movements and the conspiracy narratives associated with them, we can expect further development of the visibility and significance of entities representing them. Hence, this work can be considered opening up the field for further, more detailed research to deepen the issues raised in this publication.

As a result, this work is beneficial not only for increasing awareness of the activities of the Polish branch of anti-scientific movements but will also be very helpful for comparative studies focusing on the analysis of the structure, activities, and political connections of analogous movements in countries other than Poland. This book is also significant in the context of the recent COVID-19 pandemic and its impact on contemporary international relations and social movements, as well as the increasing polarization, radicalization of participants in political life, and new trends in extremism. The subsequent potential pandemics that scientists are warning about today may strengthen the presence of conspiracy narratives related to medicine in public discourse. Hence, a thorough study of the cases of anti-scientific movements discussed in

107 For example: University of Edinburgh, "Virus ancestry could aid bid to predict next pandemic ... " and Shafaati, "The next pandemic catastrophe ... ".

this work may benefit us if we face such a threat again. Additionally, the knowledge contained in this book's case studies of the participation of state political actors is an essential element of building our resistance to disinformation and allows us to better notice the channels of malicious dissemination of fake news. Of course, this is not the first publication discussing the topic of anti-scientific movements. However, it is a monograph that is one of the first to examine more broadly and at the same time so thoroughly the connections between political organizations and anti-scientific movements, combining the study of the narratives of these movements, their political activity and the extremist actions they undertake, additionally contributing to a better understanding of the phenomenon of home-grown terrorism.

CHAPTER 1

NWO, Big Pharma, and Electromagnetic Weapons

The Content and Origin of Conspiracy Narratives That Are the Theoretical Basis for Movements Opposing the Introduction of Fifth-Generation Telephony (5G) and Mandatory Vaccinations

In the media discourse, terms such as "electromagnetic smog," "electromagnetic field pollution," and "electrosmog" are increasingly appearing. Their effects are said to cause numerous problems for human health – migraines, cancer, and even epidemics. Even more, voices are heard in the media questioning the validity of some or all protective vaccinations. Preparations used to immunize against many diseases are accused of causing adverse reactions after vaccination and modifying human DNA or simply being ineffective. Numerous voices are talking about substances used to prepare them, such as graphene or aluminium, which are hazardous to health.

Few of these narratives are new. Most are mutations or recombinations of existing themes, often drawing on tropes and rhetorical manoeuvres with a long history. Their creators usually have a complete worldview, thanks to which they interpret new information and events to fit their existing theory. At the same time, one of the characteristic features of conspiratorial thinking is that it is self-sealing, unfalsifiable, and resistant to challenge. The lack of evidence is often treated as evidence of a mass cover-up.

On the one hand, such positions are ridiculed in traditional media and social media as conspiracy narratives. On the other hand, they gain new supporters, not only wholly anonymous but also people who influence others, such as politicians, artists, or celebrities. It is easy to read such opinions, following the comments section in press articles, video materials, or social media. There are also street demonstrations by supporters of these narratives. That is why it is essential to present the content and genesis of these views in greater detail and to indicate the most important sources on which their supporters base their narratives about dangerous electromagnetic fields and sinister vaccines.

The time of the emergence of narratives describing allegedly disastrous effects of introducing the fifth-generation telephone network is quite precisely defined. The first network of this generation was launched in 2017 as a test by

 | DOI:10.1163/9789004747852_003

Telia, Ericsson, and Intel in Tallinn and Stockholm[1] and then in London.[2] Two years later, the first phones supporting it went on sale. Therefore, the specific time and place of this technology's emergence allow us to precisely place on the timeline the emergence of narratives presenting it as a threat to the health and life of people and animals.

According to these stories, 5G is supposed to cause brain cancer, infertility, autism, and Alzheimer's disease, and to be used to control the mind, to control people through applications on their phones and microscopic chips implanted in their bodies. However, this is not the first narrative referring to alleged threats resulting from electromagnetic interactions. Doctors first spoke of "radiophobia" in 1903.[3] Then, there were concerns about power lines and microwaves in the 1970s[4] and 2G technology in the 1990s, when it was suggested that cell phone radiation could cause cancer and that this information was being suppressed. In the US, protests led to the adoption of regulations in many counties, increasing the distance of mobile infrastructure from residential areas and severely limiting the range of Wi-Fi networks due to claims that these devices cause health problems.[5]

According to proponents of the harmfulness narrative, many of the alleged threats related to 5G are very similar to those attributed by conspiracy narratives to the HAARP (High-Frequency Active Auroral Research Program), popularly called an "electromagnetic weapon." HAARP was a radio transmitter array located in Alaska and funded by the US Department of Defense in cooperation with many research universities. It was used to study the effects of various electronic devices on dynamic processes occurring in the ionosphere. It was closed in 2014. However, proponents of conspiracy theories have claimed that it was developing weapons to control the weather as well as the mind. HAARP has been accused of giving the ability to cause floods, hurri-

1 Arkadiusz Dziermański, "Intel, Telia i Ericsson uruchamiają pierwszą sieć 5G w Europie", *Antyweb*, 03.10.2017, accessed 30.08.2024, https://www.telepolis.pl/wiadomosci/wydarzenia/intel-telia-i-ericsson-uruchamiaja-pierwsza-siec-5g-w-europie.

2 Juan Pedro Tomás, "Vodafone, Ericsson carry out 5G trial in central London", *RCR Wireless News*, 21.12.2017, accessed 30.08.2024, https://www.rcrwireless.com/20171221/5g/vodafone-ericsson-carry-out-5g-trial-london-tag23.

3 John C.H. Lindberg and Denali Archer, "Radiophobia: Useful concept, or ostracising term?", *Progress in Nuclear Energy*, 149 (2022), 1–11.

4 Adam Burgess, *Cellular Phones, Public Fears, and a Culture of Precaution* (Nowy Jork: Cambridge University Press, 2003), 75–122.

5 Naomi Buck, "Could WiFi in schools be harming our kids?", *The Globe and Mail*, 11.05.2014, accessed 30.08.2024, https://www.theglobeandmail.com/life/health-and-fitness/health/is-there-such-a-thing-as-too-much-wifi.

canes, droughts and earthquakes, storms, power outages, the TWA Flight 800 crash, the so-called "Gulf War Syndrome," and chronic fatigue syndrome, as well as the ability to control the mind and cause diseases.[6]

The mentioned above conspiracy narratives related to HAARP technology have also been included in stories trying to frighten us with the dire consequences of introducing 5G. As a result, according to people who believe in them, the 5G network frequencies are similar to "electromagnetic weapons" that "destroy the neurological functioning of the body,"[7] cause mutations of viruses that are supposed to become increasingly resistant and multiply faster thanks to it, and also be used to target and attack people and lead, for example, to changes in the hormones produced by the body.[8] According to people fighting against introducing of the 5G technology, electromagnetic waves can cause over fifty different types of diseases, from cancer to changes in the psyche, leading to mass suicides and poisoning of the body, similar to arsenic poisoning.[9]

According to opponents of the introduction of 5G technology, an additional threat would be its connection with the so-called "torsion fields." This phenomenon, as they claim, is in addition to gravity, the electromagnetic force, the strong nuclear force, the weak nuclear force, a fifth force, and living organisms are adapted to receive it. According to supporters of this quasi-scientific concept, "torsion fields" are responsible for the recognition of signal particles by the receptor, antigens by the antibody, and substrate by the enzyme, and they can also be used to transmit DNA. In this way, according to people who believe in this force, it is possible to cause a cancer epidemic, send viruses directly to the human body, and perform selective killings.[10]

None of these symptoms and problems are confirmed by mainstream scientific research, and all are considered by modern science to be unrelated to

---

6 Abe Streep, "The Military's Mystery Machine: The High-frequency Active Auroral Research Program, or HAARP, has been called a missile-defense tool and a mind-control device. The truth is a bit less ominous", *Popular Science* 273, no. 1 (2008):60–63.

7 Arkadiusz Dziermański, "Krzyczą, że 5G zabija i twierdzą, że mają dowody. Podsumowanie krakowskiego zlotu przeciwników 5G", *Telepolis*, 01.03.2019, accessed 30.08.2024, https://www.telepolis.pl/artykuly/wydarzenia/siec-5g-w-polsce-szkodliwa-technologia-dowody.

8 Ibid.

9 Ibid.

10 Ibid.

5G technology.[11] At the same time, supporters of scare narratives about the harmful effects of 5G try to present various recorded, often dangerous or terrifying events as effects of introducing this technology and present them as "evidence" for the truth of their claims.

A key figure in the spread of the narrative about the alleged dangers of 5G was John Kuhles, a Dutch "ufologist" who founded the Facebook group "Stop5G" in 2018.[12] He was an active promoter of quasi-scientific conspiracy narratives as early as the early 1990s. Kuhles claims that his interest in UFOs stems from seeing an alien craft in 1990. In late 1992, Kuhles gave a lecture at the "UFO and the NWO" conference, where UFOs were linked to the conspiracy narrative of a secret world government called the New World Order.[13] In 1993, Kuhles and Australian journalist David Summers founded the Contact Network International Foundation to research and publish on UFO s and conspiracy narratives. They aired a nightly broadcast on Dutch indymedia,[14] and Kuhles also had several shows on Amsterdam cable television.[15] Still, his show was taken off the air after it was found to contain anti-Semitic and Nazi-like elements, because Kuhles reported that the Dutch Queen Juliana had met George Adamski, an ambassador from the planet Venus, in 1959, who described her Venusians as blond, blue-eyed beings with German accents whose feet left, among other things, traces of a swastika. Kuhles also believed that this might have been a new race that the Nazis wanted to breed in Neuschwabenland near Antarctica, where, according to him, a group of senior Third World War politicians had begun biological experiments there towards the end of the war.[16] Over the following years, the Dutchman continued his activities as a promoter of various conspiracy narratives and organizer of "ufological" festivals.

11 Ahmed Elzanaty, Luca Chiaraviglio, Mohamed–Slim Alouini, "5G and EMF Exposure: Misinformation, Open Questions, and Potential Solutions", *Frontiers in Communications and Networks* 2 (2021):1–11.

12 Sabrina Weiss, "5G health risks are the internet's new favorite conspiracy theory", *Wired*, 12.06.2019, accessed 30.08.2024, https://www.wired.com/story/5g–health–risks–concerns.

13 René Zwaap, "Lachen om samenzweringen is er een onderdeel van", *De Groene Amsterdammer*, 30.11.1994, accessed 30.08.2024, https://www.groene.nl/artikel/lachen–om–samenzweringen–is–er–een–onderdeel–van––2.

14 The Independent Media Center, better known as Indymedia, is an open publishing network of activist journalist collectives that report on political and social issues.

15 Ibid.

16 Ibid.

On the web pages run by Kuhles, the first theories about 5G began to emerge. One such piece of evidence for supporters of the 5G narrative was a Facebook post accusing a 5G test mast of causing mass bird deaths in the Netherlands. On November 5, 2018, conspiracy "medical" blogger Erin Elizabeth wrote about a disturbing story from the Netherlands on her blog "Health Nut News" with the headline "Hundreds of birds killed in 5G experiment in The Hague, Netherlands." Her post was shared on Facebook over 125,000 times. It was a direct copy of Kuhles' posts on the subject. In a 2018 Facebook post, he claimed that the deaths of 297 starlings in the Huijgenspark in The Hague, Netherlands, around October 19, 2018, were caused by a test of a new-generation mobile network, which Kuhles claimed took place at the same time as the birds' deaths.[17] However, local authorities have never recognized 5G as the cause of the animals' deaths. The test Kuhles describes occurred several months before the starlings' deaths (June 28, 2018). In addition, similar bird deaths have also happened in the US states of Arkansas,[18] Wisconsin,[19] and Louisiana,[20] as well as in Sweden, Japan, Thailand, and Brazil. These had nothing to do with 5G technology and occurred long before its development. Despite this, the post was widely shared on Facebook and health blogs. This story was not the only conspiracy narrative about 5G promoted by Kuhles. In another post, he suggested that the November 2018 California wildfires were a punishment by the "ruling elite" for opposing the state's 5G rollout, which was allegedly caused by a direct attack with an "energy weapon."[21] Kuhles still runs several pages that present alleged dangers of 5G, including on social media. The main one, the "Stop5G" Facebook group, now has over 32,000 members.

Tensions over 5G have increased significantly since the outbreak of the COVID-19 pandemic. According to the Global Disinformation Index, in the first four and a half months of 2019, the number of followers of the "Stop 5G" Facebook group grew from 7,000 to almost 60,000, and the size of Facebook groups on the subject grew from 0 to 50,000 followers. Meanwhile, the fol-

17 Alex Kasprak, "Did a 5G Cellular Network Test Cause Hundreds of Birds to Die?", *Snopes*, 13.11.2018, accessed 30.08.2024, https://www.snopes.com/fact-check/5g-cellular-test-birds.

18 No author, "The mystery of mass bird deaths", *BBC*, 07.01.2011, accessed 30.08.2024, https://www.bbc.com/news/world-us-canada-12135380.

19 Seth Borenstein, "Technology to blame for animal die-off panic", *NBC News*, 07.01.2011, accessed 30.08.2024, https://www.nbcnews.com/id/wbna40961721.

20 No author, "The mystery of mass bird deaths".

21 Reuters Staff, "Fact check: California wildfires were not caused by »powerful lasers«, *Reuters*, 07.08.2020, accessed 30.08.2024, http://www.reuters.com/article/uk-factcheck-california-wildfireslasers-idUSKBN25R1VH.

lowers of the 600 "Stop–5G" Twitter accounts tripled to 1.1 million in one week, while Instagram gained 20,000 followers in a week.[22] One early version of the conspiracy narrative claimed that it was no accident that 5G technology was tested in Wuhan, where the pandemic began.[23] Others claim that the "coronavirus crisis" was deliberately created to keep people at home while 5G engineers installed the technology.[24] Still others claim that "5G radiation" weakens people's immune systems, making them more susceptible to COVID-19 infection.[25] Another twist on the conspiracy theory asserts that 5G directly transmits the virus.

In turn, conspiratorial or quasi-scientific narratives about vaccinations are much older than the COVID-19 pandemic. Since the beginning of vaccinations, that is, since the end of the eighteenth century, some groups have seen them as problematic. The reasons for this reluctance varied depending on the era, but specific main trends remained constant. Smallpox vaccinations, conducted by Edward Jenner in the early nineteenth century, were already met with objections on sanitary, religious, scientific, and political grounds.

Some were put off by their children's reaction to the vaccine and how it was administered. Others, including clergy, believed that the vaccine was "unchristian." Others expressed a general distrust of medicine and Jenner's views on the spread of disease and gave completely non-medical reasons for the spread of smallpox. Finally, many people opposed vaccination because they believed it violated their freedom.[26] As more and more countries introduced mandatory vaccinations and increased their number, as well as normalized their procedures, the anti-vaccination movement continued to exist and adapting the patterns from the narratives mentioned above to the developing situation.

---

22 Chris Zappone, "Russian propaganda »very likely« stoking 5G health fears in Australia: expert claims", *The Sunday Morning Herald*, 16.09.2019, accessed 30.08.2024, https://www.smh.com.au/world/oceania/russian–propaganda–stoking–5g–health–fears–in–australia–20190916–p52rmc.html.

23 Rebecca Heilweil, "How the 5G coronavirus conspiracy theory went from fringe to mainstream", *Vox*, 24.04.2020, accessed 30.08.2024, https://www.vox.com/recode/2020/4/24/21231085/coronavirus–5g–conspiracy–theory–covid–facebook–youtube.

24 Ryan Broderick, "A Conspiracy Theory That 5G Is Causing The Coronavirus Is Spreading Alongside The Pandemic", *Buzzfeed News*, 03.04.2020, accessed 30.08.2024, https://www.buzzfeednews.com/article/ryanhatesthis/conspiracy–theory–5g–coronavirus–qanon.

25 Ingeborg Eliassen, Paulo Pena, "Real 5G issues overshadowed by Covid–19 conspiracy theories", *Investigate Europe*, 12.06.2020, accessed 30.08.2024, https://www.investigate–europe.eu/posts/5g–covid–conspiracy.

26 Dorothy Porter, Roy Porter, "The politics of prevention: Anti–vaccination and public health in 19th century England", *Medical History* 32 (1988):231–252.

During time also organizations fighting against mandatory vaccinations developed. When mandatory vaccination was introduced in the United Kingdom in the mid-19th century in subsequent legal acts (1853, 1867), the first organizations opposing it were also formed in response – the Anti-Compulsory Vaccination League and National Anti-Vaccination League.[27] There were also numerous demonstrations at that time, which, like the one in Leicester, England (1885), were able to gather almost one hundred thousand protesters.[28] Similar organizations were also established in the United States at that time.

The development of medicine did not eliminate the anti-vaccination movement. In addition, the occasional scandals related to the actual or alleged composition or side effects of some vaccines gave activists a new life and energy to act. Such high-profile cases, reinforcing the anti-vaccination message, were the controversies in the 1970s over the safety of the DTP vaccine (a combined vaccine against diphtheria, whooping cough, and tetanus), accused of causing neurological disorders in children[29] or Thimerosal – a compound containing mercury, used as a preservative in vaccines.[30] During these periods, successive waves of fear-enhancing press materials rolled through the Western media, which each time led to a decrease in vaccination rates and an increase in the incidence of certain diseases.[31]

However, the most well-known case that significantly influenced the decrease in trust in vaccinations is the study by Andrew Wakefield, published in 1998 in the scientific journal *The Lancet*. It linked vaccination with a three-component immunizing preparation against measles, mumps and rubella, enteritis, and autism. However, this study was repeatedly verified without confirming the result obtained by Wakefield, and, as it turned out during a

---

27 Robert M. Wolfe, Lisa K. Sharpe, "Anti–vaccinationists past and present", *British Medical Journal* 325, no. 7361 (2002):430–432.

28 Nadja Durbach, "They might as well brand us: Working class resistance to compulsory vaccination in Victorian England", *The Society for the Social History of Medicine* 13 (2000):45–62.

29 Donna Hilts, "TV Report On Vaccine Stirs Bitter Controversy", *The Washington Post*, 27.04.1982, accessed 30.08.2024, https://www.washingtonpost.com/archive/local/1982/04/28/tv–report–on–vaccine–stirs–bitter–controversy/80d1fc8a–1012–4732–a517–7976c86ab52d.

30 Jeffrey P. Baker, "Mercury, Vaccines, and Autism", *American Journal of Public Health*, nr 98(2), 2008, ss. 244–253.

31 Sun Kyong Lee et al., "Misinformation of COVID-19 vaccines and vaccine hesitancy", *Scientific Reports* 12 (2022), accessed 30.08.2024, https://www.nature.com/articles/s41598–022–17430–6

journalistic investigation, the group of patients examined by the English doctor was small and manipulated. Additionally, the doctor and his legal team were involved in a lawsuit against the vaccine manufacturer, and incorrect effects, such as the temporal relationship between vaccination and side effects, were misattributed.[32] The publication was withdrawn from *The Lancet* in 2010, and Wakefield was stripped of his license to practice in Great Britain. Despite this, to this day, this doctor's "research" is cited as evidence of the truth of the anti-vaccination movement's claims, and he travels and promotes his concepts. As a result, people who believe in unscientific narratives about vaccines believe that the vaccines contain aluminium, graphene, and cells from aborted fetuses,[33] that they change the DNA of the vaccinated person and that they are insufficiently studied or harmful to health,[34] and that they cause Adverse Vaccine Reactions. These allegedly include, among others, autism, epilepsy, inconsolable crying (so-called "brain screaming"), diarrhoea, fever, delay in physical and mental development, sleep disorders, swelling, vomiting, decreased immunity, regular infections, seizures, rash, paralysis, problems with muscle tone, neurological disorders, bronchitis and pneumonia, hyperactivity, apnoea, and loss of appetite.[35] However, there is no scientifically binding evidence to confirm these claims.

One of the dominant narratives about the allegedly problematic nature of vaccinations is the belief in the narrative that medical professionals are sponsored by large pharmaceutical companies (so-called "Big Pharma") and acting on their behalf are trying to make people addicted to completely unnecessary medications, services, procedures, and rehabilitation equipment. According to this conspiracy narrative, completely healthy people perceive themselves

---

32 Fiona Godlee et al., "Artykuł Wakefielda wiążący szczepionkę MMR z autyzmem był oszustwem", *Pacjent.pl*, 19.09.2018, accessed 30.08.2024, https://www.mp.pl/pacjent/szczepienia/sytuacje-szczegolne/59132,artykul-wakefielda-wiazacy-szczepionke-mmr-zautyzmem-byl-oszustwem

33 Mikołaj Pietraszewski, "»Festiwal propagandy«, »oklaski od Konfederacji«. Znowu burza po słowach Kai Godek", *Radio Zet*, 05.01.2022, accessed 30.08.2024, https://www.wiadomosci.radiozet.pl/polska/Kaja–Godek–szczepionki–na–COVID–19–to–korzystanie–ze–zbrodni–aborcji.–Komentarze–ofensywa–antyszczepionkowcow

34 Julia Theus, "Pytamy antyszczepionkowców, co by ich przekonało do szczepień przeciw COVID. Odpowiedź jest krótka: nic", *OKO.Press*, 16.04.2021, accessed 30.08.2024, https://www.oko.press/antyszczepionkowcy–co–by–ich–przekonalo–do–szczepien.

35 Astromaria, "Świadectwa rodziców, których dzieci doświadczyły problemów zdrowotnych po szczepieniach," *Jestem za, a nawet przeciw* (blog), 30.09.2014, accessed 30.08.2024, https://www.astromaria.wordpress.com/2014/09/30/swiadectwa–rodzicow–ktorych–dzieci–doswiadczyly–problemow–zdrowotnych–po–szczepieniach.

as sick or are simply driven into illness to make them permanent customers of the pharmaceutical industry.[36]

Widely shared conspiracy narratives are having the effect of changing Western societies' attitudes towards vaccines and vaccinations. As a result of scientifically unconfirmed yet numerous media reports questioning the effectiveness and safety of vaccinations, people in the US and Europe are reacting to the vaccination process with growing distrust, which results in a decrease in vaccination rates. As a result, the risk of an epidemic is increasing.[37] In such societies, people with a weak immune system, who cannot be immunized for health reasons, and newborns are particularly at risk. At the same time, the risk is growing that due to the persistence of disease outbreaks, new mutations of viruses that are more dangerous to human health will appear. Scientists indicate that due to low vaccination rates, diseases largely considered eradicated, such as polio,[38] measles,[39] whooping cough,[40] diphtheria,[41] mumps, and tuberculosis,[42] are returning.

The most severe consequences for public health were the movement's activities opposing mandatory vaccinations during the intensification of the COVID-19 pandemic. Paradoxically, the number of supporters of this move-

36 Robert Blaskiewicz, "The Big Pharma conspiracy theory", *Medical Writing* 22, no. 4 (2013):259–261.

37 Aristos Georgiou, "Anti–vax Movement Listed by World Health Organization as One of the Top 10 Health Threats for 2019", *Newsweek*, 15.01.2019, accessed 30.08.2024, https://www.newsweek.com/world-health-organization-who-un-global-health-air-pollution-anti-vaxxers-1292493.

38 Katarina Kitarovic, "Polio in a Post–Vaccine World: The Consequences of Anti–Vaccination Sentiments", *Harvard International Review*, 08.03.2023, accessed 30.08.2024, https://hir.harvard.edu/polio-in-a-post-vaccine-world-the-consequences-of-anti-vaccination-sentiments.

39 No author, "Measels", *World Health Organization*, 31.05.2023, accessed 30.08.2024, https://www.who.int/news–room/fact–sheets/detail/measles.

40 Miriam Falco, "10 infants dead in California whooping cough outbreak", *CNN*, 20.10.2010, accessed 30.08.2024, https://edition.cnn.com/2010/HEALTH/10/20/california.whooping.cough/index.html.

41 Jessica Mouzo, "Parents of diphtheria–stricken boy feel »tricked« by anti–vaccination groups", *El Pais*, 05.06.2015, accessed 30.08.2024, www.elpais.com/elpais/2015/06/05/inenglish/1433512717_575817.html.

42 Jorge L. Ortiz, "Anti–vaxxers open door for measles, mumps, other old–time diseases back from near extinction", *USA Today*, no data, accessed 30.08.2024, https://eu.usatoday.com/story/news/health/2019/03/28/anti-vaxxers-open-door-measles-mumps-old-time-diseases/3295390002.

ment increased significantly at that time,[43] what had resulted in an increase in demonstrations, pickets, blockades, and other forms of protest related to the movement, directed against mandatory vaccinations and the government's introduction of sanitary restrictions. Supporters of unscientific narratives about vaccinations did not accept the vaccination, encouraged people not to take the vaccine, demanded the lifting (and non-introduction of new) anti-pandemic restrictions,[44] banned immunization of people under their care (children and seniors), falsified data in medical systems regarding their vaccinations, and in extreme cases physically attacked vaccination points.[45] As a result, they had been infected with COVID-19 and were getting medical care in health facilities, what blocked help for other patients, including those not suffering from COVID-19.[46] As a result, when analysing the statistics on excess deaths plotted on maps and comparing them with maps showing the distribution of refusal to vaccinate, a large part of excess deaths during the intensification of the pandemic can be linked to the anti-vaccination narrative of activists of this movement.[47] At the same time, they claim that vaccines were the cause of these excess deaths.[48] They blame the governments of their countries and the pharmaceutical industry for these deaths.[49]

43 Nihal Durmaz and Engin Hengirmen, "The dramatic increase in anti-vaccine discourses during the COVID-19 pandemic: a social network analysis of Twitter", *Human Vaccines & Immunotherapeutics* 18, no. 1 (December 2022): n.pag.

44 Ms/PAP, "Warszawa. Protest antyszczepionkowców. »Zdejmijcie w końcu te obostrzenia«", *Polsat News*, 22.05.2021, accessed 30.08.2024, https://www.polsatnews.pl/wiadomosc/2021–05–22/protest–antyszczepionkowcow–w–warszawie–zdejmijcie–w–koncu–te–obostrzenia.

45 PAP, "Antyszczepionkowcy zaatakowali punkt szczepień w Grodzisku Mazowieckim. Interweniowała policja", *Rynek Zdrowia*, 26.07.2021, accessed 30.08.2024, https://www.rynekzdrowia.pl/Prawo/Antyszczepionkowcy–zaatakowali–punkt–szczepien–w–Grodzisku–Mazowieckim–Interweniowala–policja,223509,2.html.

46 Magda Nogaj, "Antyszczepionkowcy paraliżują pracę szpitala we Wrocławiu. "Nie dają się leczyć, boją się nawet termometru", *Wyborcza.pl*, 23.12.2021, accessed 30.08.2024, https://wroclaw.wyborcza.pl/wroclaw/7,35771,27939908,antyszczepionkowcy–paralizuja–prace–szpitala–we–wroclawiu–nie.html

47 Kartografia Ekstremalna, "Kiedy miesiąc temu opublikowaliśmy mapę nadmiarowych zgonów ...", *Facebook*, 18.12.2021, accessed 30.08.2024, https://www.facebook.com/153690641483461/posts/1808514839334358.

48 See further: Chris Klinsky, *Szczepionka na koronawirusa – zagrożenie dla ludzkości?* (Warszawa: Wydawnictwo św. Tomasza z Akwinu, 2021).

49 Dominika Sitnicka, "Transparent »Szczepienie czyni wolnym« to nie wypadek przy pracy. Tak Konfederacja pogrywa epidemią", *OKO.Press*, 18.12.2021, accessed 30.08.2024, https://oko.press/jak–konfederacja–politycznie–pogrywa–epidemia.

At the same time, the propagation of anti-scientific narratives, stoking fears associated with them, and strengthening the sense of uncertainty and information chaos were also used by authoritarian regimes seeking to destabilize the political situation in Western countries.[50] In this context, non-democratic countries benefit from an advantage over countries of liberal democracy, which is given to authoritarian regimes by extensive control over the sources and distribution of information. Liberal-democratic countries generally do not seek to suppress political pluralism and freedom of speech, often making these assumptions the basis of the political system. Additionally, in authoritarian countries, public media are subordinate to political power. They can act as a tool for disseminating, supporting, or amplifying conspiracy narratives, especially in the case of channels run in one of the globally popular languages, such as English.

On a global scale, the Russian Federation is particularly visible in spreading false information related to socially controversial content. The Soviet Union has already actively used disinformation activities. It was there that the term itself was also coined in the 1920s.[51] Since then, disinformation has become a vital element of the Kremlin's services, aimed at manipulating the enemy's intelligence system by placing reliable but misleading data in the information system. However, many intelligence agencies quickly began to use this tactic, including the British SIS, and the American CIA. However, disinformation remained an important tool for the USSR's services and allies. The Eastern Bloc's secret services spread false information about Western politicians as former Nazi collaborators,[52] trying to create the impression in Western societies that the CIA was involved in the assassination of US President John Fitzgerald Kennedy[53] and that the US had created the HIV virus.[54]

50 Heidi Cuda, "How Russia Joined with Anti-Vaxxers to Wage Biological Warfare by Proxy against the West", *Byline Supplement*, 18.03.2023, accessed 30.08.2024, https://www.bylinesupplement.com/p/how-russia-joined-with-anti-vaxxers.

51 Martin J. Manning, Herbert Romerstein, "Disinformation", in: *Historical Dictionary of American Propaganda*, ed. Martin J. Manning, Herbert Romerstein, (Greenwood: Bloomsbury Academic, 2004), 82–83.

52 Ladislav Bittman, *The deception game: Czechoslovak intelligence in Soviet political warfare* (Syracuse: Syracuse University Research Corp, 1972),39.

53 Max Holland, "How Moscow Undermined the Warren Commission", *Washington Decoded*, 30.03.2007, accessed 30.08.2024, https://www.washingtondecoded.com/site/2007/03/how_moscow_unde_1.html.

54 Adam Taylor, "Before »fake news«, there was Soviet »disinformation«", *The Washington Post*, 26.11.2016, accessed 30.08.2024, https://www.washingtonpost.com/news/worldviews/wp/2016/11/26/before-fake-news-there-was-soviet-disinformation.

After a period of reduced Moscow's activity in this matter in the 1990s, when Vladimir Putin took power in the Kremlin, Russia once again became active in disinformation activities, using a wide range of means – from the activity of internet trolls to propaganda and disinformation disseminated by public media such as RT and Sputnik. Moscow also began to use social media platforms such as Facebook, Twitter, TikTok, and YouTube to spread disinformation.[55] A unique role in this process is played by the state media concern Rossiya Segodnya (Russia Today), which has been established in 2013, when several of the most important Russian media companies (RIA Novosti news agency, Baltnews television, and Voice of Russia radio station) merged into one corporation. The Russian state media bundle also includes the English-language channel RT and the Sputnik news agency, which publishes articles in over thirty languages. Although RT is not formally affiliated with Rossiya Segodnya, the editor-in-chief of both entities is the same person – Margarita Simonyan. The media entities subordinate to her have been repeatedly accused by academics and Western journalists of spreading conspiracy narratives (concerning, for example, homosexuality,[56] GMO,[57] 5G[58] telephony, and

55 Media Ajir, Bethany Vailliant, "Russian Information Warfare: Implications for Deterrence Theory", *Strategic Studies Quarterly* 12, no. 3 (2018):70–89; Philip Davies and Hugo Rosemont, *Russia and Disinformation: Maskirovka* (Centre for Research and Evidence on Security Threats, 2019), accessed 13.04.2025 https://www.crestresearch.ac.uk/resources/russia-and-disinformation-maskirovka-full-report.; Marcel Schliebs and Philip N. Howard, "How Russia's Internet Research Agency Built Its Disinformation Campaign," *The Political Quarterly* 92, no. 3 (2021): 532–544, https://doi.org/10.1111/1467-923X.12690; Marek Posard, Shawn Smith, Logan Ma, and Miriam Matthews, *Russian Disinformation Efforts on Social Media* (RAND Corporation, 2022), accessed 13.04.2025 https://www.rand.org/pubs/research_reports/RR4373z2.html; Tomasz Smura, "Disinformation and Propaganda in Russia's Information Warfare. Concepts, Resources, Areas of Impact," *Athenaeum. Polish Political Science Studies* 83 (2024): 5–21 and Savvas Zannettou et al., "Disinformation Warfare: Understanding State-Sponsored Trolls on Twitter and Their Influence on the Web", *arXiv* preprint arXiv:1801.09288 (2018), accessed 13.04.2025 https://arxiv.org/pdf/1801.09288.

56 Marcin Antosiewicz, "Straszyła LGBT+ i imigrantami. Wspierała antyszczepionkowców i płaskoziemców. Tak przez lata działała rosyjska dezinformacja w Polsce", *Newsweek*, 15.12.2019, accessed 30.08.2024, https://www.newsweek.pl/swiat/rosyjska–dezinformacja–w–polsce–straszyla–lgbt–wspierala–antyszczepionkowcow/p87g1w5.

57 Shawn F Dorius, Carolyn J Lawrence-Dill, "Sowing the seeds of skepticism: Russian state news and anti-GMO sentiment", *GM crops & food* 9, no. 2 (2018):53–58.

58 Mark Lynas, "Anti–vaxxers and Russia behind viral 5G COVID conspiracy theory", *Alliance for Science*, 08.04.2020, accessed 30.08.2024, https://allianceforscience.org/blog/2020/04/anti-vaxxers-and-russia-behind-viral-5g-covid-conspiracy-theory.

vaccines),[59] disinformation, duplicating fake news, and promoting one-sidedly pro-Russian journalism. In 2012, Simonyan said that RT was needed "for much the same reason that the country needs a Ministry of Defense" and RT can "wage information warfare against the entire Western world" using "information as weapons." According to her, RT's strategic goal is to "gain" and "expand its audience" to use access to that audience at "critical moments"[60] in Russia favour.

In effect, in 2017, Facebook (now Meta) reported that as many as 126 million of its users had seen content from Russian disinformation campaigns on its platform.[61] Twitter, meanwhile, reported that it had detected 36,000 Russian bots spreading tweets related to the 2016 US elections.[62] Russian entities particularly active in spreading disinformation were the Internet Research Agency (ABI) and the Russian Institute for Strategic Studies (RISS). ABI spread posts and comments criticizing Alexei Navalny and his sponsors, the Russian opposition, the foreign policy of Ukraine and the United States, and leading politicians from these countries, and praising Vladimir Putin, Syrian dictator Bashar al-Assad, and the policies of the Russian Federation, Syria, Iran, and Belarus. ABI also used internet trolls to undermine trust in American political and media institutions and portray some politicians as incompetent.[63] RISS promoted, among other things, anti-Ukrainian, anti-LGBT, anti-abortion, anti-divorce, and critical views of the idea of sex education and attempted to influence the outcome of the 2016 US presidential election.[64]

---

59 Jonathan Vanian, "Russian disinformation campaigns are trying to sow distrust of COVID vaccines, study finds", *Fortune*, 23.07.2021, accessed 30.08.2024, http://www.fortune.com/2021/07/23/russian-disinformation-campaigns-are-trying-to-sow-distrust-of-covid-vaccines-study-finds.

60 Ben Nimmo, "Question That: RT's Military Mission", *Medium*, 08.01.2018, accessed 30.08.2024, https://medium.com/dfrlab/question–that–rts–military–mission–4c4bd9f72c88.

61 David Ingram, "Facebook says 126 million Americans may have seen Russia–linked political posts", *Reuters*, 30.10.2017, accessed 30.08.2024, http://www.reuters.com/article/us-usa-trump-russia-socialmedia-idUSKBN1CZ2OI.

62 Scott Neuman, "Russia Using Disinformation To »Sow Discord In West«, Britain's Prime Minister Says", *National Public Radio*, 14.11.2017, accessed 30.08.2024, http://www.npr.org/sections/thetwo-way/2017/11/14/564013066/russia-using-disinformation-to-sow-discord-in-west-britains-prime-minister-says.

63 Jarred Prier, "Commanding the Trend: Social Media as Information Warfare", *Strategic Studies Quarterly* 11, no. 4 (2017):50–85.

64 Grzegorz Rzeczkowski, "Służba Putina do zadań specjalnych zatruwa Polskę. Tak działa RISI", *Wyborcza.pl*, 20.04.2022, accessed 30.08.2024, https://www.wyborcza.pl/7,75398,28351668,sluzba–putina–do–zadan–specjalnych–zatruwa–polske–tak–dziala.html.

At the same time, however, assessing their actual impact on the democratic process in the United States is complex. Linvill and Warren researched the effects of ABI activities on the Twitter platform during the 2016 US presidential election. They examined primarily activities aimed at manipulating American public opinion and influencing the electoral process. Linvill and Warren identified five main types of ABI-run Twitter accounts, namely: accounts posing as right-wing users, supporting Donald Trump and criticising Hillary Clinton; accounts posing as liberal and left-wing activists, often promoting extreme left-wing views; accounts posing as independent news sources, disseminating both true and false information; accounts creating and promoting specific hashtags to increase the reach of particular narratives; accounts spreading disinformation and conspiracy theories, aimed at instilling fear and uncertainty. Different tactics have been used depending on the target group. Fake "Right-wing" accounts focused on mobilising traditionalistic voters by promoting pro-Russian and pro-Trump content. Accounts posing as liberal and left-wing activists sought to discourage such voters from voting for Hillary Clinton by promoting divisions among Democratic Party supporters. Content posted by ABI-affiliated accounts could reach millions of American Twitter users. However, compared to the overall amount of political content on the platform, the share of this content was relatively small. The study found no significant association between exposure to Russian influence campaign content and changes in political attitudes, polarization, or voting behaviour among US Twitter users.[65]

In their subsequent work, Linvill and Warren also described how the ABI coordinated its information operations, engaging social media users and building networks of influence. In "Engaging with Others: How the IRA Coordinated Information Operation Made Friends," Linvill and Warren analysed the ABI Twitter activity from 2015 to 2017. Firstly, the ABI replied foremost to posts by prominent users to gain visibility and credibility in some communities. Later, the ABI accounts published new content and added the external retweets, building a reputation as a news source and increasing their following. Then, the ABI focused on retweeting of external accounts ideologically aligned with the ABI narrative. By retweeting content from less popular,

65 Darren Linvill and Patrick Warren, "Troll Factories: Manufacturing Specialized Disinformation on Twitter", *Political Communication*, February 37, no 4 (2020): 1–21; see also Darren Linvill, Patrick Warren, and William Budak, "Talking to Trolls – How Users Respond to a Coordinated Information Operation," *Journal of Computer-Mediated Communication* 27, no. 1 (2021).

unverified accounts, the ABI sought to increase its reach and influence, which could lead to the organic spread of desired narratives.[66]

In their work, "Writers of the Storm: Who's Behind the Ongoing Production of Pro-Russian False Narratives," they analysed a disinformation campaign called Storm-1516 (running since August 2023), which. The campaign intensified in time of the US elections, focusing on spreading deep-fakes and fake news to discredit specific candidates, and achieved several viral successes in spreading pro-Kremlin narratives, like smearing Minnesota Governor Tim Walz. Storm-1516 used a network of websites and social media accounts working together to strengthen the disinformation and make it appear authentic. This campaign aimed to undermine trust in democratic institutions in the United States by spreading false narratives and causing social division.[67]

Even before the pandemic, bots and trolls controlled by Russia amplified disputes about vaccines, supporting both pro- and anti-vaccination sides.[68] The aim was to exploit and deepen existing social divisions to weaken public understanding of the importance of vaccination.[69] The COVID-19 pandemic has also created vocal and active political groups used in Russian disinformation activity.[70] American scientists from George Washington University examined a random sample of 1 793 690 tweets[71] (published between 2014 and 2017). It turned out that 93% of the published information was generated by accounts that could not be definitively categorised as bots or human users but exhibited behaviours suggestive of coordinated influence operations. As was

---

66 Darren Linvill and Patrick Warren, "Engaging with Others: How the IRA Coordinated Information Operation Made Friends," *Harvard Kennedy School (HKS) Misinformation Review*, 14.02,2020, accessed 30.03.2025 https://misinforeview.hks.harvard.edu/article/engaging-ira-coordinated-information-operation/.

67 Darren Linvill, Patrick Warren et al., *Writers of the Storm: Who's Behind the Ongoing Production of Pro-Russian False Narratives*, Clemson University Media Forensics Hub, October 2024, accessed 30.03.2025 https://open.clemson.edu/mfh_ci_reports/10.

68 Małgorzata Fraser, "Eksperci: rosyjska dezinformacja antyszczepionkowa rośnie w siłę wraz z wariantem Delta", *CyberDefence 24*, 09.08.2021, accessed 30.08.2024, https://www.cyberdefence24.pl/eksperci–rosyjska–dezinformacja–antyszczepionkowa–rosnie–w–sile–wraz–z–wariantem–delta.

69 Wesley R. Moy and Kacper Gradon, "COVID-19 Effects and Russian Disinformation Campaigns", *Homeland Security Affairs Journal* Special COVID-19 Issue (2020).

70 Isobel Cockerell, "Why anti–vaccine groups are running with Putin's narrative", *Coda*, 01.04.2022, accessed 30.08.2024, https://www.codastory.com/newsletters/putin–anti–vaxxers.

71 David A. Broniatowski et al., "Weaponized Health Communication: Twitter Bots and Russian Trolls Amplify the Vaccine Debate", *American Journal of Public Health* 108 (2018): 1378–1384.

pointed out, accounts they identified as "Russian trolls" published both pro-vaccine and anti-vaccine content. The idea here was not to promote a particular dogma but to spread disinformation, polarise society, and undermine trust in public institutions. The purpose was to build the impression that the vaccination discussion was controversial and divisive, even when the scientific data suggested otherwise, and so-called "Content polluters" and other bots concentrated on spreading anti-vaccine narratives, boosting conspiracy theories and enabling disinformation, and questioning the effectiveness of vaccines. Researchers found that Russian trolls often used the hashtag "#VaccinateUS" to post politically charged, provocative posts that linked the topic of vaccination to other social controversies (e.g., civil liberties, distrust of government, racism). In effect the activities of trolls and bots helped to manufacture fake balance in the public discourse, strengthening anti-scientific narratives and threatening public health by reducing trust in vaccinations.[72]

US government institutions have also pointed to Russia's apparent involvement in propagating disinformation and conspiracy narratives about vaccines and fifth-generation telephony. The 2020 US Department of State report, "Pillars of Russia's Disinformation and Propaganda Ecosystem," analysed the mechanisms and structures through which Russia conducts disinformation campaigns on a global scale, including in the US. It identifies the goals of Russian disinformation as undermining trust in democratic institutions, NATO, the European Union, and public health systems; breaking up the unity of the West by strengthening social divisions (including in the US); and enhancing Russia's geopolitical position without the need for conventional military confrontation. It identifies the main channels of Russia's activity in this matter, including Government-Sponsored Media (RT, Sputnik, and Ruptly), which spread conspiracy theories and narratives that undermine trust in Western governments, including the alleged dangers of vaccines and 5G technology; Narrative Proxy Websites (Global Research, Strategic Culture Foundation, SouthFront.) – websites and blogs that appear to be independent sources but replicate Russian narratives; Social Media Amplification – bots, trolls and fake accounts used to spread disinformation, polarise public debates and promote anti-vaccination and anti-5G narratives; Information Manipulators, intentionally mixing facts with lies or information taken out of context too, for example, distort scientific research on vaccinations and their alleged side

72 Caitlin O'Kane, "Russian trolls fueled anti-vaccination debate in U.S. by spreading misinformation on Twitter, study finds", *CBS News*, 31.05.2019, accessed 30.08.2024, https://www.cbsnews.com/news/anti-vax-movement-russian-trolls-fueled-anti-vaccination-debate-in-us-by-spreading-misinformation-twitter-study.

effects; and the use of Sympathetic Voices & "Useful Idiots," that is politicians, journalists, and activists who unwittingly (or intentionally) replicate Russian propaganda.[73]

Also the October 2020 US Department of Homeland Security "Homeland Threat Assessment" identified Russia as a primary actor conducting disinformation campaigns in the United States, including those related to public health and technology. According to the conclusions of the document, Russia, China, and Iran engaged in coordinated efforts to influence public opinion in the United States, manipulated political debate, undermined trust in democratic institutions, and boost confusion during the presidential election period. Russian actions were the most aggressive, including the spread of disinformation about vaccines, 5G technology, and the COVID-19 pandemic. DHS indicated that some states (especially Russia and China) have exploited the pandemic to spread disinformation in order to weaken the US response and polarize society (e.g., on the issue of masks, lockdowns, and vaccinations).[74]

Also, a July 16, 2020 alert issued by the Cybersecurity and Infrastructure Security Agency (CISA), in cooperation with the UK's National Cyber Security Centre (NCSC), Canada's Communications Security Establishment (CSE), and the National Security Agency (NSA), titled "Malicious Activity Targeting COVID-19 Research, Vaccine Development," discussed cyber activities during the COVID-19 pandemic that targeted institutions involved in research on the virus and vaccine development (research institutions, pharmaceutical companies, universities, and government organizations involved in the development of COVID-19 vaccines and treatments) and identified groups sponsored by Russia (as well as China, Iran, and North Korea) as perpetrators.[75]

Some findings of the above reports were confirmed by the research by Zannettou et al. (the activities of ABI and trolls, the reinforcement of chaos in public discourse as a result of their actions, coordinated and aimed at manip-

73 U.S. Department of State, *GEC Special Report: Pillars of Russia's Disinformation and Propaganda Ecosystem*, U.S. Department of State, 04.08.2020, archived March 31, 2021, accessed 13.04.2025, https://web.archive.org/web/20210331233359/https://www.state.gov/wp-content/uploads/2020/08/Pillars-of-Russia%E2%80%99s-Disinformation-and-Propaganda-Ecosystem_08-04-20.pdf.

74 U.S. Department of Homeland Security, *Homeland Threat Assessment*, Washington, DC: Department of Homeland Security (October 2020), accessed 30.03.2025 https://www.dhs.gov/sites/default/files/publications/2020_10_06_homeland-threat-assessment.pdf.

75 Cybersecurity and Infrastructure Security Agency, *Malicious Activity Targeting COVID-19 Research, Vaccine Development*, 16.07.2020, accessed 30.03.2025 https://www.cisa.gov/news-events/alerts/2020/07/16/malicious-activity-targeting-covid-19-research-vaccine-development.

ulating public opinion). However, they noted the limited influence of the trolls themselves, indicating a much more significant role of the Russian public media.[76] The complexity of the Russians' way of influencing the vaccine debate in the US is demonstrated by research published in *The American Journal of Public Health*. The researchers analysed 2.82 million tweets published by 2,689 IRA-linked accounts between 2015 and 2017. They used unsupervised machine learning and network analysis to identify profiled fictitious individuals (with fabricated American identities) – accounts that consistently shared similarly themed content. The messages they created were aimed at different segments of American society, with other tones.[77]

A similar phenomenon is occurring in Poland. Studies of the anti-vaccination movement's message have found a noticeable overrepresentation of anonymous users in comments under videos about vaccination on YouTube[78] and Facebook.[79] Similar connections between the spread of anti-vaccination narratives through Russian disinformation channels have also been described in Australia,[80] the USA,[81] Scotland,[82] Germany,[83] Austria,[84]

76 Zannettou et al., "Disinformation Warfare," 2019.

77 Dror Walter, Yotam Ophir, and Kathleen Hall Jamieson, "Russian Twitter Accounts and the Partisan Polarization of Vaccine Discourse, 2015–2017," *American Journal of Public Health* 110, no. 5 (2020): 718–24.

78 Maciej Piasecki, "Rosyjskie trolle nakręcają antyszczepionkową histerię w Polsce? Wysyp anonimowych kont", *OKO.Press*, 31.01.2018, accessed 30.08.2024, https://www.oko.press/rosyjskie-trolle-nakrecaja-histerie-antyszczepionkowa-w-polsce-wysyp-anonimowych-kont.

79 Anna Mierzyńska, "Rosyjski troll sieje chaos na polskim Facebooku. Jest wśród antyszczepionkowców", *OKO.Press*, 14.10.2018, accessed 30.08.2024, https://www.oko.press/rosyjski-troll-sieje-chaos-na-polskim-facebooku-jest-wsrod-antyszczepionkowcow.

80 Josh Butler, Sarah Martin, "Australian online anti-vaccine groups switch to Putin praise and Ukraine conspiracies", *The Guardian*, 01.03.2022, accessed 30.08.2024, https://www.theguardian.com/australia-news/2022/mar/02/australias-anti-vaccine-groups-switch-focus-to-putin-praise-and-ukraine-conspiracies.

81 Brian Contreras, Wendy Lee, "Putin targets lots of Americans with disinformation. One example? Anti-vaccine groups", *Los Angeles Times*, 25.02.2022, accessed 30.08.2024, https://www.latimes.com/business/technology/story/2022-02-25/heres-what-putins-disinformation-war-looks-like-on-the-internet.

82 The Ferret, "Pro-Russian propaganda spreading in Scots anti-vax group", *The Ferret*, 20.03.2022, accessed 30.08.2024, https://www.theferret.scot/pro-russia-propaganda-anti-vaccine-white-rose.

83 Bret Schafer, "RT Deutsch Finds a Home with Anti-Vaccination Skeptics in Germany", *Alliance for Securing Democracy*, 05.10.2021, accessed 30.08.2024, https://securingdemocracy.gmfus.org/rt-deutsch-youtube-antivaccination-germany.

84 Aleksandar Brezar, "From anti-vax to pro-Putin, how conspiracy theorists are backing Russia's war in Ukraine", *Euronews*, 13.05.22, accessed 30.08.2024, https://www.euro

and Italy.[85] In these and many other countries, anti-vaccination groups have been used to share and spread anti-Ukrainian, pro-Russian, and different views supporting the Kremlin's foreign policy.

EU institutions have also spoken out about Russia's negative influence on public discourse on the topics of the vaccines, COVID-19, and 5G. In June 2020, the European External Action Service (EEAS) and the European Commission published a joint communication entitled "Tackling COVID-19 Disinformation – Getting the Facts Right," which analysed the spread of fake news and disinformation related to the COVID-19 pandemic. Russia and China were explicitly identified as the source or reinforcement of their presence in public discourse. These actions aimed to weaken social cohesion and undermine trust in public institutions.[86] Also in a special separate report by the European External Action Service (December 2020–April 2021), the institution analysed narratives related to the pandemic, pointing to Russia's disinformation activities and this state support for the conspiracy theories. In the period under review, the Russian side not only promoted its vaccine (Sputnik V) but also undermined trust in Western preparations, using state-controlled media and social media for both purposes. They also promoted conspiracy theories linking the pandemic with biological warfare and 5G technology.[87] A good summary of the analyses of EU institutions on the Russian influence on information about vaccines is the report by the Center for European Policy Analysis (CEPA), "Jabbed in the Back: Mapping Russian and Chinese Information Operations During the COVID-19 Pandemic." It analyses in detail Russia's strategy in the matter of building the image of vaccines and the pandemic, pointing to the aggressive promotion of its own Sputnik V vaccine, combined with the dissemination of doubts about the effectiveness and safety of Western prepa-

---

news.com/my–europe/2022/05/13/from–anti–vax–to–pro–putin–how–conspiracy–theorists–are–backing–russia–s–war–in–ukraine.

85 Hannah Roberts, "Italy and Russia: A love affair that hasn't quite ended", *Politico*, 06.04.2022, accessed 30.08.2024, https://www.politico.eu/article/italy–russia–love–affair–no–end.

86 No author, "Tackling COVID-19 Disinformation: Getting the Facts Right", *European Commission and European External Action Service*, 10.06.2020, accessed 30.03.2025 https://commission.europa.eu/strategy-and-policy/coronavirus-response/fighting-disinformation/tackling-coronavirus-disinformation_en.

87 No author, "EEAS SPECIAL REPORT UPDATE: Short Assessment of Narratives and Disinformation Around the COVID-19 Pandemic (UPDATE DECEMBER 2020 – APRIL 2021), *European External Action Service*, 28.04.2021, https://euvsdisinfo.eu/uploads/2021/04/EEAS-Special-Report-Covid-19-vaccine-related-disinformation-6.pdf.

rations such as Pfizer and AstraZeneca using social media, influencers, memes and animations.[88]

The Australian Senate Committee on Foreign Interference through Social Media (2023) linked Russia to the spread of narratives about the harmfulness of vaccines and 5G. It also cited foreign actors (including Russia) using the pandemic as an opportunity to spread disinformation aimed at deepening social divisions and undermining trust in scientific and public health institutions, as well as their spreading of false information about COVID-19, vaccines and treatments, amplifying anti-vaccination narratives, supporting local movements opposing mandatory vaccinations in Australia, and linking conspiracy narratives about 5G to COVID-19, including the alleged "induction of coronavirus symptoms by 5G waves."[89] This situation is confirmed by the Australian Strategic Policy Institute (ASPI) report "COVID-19 Disinformation and social media manipulation: pro-Russian Vaccine Politics Drives New Disinformation Narratives" from August 2020, which analysed how Russian vaccine disinformation is penetrating the international information ecosystem, including Australian anti-vaccine groups on Facebook. It describes, for example, a case in which a message published on July 17, 2020, on the website of the so-called Luhansk People's Republic, was infiltrated into these groups. According to this fake news, American virologists conducted COVID-19 vaccine tests on Ukrainian volunteers in Kharkiv (an area controlled by the Ukrainian government), including soldiers, five of whom allegedly died from it.[90]

Canadian public institutions have also pointed to Russia's involvement in disinformation campaigns aimed at manipulating public opinion in the country. It is mentioned in the report of The Standing Committee on Public Safety and National Security,[91] the report of the Canadian Security Intelligence Ser-

---

88 Ben Dubow, Edward Lucas and Jake Morris, "Jabbed in the Back: Mapping Russian and Chinese Information Operations during the COVID-19 Pandemic", *Center for European Policy Analysis*, 02.12.2021, accessed 30.03.2025 https://cepa.org/comprehensive-reports/jabbed-in-the-back-mapping-russian-and-chinese-information-operations-during-the-covid-19-pandemic.

89 Senate Standing Committees on Finance and Public Administration, *Foreign Interference through Social Media*, Parliament of Australia, August 2023, accessed 30.03.2025 https://www.aph.gov.au/Parliamentary_Business/Committees/Senate/Foreign_Interference_through_Social_Media/ForeignInterference47/Report.

90 Elise Thomas, Albert Zhang and Emilia Currey, *COVID-19 Disinformation and Social Media Manipulation: Pro-Russian Vaccine Politics Drives New Disinformation Narratives,* International Cyber Policy Centre, 2020, accessed 30.03.2025 https://apo.org.au/node/307763.

91 The Standing Committee on Public Safety and National Security, *Up to the Task: Strengthening Canada's Security Posture in Relation to Russia* (Ottawa: House of Com-

vice,[92] and the Global Affairs Canada Statement.[93] However, while these documents do not mention Russia's role in disinformation related to vaccines and the pandemic, the Report of the Canadian Security Intelligence Service and Report of the National Security and Intelligence Committee describe it directly. The first document indicated that Russian state media and affiliated internet platforms were spreading conspiracy theories and false information about the effectiveness and safety of vaccinations using social media, often through anonymous accounts and bots.[94] The second document accused Russia, China, and Iran of spreading disinformation about COVID-19, and Russia itself of amplifying disinformation about doubts about the effectiveness and safety of vaccines developed in Western countries (false information about side effects and questioning the results of clinical trials).

The German Ministry of the Interior (BMI) also pointed to the governments of Russia and China as being responsible for spreading fake news and misinformation about vaccines developed in Germany, the United States, and Great Britain.[95] Several scientific studies have also pointed to the links between narratives questioning vaccines in Germany and Russian narratives or Russia itself. For example, Jarynowski found that as many as 62% of users with pro-Kremlin sympathies engaged in anti-vaccination discussions. They were also 51 times more likely to participate in both anti-lockdown and anti-vaccination groups compared to anti-Kremlin users. This result indicates a very high correlation between pro-Russian and anti-vaccination views in the studied group of German Twitter users. Such results suggest that German Twitter users supporting the Kremlin may play a significant role in spreading disinformation

---

mons Canada, 2023), accessed 13.04.2025 https://www.ourcommons.ca/DocumentViewer/en/44-1/SECU/report-12.

92 Canadian Security Intelligence Service, *Who Said What? The Security Challenges of Modern Disinformation* (Ottawa: CSIS, 2017), accessed 13.04.2025 https://www.canada.ca/content/dam/csis-scrs/documents/publications/disinformation_post-report_eng.pdf.

93 Global Affairs Canada, Foreign Interference Threats to Canada's Democratic Process (Ottawa: Government of Canada, 2021) accessed 13.04.2025 https://www.canada.ca/en/security-intelligence-service/corporate/publications/foreign-interference-threat-to-canadas-democratic-process.html.

94 National Security and Intelligence Committee of Parliamentarians (NSICOP), *Annual Report 2020* (Ottawa: Government of Canada, April 2021) accessed 30.03.2025 https://www.nsicop-cpsnr.ca/announcements/ann-2021-04-09/ann-2021-04-09-en.html.

95 Federal Ministry of the Interior, Building and Community (BMI), "Disinformation Related to the Coronavirus", *Federal Ministry of the Interior, Building and Community*, no data, accessed 10.10.2025 https://www.bmi.bund.de/SharedDocs/schwerpunkte/EN/disinformation/disinformation-related-to-the-coronavirus.html.

and conspiracy theories about the COVID-19 pandemic.[96] A study of over 400,000 posts from over 1,000 users on platforms such as Facebook, Twitter, Instagram, and Telegram conducted by the Institute for Strategic Dialogue between December 21, 2020 and April 5, 2021 indicated a similar correlation. Leaders of movements such as the "Querdenker" and far-right extremists regularly supported each other, creating a sense of scientific legitimacy for their views while promoting narratives about deaths after vaccination, health damage caused by vaccines, fears of forced vaccination and discrediting experts.[97]

Ukraine was a country where Russian disinformation played a critical role. Firstly, Russia was directly involved in the conflict with that country and used it to tip the scales in its favour in this war. Secondly, it was significantly facilitated by the widespread knowledge of Russian in Ukraine and the numerous contacts between Ukrainians and Russians stemming from the period of the existence of the Soviet Union. As a result, Russian disinformation, including that related to vaccinations, was noticeable in Ukraine. As a culturally close country and a neighbour of Poland, this context is essential for a better view of the Polish landscape. Ukrainian Centre for Strategic Communication published a report describing a network of media, including channels on Telegram controlled or associated with Russia, spreading false information about the allegedly harmful effects of vaccinations. The same channels in other regions spread attacks on opponents of vaccinations, but they did so in areas occupied by Russia, such as Crimea or Donbas. Disinformation was amplified by bots, fake accounts, and automated commenting systems, creating a surge in anti-vaccine narratives, including one Orthodox church that organised an anti-vaccine protest in Kyiv on October 19, 2021.[98] The authors of the 2021 study reached similar conclusions. They point out that Russian disinformation campaigns in Ukraine included the promotion of anti-vaccination content and undermining trust in the health measures taken by the government. The method of such action was spreading fake news, which undermined the

---

96 Andrzej Jarynowski, "Pro-Kremlin German Twitter Users Are More Likely to Be Involved in both Anti-Lockdown and Anti-Vaccine Discourse Than Anti-Kremlin Users", preprint, *Social Science Research Network*, 01.04.2022, accessed 13.04.2025 https://papers.ssrn.com/sol3/papers.cfm?abstract_id=4079045.

97 Hannah Winter et al., "Disinformation Overdose: A Study of the Crisis of Trust among Vaccine Sceptics and Anti-Vaxxers", *Institute for Strategic Dialogue*, 06.06.2021, accessed 13.04.2025 https://www.isdglobal.org/wp-content/uploads/2021/07/Disinformation-Overdose3.pdf

98 National Security and Defense Council of Ukraine. *Report by the Center for Countering Disinformation: Anti-Vaccination Propaganda* (Kyiv: NSDC, 2021) accessed 13.04.2025 https://www.rnbo.gov.ua/files/2021/

validity of preventive measures, including vaccines, to increase information chaos and reduce trust in the authorities.[99] Keegan draws similar conclusions, comparing the cases of Serbia and Ukraine and pointing out that in the case of Ukraine, Russian portals such as Ukraina.ru spread a wide range of contradictory narratives, mixing information about vaccinations with geopolitical topics and criticism of the Ukrainian authorities. According to the author, such actions aimed to sow confusion and undermine public trust.[100] Fursai and Danylenko reach similar conclusions.[101]

The case of 5G was no different. Since 2018, the American branch of RT has reported in an alarmist tone dozens of times about alleged threats resulting from the introduction of 5G technology, using terms such as "5G apocalypse" or "5G dangerous experiment on humanity" that "5G could kill you" and suggesting that the technology could cause brain cancer, infertility, autism, heart cancer, Alzheimer's disease, nosebleeds, and other health problems.[102] The programmes have featured slogans such as: "How to Survive the Danger of 5G," "5G Tech is a Crime Under International Law," "Complete Madness: Telecom Industry Ignores the Dangers of 5G," "Could 5G Expose More Children to Cancer?"[103] All mobile phones use radio waves. RT America, however, called the signals "radiation," associating them with powerful rays at the other end of the electromagnetic spectrum, such as X-rays and ultraviolet rays, which, in high doses, can damage DNA and cause cancer. RT also selectively highlighted the most sensational claims, giving a few fringe opponents of wireless technology a new very visible forum. The hosts reported that doctors, scientists, and activists were calling for a ban on the technology. RT programmes featured primarily opponents of 5G, including

---

99 Sonny Patel et al., "The Landscape of Disinformation on Health Crisis Communication During the COVID-19 Pandemic in Ukraine: Hybrid Warfare Tactics, Fake Media News and Review of Evidence." *Journal of Science Communication* 19, no. 5 (2020): A02.

100 Katrina Keegan, "Clarity for Friends, Confusion for Foes: Russian Vaccine Propaganda in Ukraine and Serbia," *Harvard Kennedy School Misinformation Review* 3, no. 3 (25.04.2022), accessed 13.04.2025, https://misinforeview.hks.harvard.edu/article/clarity-for-friends-confusion-for-foes-russian-vaccine-propaganda-in-ukraine-and-serbia.

101 Oleksandra Fursai and Serhiy Danylenko, "»Vaccinodemic« as a component of the global hybrid conflict between democracy and autocracy: the case of Ukraine, *Rocznik Instytutu Europy Środkowo-Wschodniej* 20, vol 2, (2022), 19–45.

102 William J. Broad, "Your 5G Phone Won't Hurt You. But Russia Wants You to Think Otherwise", *The New York Times*, 12.05.2019, accessed 30.03.2025 http://www.nytimes.com/2019/05/12/science/5g–phone–safety–health–russia.html.

103 Ibid.

those who have earlier accused cell phones, power lines, and other everyday sources of electromagnetic waves of worsening health and causing disease. Most of this people papers appeared not in reputable scientific journals but in obscure reports, publications, and self-published books. These "experts" also tend to cite each other's research or their own papers.[104] Russia similarly exploited the social media accounts of these experts. One example is Oleg A. Grigoryev, a self-proclaimed "radiobiologist" and chairman of the Russian National Committee on Non–Ionizing Radiation Protection (RusCNIRP). Despite having only 1,197 followers on Twitter (currently "X"),[105] his tweets were retweeted by accounts and link to frightening stories about 5G "radiation." Hundreds of blogs and websites replicated such alarmist reports, rarely, if ever, noting their Russian origins.

The American Office of the Director of National Intelligence described RT in a 2017 report as "the Kremlin's primary international propaganda channel." Experts of this institution say Moscow's goal is to destabilize the West by undermining trust in democratic leaders, institutions, and political life. In their opinion Russian government is trying to amplify and intensify the debate between pro- and anti-vaccine activists, 5G supporters, and opponents, contributing to the illusion of a large number of conflicting opinions on a given issue while at the same time increasing in the Western societal conflicts and distrust.[106] In effect of this approach RT gives marginal groups a megaphone and visibility within a false pluralism. Previous campaigns have aimed to attack fracking, vaccinations, and genetically modified organisms. One programme called GMO tomatoes a "beautiful poison."[107] The result is increasing political polarization, stronger emotions manifesting in public discourse, and the paralysis of state action, for example, in combating the pandemic.[108]

The issue of 5G technology and related conspiracy narratives were also widely indicated by reports of public institutions, think tanks, and non-governmental organisations in the US and the European Union as a space of

---

104 Ibid.

105 Oleg A. Grigoriev profile in "X" see: https://twitter.com/O_Grigoriev.

106 Katharine Kirk, "How Russia Sows Confusion in the U.S. Vaccine Debate", *Foreign Policy*, 09.04.2019, accessed: 13.04.2025. http://www.foreignpolicy.com/2019/04/09/in–the–united–states–russian–trolls–are–peddling–measles–disinformation–on–twitter.

107 William J. Broad, "Your 5G Phone Won't Hurt You. But Russia Wants You to Think Otherwise", *The New York Times*, 12.05.2019, accessed 30.08.2024, http://www.nytimes.com/2019/05/12/science/5g–phone–safety–health–russia.html.

108 No author, "How Russia Influences Vaccination", *Center for Strategic Communication*, 12.11.2021, accessed 30.08.2024, http://www.spravdi.gov.ua/en/how–russia–influences–vaccination.

influence for Russian disinformation. According to these documents, the mechanism of this action was very similar to that used by Russia in the context of vaccines. Russian efforts to undermine confidence in the US rollout of 5G technology were underway by 2019 and involved several conduits of disinformation, including Russian state-sponsored English-language news outlets, English-language reports from Russian state-sponsored "think tanks," and a covert social-media campaign. The principal benefits of Russian disinformation techniques are the ability to craft different – and even opposing – narratives to appeal to various target audiences and take advantage of the "media multiplier effect," the strengthened ability to influence discourse by simultaneously deploying disinformation on multiple media platforms.[109]

Russia's links to 5G disinformation were described in a study by Sten Hansson and his colleagues from the University of Tartu. According to the researchers, Russia, as a state, is engaging in the global debate about 5G through strategic narratives that aim to shape public opinion worldwide. Strategic narratives are a set of stories that states and organisations use to influence the perceptions and attitudes of others. In the case of 5G, Russia is presenting these technologies not only as a threat but also as a tool used by other states, particularly the United States, to harm other countries. 5G was portrayed by Russian media and Russian controlled social accounts as a dangerous technology. These narratives often included conspiracy narratives suggesting that 5G is not only a tool for mass surveillance but also has serious health consequences. Russian propaganda agencies are using social media to spread disinformation about 5G, often using internet trolls and bots to help spread manipulated information quickly.[110] Russian influence on 5G-related disinformation is also indicated by studies by Meese et al.[111] and Ventsel et al.[112]

---

109 Network Contagion Research Institute (NCRI), *The Future of Disinformation Operations: Emerging Narratives and Trends in Online Manipulation*, NCRI February 2024, accessed 13.04.2025 https://networkcontagion.us/reports/the-future-of-disinformation-operations/.

110 Sten Hansson et al., "Discourses of Blame in Strategic Narratives: The Case of Russia's 5G Stories," *European Security* 32, no 1(2022),62–84. doi:10.1080/09662839.2022.2057188.

111 Meese, James et al. "COVID-19, 5G conspiracies and infrastructural futures." *Media International Australia* 177, no 1 (2020): 30–46. doi:10.1177/1329878X20952165.

112 Andreas Ventsel, Mari-Liis Madisson, and Sten Hansson, "Russia's Strategic Blame Narratives: Comparative Analysis of Domestic and International Media Coverage about 5G," in *Disinformation and Digital Media as a Challenge for Democracy*, ed. Michael Kramer et al. (Cham: Springer, 2021), https://doi.org/10.1007/978-3-030-73955-3_14.

The dissemination of 5G conspiracy narratives by Russia or its affiliated media is also indicated by reports from the Institute for Strategic Dialogue,[113] Institut Montaigne and Institute For Strategic Dialogue,[114] US Department of State,[115] an expert hub operating under the auspices of NATO and the EU European Centre of Excellence for Countering Hybrid Threats[116] European External Action Service[117] and reports of the EUvsDisinfo – East Stratcom Task Force, a team of experts with a background mainly in communications, journalism, social sciences and Russian studies, and a part of the EU's diplomatic service.[118]

In response to the growing disinformation from Russia and its public media, the European Union and NATO have created special units to analyse and debunk fabricated messages produced on the Kremlin's orders.[119] Access to RT was blocked in Ukraine in 2014 following Russia's annexation of Crimea. Latvia and Lithuania implemented similar blocks in 2020. Germany blocked RT in February 2022. Following the Russian invasion of Ukraine in 2022, Poland, then the entire European Union and Canada, announced they were formally blocking RT as well. Social media sites followed, blocking external links to RT's website and restricting access to its content. Microsoft removed RT from its app store and lowered its search rankings on Bing, while Apple

113 Institute for Strategic Dialogue, *Behind the Curtain: Unveiling Pro-Kremlin Ecosystems in Europe* (London: ISD, January 2025), accessed 13.04.2025 https://www.isdglobal.org/wp-content/uploads/2025/01/Behind-the-curtain-Unveiling-pro-Kremlin-ecosystems-in-Europe.pdf.

114 Institut Montaigne, *Information Manipulations Around Covid-19: France Under Attack*, Institut Montaigne, Policy Paper, June 2021, accesed 13.04.2025 https://www.institutmontaigne.org/ressources/pdfs/publications/information-manipulations-around-covid-19-france-under-attack-policy-paper_0.pdf.

115 U.S. Department of State, *Pillars of Russia's Disinformation and Propaganda Ecosystem ...*

116 Hybrid CoE, (European Centre of Excellence for Countering Hybrid Threats), *European Centre of Excellence for Countering Hybrid Threats*, Working Paper 29, March 2024, accessed 13.04.2025, https://www.hybridcoe.fi/wp-content/uploads/2024/03/20240306-Hybrid-CoE-Working-Paper-29-The-impact-of-Kremlin-disinformation-WEB.pdf.

117 European External Action Service (EEAS), "EEAS Special Report Update: Short Assessment of Narratives and Disinformation around the COVID-19 Pandemic", EEAS June 2020, accessed 13.04.2025, https://www.eeas.europa.eu/delegations/georgia/eeas-special-report-update-short-assessment-narratives-and-disinformation-around-covid-19-pandemic_en.

118 EUvsDisinfo, "Figure of the Week: 5G," *EUvsDisinfo*, 12.04.2021, accessed 13.04.2025 https://euvsdisinfo.eu/figure-of-the-week-5g.

119 Neil MacFarquhar, "A Powerful Russian Weapon: The Spread of False Stories", *The New York Times*, 29.08.2016, accessed 30.08.2024, http://www.nytimes.com/2016/08/29/world/europe/russia–sweden–disinformation.html.

removed the RT app from all countries except Russia. Following the Russian invasion of Ukraine in late February 2022, European Commission President Ursula von der Leyen announced a block on Sputnik in the whole European Union. Facebook, Instagram, TikTok, and YouTube have blocked the news agency's content for European Union users. Also Reddit has blocked outgoing links to its websites, and Microsoft and Apple Inc. have removed Sputnik apps from the Microsoft Store and App Store.

CHAPTER 2

The 5G Files

Poland's Anti-5G Movement: Activists, Major Organizations, and "Experts"

Conspiracy narratives about 5G unite people from very different parts of the political spectrum. On the one hand, they attract the right-wingers, who see new telecommunication technologies as part of an attack of the government on individual freedom. On the other hand, these narratives are also close to the anti-vaccination community and leftists who distrust big corporations. You can also find libertarians, Neo-fascists, and ultra-conservatives among the supporters of conspiracy narratives about 5G. However, among the hardcore supporters, belief in the narrative about the harmfulness of 5G is so central that it is a crucial element of their faith. Any political connections they have are entirely secondary, ad hoc, and dependent of the support by the particular political milieu of their central axis of interest: narrative about harmfulness of 5G.

In Poland, the most famous anti-5G activist is Ewa Pawela. Her views combine technophobic, anti-American, anti-vaccination, anti-Semitic, and Russian-inspired themes. She claims to be a victim of what she calls "electronic mind control." She believes that governments, in concert with corporations, are using actions against her and many other people to influence their decisions, choices, and health and even to induce or encourage suicide. In her opinion, this is done using modern electronic and telecommunications technologies.[1]

The US government plays an essential and demonic role in her story about 5G. Pawela combines her vision of new telecommunications technologies with the conspiratorial narrative about the so-called "electromagnetic weapon," or HAARP. In her opinion, it was used to trigger climatic phenomena, while simultaneously influencing the moods and thoughts of entire populations.[2] In her opinion, the US Army has such a tool and has been conducting experiments on humans since 1995. Pawela warns against American soldiers who, on

1 Rafał Mossakowski, "Ewa Pawela: Kontrola umysłu trafiła pod strzechy – aktualne ciekawe książki i ważne teksty artykułów", *Youtube*, 12.09.2023, accessed 06.09.2024, https://www.youtube.com/watch?v=HDybkiodF-4.

2 Arkadiusz Dziermański, "Krzyczą, że 5G zabija i twierdzą, że mają dowody. Podsumowanie krakowskiego zlotu przeciwników 5G", *Telepolis*, 01.03.2019, accessed 06.09.2024,

 | DOI:10.1163/9789004747852_004

bases outside the US, are allegedly en masse kidnapping children, taking part in satanic rituals, and drinking their blood.[3] According to Pawela, the introduction of 5G technology is intended to keep every person "on a leash." After the Internet of Things introduction, there will be a short transition period. After that, the era of the Internet of People will come: "A person without a chip and an electronic access code will not be able to use any device in our environment. Then the prophecy from the apocalypse about the mark of the beast may come true," Pawela says in her "lectures."[4]

Pawela intertwines her technophobic stories with anti-Semitic narratives, finding Jewish influences at the foundation of new, dangerous technologies. She considers Dr Naftali Berg, "long-time director of the Department of Scientific Research in the US Defense Department and advisor to several successive US presidents," to be the "father and greatest promoter of this sinister installation," which she believes gives the United States unquestionable hegemony in the world, "while also being a geophysical, climatic, and psychotronic weapon." Pawela believes that Dr Berg was "a rabbi, a member of the Chabad Lubavitch Hasidic movement, called by his coreligionists the pride or super weapon of Chabad" and that, in her opinion, "we owe him the introduction of the principle of the so-called dual use of military technologies, i.e., the possibility of using them also against civilians in times of peace."[5]

Pawela's struggle against 5G is combined with anti-vaccination views and a conspiratorial narrative about depopulation. She believes that 5G, together with mandatory vaccinations, will lead to another pandemic and result in the depopulation of the Earth. In her opinion, the vaccine introduces nanoparticles into the body, which, using low-frequency waves sent, among others, by Wi-Fi, enable the control of the human mind and behaviour. Ultimately, 5G technology will cause the disappearance of life on Earth within three generations, and the whole process started in 1973 during a conference held in Warsaw,[6] where it was "predicted" that Poland would be destroyed by microwaves.[7] Ewa Pawela also talks about the 17 quintillion dollars that telecommu-

---

https://www.telepolis.pl/artykuly/wydarzenia/siec-5g-w-polsce-szkodliwa-technologia-dowody.

3 Jerzy Targalski, "Archipelag szurów (2)", *Jerzy Targalski*, za: https://jozefdarski.pl/13594-archipelag-szurow-2.

4 Dziermański, "Krzyczą, że 5G zabija".

5 Ewa Pawela, "5G narzędzie masowej kontroli umysłu – wykład", *Stowarzyszenie Stop Zet*, 04.03.2019, accessed 06.09.2024, https://stopzet.pl/2019/03/04/5g-narzedzie-masowej-kontroli-umyslu-wyklad.

6 Targalski, "Archipelag szurów".

7 Dziermański, "Krzyczą, że 5G zabija".

nications companies will divide among themselves after introducing the 5G network.[8] Pawela claims that COVID-19 is a code transmitted by the New World Order[9] through 5G antennas, which our genes receive to replicate the virus genome under its influence, controlling our neurons and turning us into zombies obedient to the world government by controlling our minds.[10]

There is also a Russian thread in her views. In her lectures, Pawela refers to experts from that country, including biologist Igor Nikulin. He told RT that the coronavirus is a biological weapon created by the US or Israel to reduce the Earth's population by 90 percent. Nikulin also said the same thing in the main news program of the All-Russian State Television and Radio Company "Vremya".[11]

Pawela often collaborates with Witold Hake. He is a historian by education, but his main occupation is the fight against "mind control." He also believes that he is a so-called "target," a person attacked, for example, with electric beams because "he knows too much." During his speeches, he shows photos of wounds on his body, alleged effects of this attacks.[12] He also cites Russian sources in his public statements supporting conspiratorial narratives about new telecommunications technologies, such as an interview published by the newspaper *Konsomolskaya Pravda* with KGB General Boris Ratnikov,[13] which was supposed to confirm information about the use of psychotronic weapons by superpowers, etc., experiments on humans, similar to the one of which Hake himself claims to be a "victim".[14]

Another Polish "5G whistleblower" is Hanna Kazahari, who has lived in Tokyo since 1991. She works in the IT industry. Kazahari claims that the COVID-19 virus does not exist. In her opinion, the world government is turning up the 5G antennas so that they modify oxygen, making it incapable of being absorbed by haemoglobin. As a result, people suffocate and die. According to

---

8 Ibid.

9 New World Order, the kind of world government that many supporters of conspiracy narratives believe in.

10 Ewa Pawela, "Kamienie Milowe NWO (WTC 9/11)", *CDA*, 11.09.2023, accessed 06.09.2024, https://www.cda.pl/video/17032762f0.

11 Anna Mierzyńska, "Epidemia dezinformacji w Polsce. Część fałszywek wskazuje na aktywność Rosji", *OKO.Press*, 14.03.2020, accessed 06.09.2024, https://oko.press/koronawirus–wywolal–dezinformacyjna–powodz.

12 Ewa Pawela and Witold Hake, "Międzynarodowy Dzień Targeta w Polsce", *CDA*, 29.08.2022, accessed 06.09.2024, https://ebd.cda.pl//620x368/1184288056/vfilm.

13 Светлана Кузина, "Психотронное оружие все–таки существует?", *KP*, 09.08.2007, accessed 06.09.2024, https://www.kp.ru/daily/23948.4/71317.

14 Witold Hake, "Generał KGB i agent Mosadu o mind control – Witold Hake (25)", *YouTube*, 11.12.2020, accessed 06.09.2024, https://www.youtube.com/watch?v=IX5YjIjuyWw.

Kazahari, the pandemic is a "typical cover-up," and the goal is top-down depopulation. Her views are a mixture of various conspiracy narratives held together by an anti-Semitic component. Kazahari believes that the Illuminati, the Rothschilds, the Jews, and the Freemasons are working together to build a global New World Order. In her narrative about COVID-19, she attributes the pandemic to the actions of Zionists and the Rothschild and Rockefeller families, additionally identifying them with members of the World Health Organization (WHO), who "want to control all the people of the world, want to have people on chains, in cages, slaves. They want to turn the whole world into a big, beautiful concentration camp."[15] In her view, fascism was German national socialism originating from the ideology of the Rothschilds and Rockefellers, just like communism, Marxism, and Leninism, and all of these concepts were supported and financed by them. "All of these total absurdities, gulags, collective farms were the idea of the Rothschilds, the Rockefellers [...] – so that we would be like cattle led to slaughter. Concentration camps were their idea. They got bored, and for a few years, they made concentration camps for Muslims for Palestinians. Now it's time for Europe, the United States, and Japan."[16] In her opinion, the Fourth Reich is currently being introduced by Bill Gates and Zionists.

Kazahari's narratives of a Jewish conspiracy are combined with other anti-Semitic motifs, additionally mixed with narratives typical of the American QAnon movement. Kazahari believes that currently, rituals of drawing and consuming baby blood are taking place among global elites, including Hillary Clinton, referring to the so-called "pizzagate," a conspiracy narrative spread by the QAnon movement, and adding her thread to it, "a cookbook written by Dr. Mengele together with Hillary Clinton, where people are cut up".[17]

Hake and Pawela worked closely with the first Polish organization, initially fighting against "electromagnetic radiation" and later 5G, the STOPZET STOP Organized Electronic Torture Association. Until 2015, it operated as a Facebook group and brought together people complaining about alleged "mind theft" and "remote poisoning." The group is full of posts like this one by A.K., which describes the alleged torture to which she is subjected by an unknown group that persecutes her. A.K. writes: "All day, especially since noon, and now at night, they poison me with heavy metals. They approached from the

15 CEPowiśle, "Hanna Kazahari (Tokio): Plandemia – zaplanowana przez globalistów i syjonistów fałszywa flaga wirusa", *BitChute*, 08.04.2020, accessed 06.09.2024, https://www.bitchute.com/video/ccEa9MRbElOQ.

16 Ibid.

17 Ibid.

field and behind the bushes when I was in the forest. As soon as they approached, I immediately felt sick; my head hurt, my intimate organs, and I felt inside what they were putting in me. There was a moment of peace when I moved away from them, but then they drove up. You could hear one of them had a motorbike. He was driving behind me. They attack for so many hours that you can't do anything. I feel sick all the time. That's how it is. You can't even throw it up to make it easier ... Lately, my hair has been falling out a lot, and they are dry and thin. They also continue to add nanoparticles to my food in stores. No matter what store I go to, they do it. And they are particularly keen on meat. Recently, I bought meat in a store full of white particles. They really want me not to eat meat ... They also continue to hide bread from me. Every time I go to the store, I hear: There's no more bread. They only leave some toast leftovers, sprinkled with nanoparticles" (original spelling).[18] These symptoms were often identical to those described by people suffering from disorders belonging to the schizophrenia spectrum, such as the feeling of being controlled, watched, and tortured by invisible beings, being the victim of invisible rays that cause pain or physical injuries, the belief in the existence of a conspiracy directed against such a person, a sense of exceptionalism and a sense of mission.[19]

While STOPZET mainly deals with "victims" of "electronic torture," another pioneer of the movement struggling against introduction of the 5G network, the Polish Association for Combating Electrocontamination "Right to Life," is fighting with alleged perpetrators of the contamination. It has been headed by Zbigniew Gelzok since 2006. This former miner also worked five years as an assembler installing mobile phone base stations. He claimed that he lost his health at work, developing dangerous changes in his spine, gonads, and nervous system, as well as headaches and insomnia. He was convinced that these were symptoms of microwave disease, which he had contracted while installing the telecommunication base stations, but he had trouble proving it. The Institute of Occupational Medicine in Sosnowiec refused to issue him a certificate of occupational disease, so he appealed to the court. The court ordered a medical explanation of the origin of such symptoms. Meanwhile, in 2002, by a regulation of the Council of Ministers, microwave disease was removed from the list of occupational diseases of civilians, which prevented him from obtaining compensation from the telecommunications company.

---

18 Comment in the Stop Zet Association group on Facebook (name anonymized), https://www.facebook.com/groups/stopzet/posts/2107054576106179.

19 Sylwia Rozbicka, "Schizofrenia – przyczyny, rodzaje, terapia", *Mind Health*, no data, accessed 06.09.2024, https://mindhealth.pl/co-leczymy/schizofrenia.

As a result of this dispute, he began to support colleagues from the mobile phone industry who were suing their employers. In 2006, he established the National Association for Combating Electrocontamination "Right to Life" in Rzeszów. The impetus for its establishment came from a doctor near Rzeszów, who had a mobile phone base station built next to her house. She and Gelzok challenged the location and environmental decision, and the mayor, who initially supported erecting the mast, sided with them.[20] Gelzok deals with legal and technical issues related to mobile telephony and fights telecommunications companies in the Administrative Courts, accusing them of numerous abuses and frauds. "Right to Life" and his second association, the Social Initiative "Law Every Day," have won over 90 cases.[21] "Right to Life" eventually became the central organization of people fighting 5G, and Tobiasz Żuchiewicz, Rafik Ennaoui, and the mayor of Ciechocinek, Jarosław Jucewicz, who were active in the anti-5G protests, began operating under its banner. In 2020, the Rybnik prosecutor's office charged Gelzok with insulting public officials – including, among others, Minister of Digitization Marek Zagórski and former Deputy Minister of Digitization Wanda Buk, because in many posts he published on the social networking site Facebook, he accused them of accepting millions of zlotys in bribes from "telecoms," operating in an organized criminal group, committing numerous frauds and lies, and writing laws "dictated by telecoms."[22]

Another link in the network of anti-5G activist organizations is "My Wolni Ludzie," led by Jadwiga Łopata and her partner Julian Rose. Łopata comes from the ecological movement and runs an environmental farm in Stryszów near Wadowice in Lesser Poland. The farm was previously managed by Richard F. Gauthier,[23] who came from the Microvita Research Institute, part of the Ananda Marga (Path of Happiness) religious movement founded by Prabhat Ranjan Sarkar. This group is a religious and social organization that offers meditation and yoga instruction, promoting a system of spiritual practice that is a synthesis of Vedic and tantric philosophy, condemning material-

20 Grażyna Ostropolska, "Prawnik po zawodówce", *Nowości. Dziennik Toruński*, 30.03.2013, accessed 06.09.2024, https://nowosci.com.pl/prawnik-po-zawodowce/ar/10909193.

21 Zbigniew Gelzok, "Zbigniew Gelzok Geniusza Historia Prawdziwa", *Prawo do Życia*, no data, accessed 06.09.2024, https://web.archive.org/web/20161103183404/http://prawodozycia.pl/services/zbigniew-gelzok-historia-prawdziwa/.

22 Zbigniew Podjadek, "Anatomia polskiego ruchu anty5G. Kim są ludzie, którzy walczą z budową sieci nowej generacji?", *Wyborcza.pl*. 16.09.2020, accessed 06.09.2024, https://wyborcza.biz/biznes/7,177150,26290907,anatomia-polskiego-ruchu-anty5g-kim-sa-ludzie-ktorzy-walcza.html.

23 Also writing under the pseudonym "Richard Richardson".

ism and capitalism, describing the universe as the result of consciousness entering into slavery of its nature.[24] Łopata herself ran another farm of that group – the 47-hectare Ananda Putta Bhumi (Pure Thoughts Place) in the village of Głębock, in the Podgórzyn commune, near Jelenia Góra. She also organized the Neohumanist Ecology Festivals there since 1991, which were part of the activities of this religious group in Poland.[25] She also tried to create the Institute of Transforming Ecology in the Jelenia Góra Valley and conducted proselytizing campaigns. She continued her ecology-related activities after moving to Stryszów and leaving the Path of Happiness. She was the main initiator and co-founder of the GMO-Free Poland Coalition and the European Centre for Organic Agriculture and Tourism – ECEAT, which is, in practice, an association of several agritourism eco-farms. Łopata runs a small, educational, traditional organic farm called Eco-center of ICPPC. She was awarded the Goldman Prize and was decorated, among others, with the Polish Golden Cross of Merit and the title of "Meritorious for the Stryszów Commune."

Her main collaborator and life partner, Sir Julian Rose, is an organic farmer and English baronet. In the 1960s, he was an active participant in the counter-culture movement. He studied in London at the Royal Academy of Drama and worked as an actor and director in a regional theatre. In 1971, he left for the USA, where he joined the experimental group Players Theatre of New England in Boston and joined the hippie movement, exploring new age and organic farming issues. He returned to his family estate in 1983 and devoted himself entirely to organic farming. In 1984, he joined the board of the English organic farmers' association Soil Association. In the 1990s, together with Łopata, he began to create the International Coalition to Protect the Polish Countryside (ICPPC).[26] In June 2002, the ICPPC headquarters in Stryszów was visited by the future King of England, Charles III. Rose and Łopata are running a joint campaign, "GMO-free Poland," to block the introduction of genetically modified crops into Poland.[27]

---

24 See further: Richard Richardson, *The Microvita Revolution* (Calcutta, Microvita Research Institute, 1990).

25 No author "ZAPOWIEDZI. V Festiwal Ekologii Neohumanistycznej. 20.–24.7.95 Głębock", *Zielone Brygady*, no data, accessed 06.09.2024, http://www.zb.eco.pl/zb/72/zapowie1.htm.

26 Watkinsbooks, "Julian Rose: Organic Farming and Green Wisdom", *YouTube*, 23.02.2014, accessed 06.09.2024, https://youtu.be/hFaWwQIWd_4.

27 Julian Rose, "Polska do wzięcia?", *Nowy Obywatel*, 07.05.2012, accessed 06.09.2024, https://nowyobywatel.pl/2012/07/02/polska-do-wziecia.

However, his interests do not end with ecology and agriculture. He is a supporter of Polexit and has also published in a magazine sympathetic to Russian nationalist narratives like "Polityka Polska,"[28] "Nowy Obywatel,"[29] and on a portal belonging to the Neo-fascist party National Rebirth of Poland.[30] His texts, similar to those of Pawela and Kazahari, also contain numerous anti-Semitic motifs. He believes that "the Rothschild, Rockefeller, Soros dynasties, with their strong links to the Project for the New American Century, the Trilateral Commission and the Bilderberg Group, maintain an uninterrupted influence on the development of Israeli policy" (…) "This cabal has always planned to destroy the sense of unity of the Middle East and replace it with a form of military dictatorship, sponsored and directed by supporters of the totalitarian New World Order," Rose states in a text for the WNET radio portal.[31] He believes that the NWO and Israel are direct extensions of the Third Reich, with its openly declared intentions of genocide and eugenics, and also praises Russia's foreign policy. He considers the sanctions against this country to be the result of fabricated evidence and part of a long-standing plan created by the Trilateral Commission, the CIA, the Project for the New American Century, the Institute of Foreign Affairs, Chatham House (the Royal Institute of International Affairs based in London) and other similar organizations associated with NATO/USA, which aim to gain unquestionable and dominant power over the entire planet – the position of rulers of the New World Order.[32] He also believes that chemtrails are being sprayed over Poland, which causes the suppression of brain cell activity so that citizens become "passive."[33]

Several scientists also publicly expressed their opposition to 5G technology. These people speak about communication technologies, even though their

---

28 Admin. "Warto przeczytać: »Polityka Polska« – nowy miesięcznik." *Geopolityka.net*, 29.05.2015, accessed 06.09.2024, https://geopolityka.net/polityka-polska-nowy-miesiecznik.

29 Julian Rose, "Polska do wzięcia?".

30 Julian Rose, "Polska do wzięcia?", Nacjonalista.pl, 23.05.2012, accessed 06.09.2024, https://www.nacjonalista.pl/2012/05/23/julian-rose-polska-do-wziecia.

31 Julian Rose, "Izraelskie zbrodnie przeciwko ludzkości. Podział w sercu konfliktu izraelsko-palestyńskiego", WNet.pl, 14.07.2018, accessed 06.09.2024, https://wnet.fm/kurier/elita-z-krwi-zawsze-u-steru-bezposrednie-przedluzenie-iii-rzeszy-wraz-z-jej-zamiarami-ludobojstwa-i-eugeniki.

32 Julian Rose, "NATO – prowokator nieustającej wojny", *Zmieniając kurs na życie*, 03.08.2016, accessed 06.09.2024, https://renesans21.pl/category/artykuly.

33 Julian Rose, "Cicha broń masowego rażenia", *Zmieniając kurs na życie*, 28.11.2018, accessed 06.09.2024, https://renesans21.pl/category/artykuly.

academic titles concern completely different research fields or are pursuing entirely hobbyist scientific fields outside the organized structures of the academy.

One such person is Diana Wojtkowiak. She claims to be a researcher of "torsion field physics," "homeopathy," "radionics," and "physical therapy." However, she obtained her academic title based on a dissertation in microbiology (topic: "ATP-dependent proteases of Escherichia coli hydrolyzing replicative phage protein λO"). Her main topic of public appearances is the extra-scientific torsion field theory (TPT). This idea originated in Russia. Secret research on "torsion fields" in the mid-1980s was conducted by Soviet scientists under the supervision of the KGB, operating in the State Committee of the Council of Ministers of the USSR for Science and Technology. However, in July 1991, the Committee on Science and Technology under the Supreme Soviet of the USSR declared the concept pseudoscientific, and the research programme was closed. The initiators of the research, Gennady Shipov, and Anatoly Akimov, did not accept the decision and independently continued their research under the auspices of the Russian Union of Gravitational Sciences and the Scientific Center for Plasma Physics in Moscow. Despite their serious-sounding names, these entities exist outside the official mainstream of Russian science.[34]

Supporters of the idea of "torsion fields" derive their concept from the Einstein-Cartan theory and some solutions of James Clerk Maxwell's equations[35] and believe that torsion fields are an explanation for homeopathy, teleportation, tunnels to parallel universes, chakras, dowsing, orgonites, chemtrails, and even magic. Shipov even claims to know the "formula for the existence of God," identifying it with a level of reality that he calls "Absolute NOTHING."[36] French mathematician Élie Joseph Cartan used this term to describe a hypothetical physical field that is created as a result of the spinning of space. For Szypov and Akimov, however, a torsion field is a type of matter that carries information instead of energy. Left-handed and right-handed "vortices of Absolute Nothing" saturate the vacuum with information, and its source can be any physical body and can be artificially created.[37]

34 Ewa Korzeniewska et al., "Pola torsyjne – przykład pseudonaukowej koncepcji w fizyce", *Przegląd Elektrotechniczny* 1 (2021):196–199.

35 Anatoly E. Akimov, and Gennady I. Shipov, "Torsion fields and experimental manifestation?", *New Energy News* (2017): 11–14.

36 Korzeniewska et al., "Pola torsyjne".

37 Ibid.

According to Akimov, torsion fields can be used in transmission using the fifth-generation telephone network, which is supposed to be an "explanation" of the threat resulting from the use of this technology. According to supporters of this concept, the network carries distorted information, which is supposed to affect the state of human DNA. This alleged "property" also links this concept with conspiracy narratives about the COVID-19 virus and vaccines. According to Wojtkowiak, radio waves independently stimulate two signal paths in biocells, causing SARS-CoV to infect them more aggressively.[38] The idea of torsion fields was also propagated by "experts" of the anti-vaccination movement (Jerzy Zięba[39] and Jerzy Jaśkowski).[40] In May 2020, Wojtkowiak decided to file a report "of committing a crime bearing the characteristics of a crime of genocide,"[41] accusing the Minister of Health in Mateusz Morawiecki's government, Łukasz Szumowski, who signed a regulation on permissible levels of electromagnetic fields in the environment, which in her opinion will cause "higher levels of microwave intensity" to increase the "virulence of the virus" and cause "higher patient mortality".[42]

Wojtkowiak's connections with Russia do not end with the sources of her theory. The scientist's letter to President Andrzej Duda was published in July 2019 by the nationalist "Warszawska Gazeta,"[43] which promotes Russian and anti-Semitic narratives. Experts have indicated that this medium is one of the elements of the Russian disinformation network.[44]

The primary "expert" organization in the matter of "electrosmog" and "EMF pollution" is considered to be the James Clerk Maxwell Electromagnetic

---

38 Porozmawiajmy.tv, "Zabójczy wpływ sieci 5G na pola torsyjne organizmów żywych – dr Diana Wojtkowiak", *YouTube*, 10.03.2019, accessed 06.09.2024, https://www.youtube.com/watch?v=Qn6JKMrLN5g.

39 Moment w Atmosferze, "Jerzy Zięba – Pola torsyjne (14 grudnia 2023)", *YouTube*, 15.12.2023., https://youtu.be/yukyZ4tpRbU.

40 Leszek Korolkiewicz, "Elektrosmog – puszka Pandory?", *Lepsze Zdrowie* (blog), no data, accessed 06.09.2024, http://www.lepszezdrowie.info/elektrosmog.

41 Diana Wojtkowiak, "Zawiadomienie o popełnieniu przestępstwa zbrodni ludobójstwa przez Szumowskiego!", *Głos Gminny,* 20.05.2020, accessed 06.09.2024, https://web.archive.org/web/20201117124610/https://www.glosgminny.pl/zawiadomienie-o-popelnieniu-przestepstwa-zbrodni-ludobojstwa-przez-szumowskiego.

42 Ibid.

43 Diana Wojtkowiak, "Petycja w sprawie procedowania megaustawy UD192", 12.07.2019, accessed 06.09.2024, www.docplayer.pl/146010826-W-p-prezydent-rzeczypospolitej-polskiej-andrzej-duda.html.

44 Mateusz Bajtek, "Na wojnie z kłamstwem", *Interia.pl,* 15.01.2018, accessed 06.09.2024, https://wydarzenia.interia.pl/zagranica/news-na-wojnie-z-klamstwem,nId,2508561#utm.

Research Institute. Despite its name, it is not a research institution but a non-governmental organization. Jerzy Gremba serves as its president. He defended his doctorate when he was already in his fifties. Previously, he had run an electronic services company in Kielce called "Jim Elektronik" for years and repaired household appliances. His doctoral thesis was concerned not with telephony but with interference with the reception of satellite television signals. His function reinforces his "scientific" authority as a court expert in the field of electronics, radio communications (analogue and digital wireless systems for transmitting and receiving video, audio, and data signals, terrestrial and satellite), and mobile telecommunications (mobile telephone systems and other systems based on wireless communication). However, he does not work at any university or publish works considered scientific.

The "Institute" itself is registered in Kraków at the address of Barbara Gałdzińska–Calik (board member of the James Clerk Maxwell Institute of Electromagnetic Research). She graduated from the Kraków University of Economics. She later obtained a doctorate in economics (international accounting focusing on the Chinese market, the subject of her doctoral dissertation: *Changes in financial accounting systems in the globalization of the world economy*) at the Jagiellonian University. She also completed postgraduate studies in environmental protection at the Kraków University of Technology. She is the co-author of only one scientific article on the study of magnetic radiation, published in the journal "Przegląd Lekarski."[45]

Her fight against telecommunications technologies and their alleged harmfulness began in 2006 when she moved to the Bielany housing estate in Kraków. After a miscarriage, she decided that the base station located near her house was to blame. She decided that her daughter was an "electrosensitive child" and began actions aimed at stopping the introduction of 5G technology in Kraków. In her opinion, due to the high density of transmitters, the population began to fall ill and die en masse. The protests of residents organized by her led to a court order to move the nearby transmitters.[46] In 2018, she even received the "Lajkonik" award for social activity "for unprecedented involvement in matters related to building awareness of the harmfulness of electro-

45 Jolanta Kaszuba-Zwoińska, Jerzy Gremba, Barbara Gałdzińska-Calik, Karolina Wójcik-Piotrowicz, Piotr J. Thor, "Electromagnetic field induced biological effects in humans", *Przegląd Lekarski* 72, no. 11 (2015):636–641.

46 Małgorzata Jankowska, "Blaski i cienie 5G, czyli jak ucywilizować cyfrową cywilizację (1)", *Instytut Spraw Obywatelskich*, 20.12.2019, accessed 06.09.2024, https://instytutsprawobywatelskich.pl/blaski-i-cienie-5g-czyli-jak-ucywilizowac-cyfrowa-cywilizacje-1.

magnetic radiation."[47] She also ran in local elections from the lists of the local committee Wspólnota Małopolska.

The last member of the "Institute" board is Paweł Wypychowski, a graduate of electronics and telecommunications at the Faculty of Electrical Engineering of the Szczecin University of Technology (currently the West Pomeranian University of Technology). In 1990–95, he was one of Poland's first commercial internet service providers. Then, he worked as an IT consultant. He is a lecturer at the West Pomeranian Business School. He, in turn, is the primary Polish seller of equipment that is supposed to protect against the harmful effects of 5G. Because the fight against 5G is also a profitable business activity. Along with the growing group of people afraid of this technology, many business people, inventors, and service providers who want to alleviate this fear with their products or activities have appeared. Advertisements for services protecting against "electromagnetic radiation" have also appeared in Poland.[48] More original products have also appeared on sale, such as an American anti-5G cream, supposed to protect against "radiation," which turned out to be a mixture of Vaseline and sunscreen.[49] In Poland, you can buy an "Anti-5G de-Radiator," which is supposed to absorb electromagnetic radiation, promote regeneration, and remove fatigue, "personal de-radiators," which additionally emit "healing light" basing their alleged function "on the principle of emission of waves with the opposite phase, transformation, and impregnation – which in effect neutralizes the effects of harmful geopathic radiation and has a healing and strengthening effect, adds energy and life forces, strengthens the immune system." There are also gadgets for your wallet or pocket on the market that supposedly stop 5G "radiation," emitting "Schuman resonance waves" and "using the Fibonacci sequence" and an "anti-torsion field generator," which is supposed to "eliminate harmful torsion fields and geopathic radiation of electrical devices, starting from an ordinary light bulb."[50]

47 Ibid.

48 Tomasz Domański, "Rozdają ulotki, na których straszą 5G. Za pieniądze zabezpieczą twoje mieszkanie", *Spider's Web*, 08.05.2020, accessed 06.09.2024, https://spidersweb.pl/2020/05/alert-5g-promieniowanie-badania-skutki.html.

49 Marcin Kaczmarek, "Krem anty-5G. Miliony dolarów zysku ze sprzedaży", *ITBiznes*, 29.05.2020, accessed 06.09.2024, https://itbiznes.pl/technologie/krem-anty-5g-miliony-dolarow-zysku-ze-sprzedazy.

50 Tomasz Kurzak, "Gadżety chroniące przed 5G. Oszuści wykorzystują obawy przed nową techniką", *Komputer Świat*, 30.05.2020, accessed 06.09.2024, https://www.komputerswiat.pl/artykuly/redakcyjne/gadzety-chroniace-przed-5g-oszusci-wykorzystuja-obawy-przed-nowa-technika/l4d7wwn#slajd-2.

The first companies specializing in products to stop "harmful 5G radiation" are also appearing on the market. Initially, these were businesses located in Western Europe. British company BioShield Distribution sells the 5GBioShield, which it claims "protects your home and family, using a wearable holographic nano-layer catalyst that can be worn or placed near your smartphone or any other device that emits electrical energy, radiation or electromagnetic fields. (...) Using a quantum oscillation process, the 5GBioShield USB dongle balances and re-harmonizes the disturbing frequencies resulting from electrical fog induced by devices such as laptops, wireless wifi cards, tablets, etc." According to tests by Pen Test Partners, this is an ordinary USB pen that sells in China for about six dollars. BioShield Distribution, meanwhile, prices it at $346.[51] Protective clothing, bedding, underwear, hats, phone covers, and sleeping canopies that supposedly protect against 5G are offered by Swiss company Wavesafe.[52]

Poles are also setting up similar businesses. Youtuber Michał Podhajski, who introduces himself as "Vincenty Docent," "awakening activist," "artist," and "mystic," shares mystical theories on his channel on the portal and reports on his experience with consuming psychoactive substances[53] in 2019 on the crowdfunding platform Patronite announced a fundraising campaign for the production of "a-hat." In his opinion, this object is "a specialist protective hat that effectively blocks almost 98% of microwave radiation from GSM, 3G, 4G, 5G transmitters, Wi-Fi routers, laptops, and mobile phones" and also "works wonderfully during sleep, because it gives our brain a many-hour break from microwaves."[54] He collected the sum, and the hat went into production and was sold for PLN 199 per piece (approximately 50 Euro).

Another YouTuber, Sebastian Szaman (Szaman 21 Wieku), runs a similar business, with 165 thousand subscribers on his channel.[55] On his website, he

---

51 Paweł Czajkowski, "Boisz się 5G? Kup do ochrony pendrive z »kwantową technologią« za 346 USD", *IT Hardware*, 29.05.2020, accessed 06.09.2024, https://ithardware.pl/aktualnosci/boisz_sie_5g_kup_do_ochrony_pendrive_z_quot_kwantowa_technologia_quot_za_346_usd-12352.html.

52 Aleksandra Ptak-Iglewska, "Sklep sprzedaje bieliznę, która ochroni przed 5G", *Rzeczpospolita*, 09.07.2019, accessed 06.09.2024, https://cyfrowa.rp.pl/telekomunikacja/art18197041-sklep-sprzedaje-bielizne-ktora-ochroni-przed-5g.

53 Vincenty Docent, "MISTYK W ROLLS ROYCE ★ MOJA HISTORIA – Sport, Plastyka, Psylocybina, Przebudzenie, Port Kosmiczny ★", *YouTube*, 20.09.2023, accessed 06.09.2024, https://youtu.be/SCDkttOojGs.

54 Vincenty Docent, "»Jeżeli to co zrobię pomoże chociaż jednej osobie, to warto to zrobić«", *Patronite.pl*, no data, accessed 06.09.2024, https://patronite.pl/Vincenty.

55 User "Sebastian Szaman 21 Wieku" channel on YouTube, https://www.youtube.com/channel/UCITOhCiRg6bre4pEUEnUcaw.

sells water and air ionizers, "space pillows," "de-radiators" and "ozonators," as well as, importantly in the context of 5G, mats, which he claims "protect against all types of radiation, both magnetic, electrical and radio" including 2G GSM/GPRS, 3G UMTS/HSPA, 4G LTE, 5G), Wi-Fi routers, digital TV, radars and "electric fields generated by cables and electrical and electronic devices."[56]

Wypychowski, an "expert" at the James Clerk Maxwell Institute of Electromagnetic Research and a member of its board, also founded the company EMF3 – Electromagnetic Safety, whose goal is to provide devices that supposedly protect against electromagnetic fields (EMF). Through it, it sells various products, from paint through curtains to shorts, which are supposed to protect against harmful effects of not only 5G but also GSM, UMTS, LTE, Wi-Fi, Bluetooth, DECT phones, radars, and millimetre wave scanners used at airports.[57]

However, the profits from the fight against 5G do not end with funds obtained for questionable quality products from individual customers. Activists and their organizations also receive public funds. The fight against "electrosmog" (i.e., 5G) has already cost the Krakow City Office PLN 1.133 million in 2016–2018, and "experts" have tried to obtain another PLN 6 million from the city budget.[58] The position of Social Advisor to the Mayor of Krakow for Bioelectromagnetic Issues has also been established in Krakow, initially held by Gałdzińska-Calik. During its two-year period of operation, it earned PLN 260000 (approximately 60000 Euro), granted entirely on a discretionary basis, because any regulations or laws do not regulate the activity of the Social Advisor.[59] Another "expert" from the James Clerk Maxwell Gremb Institute of Electromagnetic Research also co-created the "Programme for Environmental Protection against electromagnetic fields for 2018–2022" in Krakow, earning PLN 30000 (approximately 7000 Euro). Other experts from the institute earned another PLN 70000 (approximately 17000 Euro).[60]

The Krakow City Office also purchased specialist equipment for measuring the intensity of electromagnetic fields, i.e., EME Spy 200 experimenters, usu-

56 No author, "Co polecam", *Szaman 21 Wieku*, no data, accessed 06.09.2024, https://szaman21wieku.com/category/co-polecamy-testy.

57 No author, "Ekranowanie pól elektromagnetycznych", *EMF Busters*, no data, accessed 06.09.2024, https://emfbusters.pl.

58 Arkadiusz Dziermański, "Działalność krakowskich ekspertów ds PEM kosztowała miasto ponad milion złotych. W planach było dużo więcej", *What Next*, 31.05.2020, accessed 06.09.2024, https://whatnext.pl/krakow-przeciwnicy-sieci-komorkowych-wydatki-elektrosmog.

59 Ibid.

60 Ibid.

ally used by installers of telecommunications and electrical networks. The City Office spent PLN 594314,17 (approximately 140000 Euro) on them and the dedicated software. Residents rented them 180 times, which gives a cost of approximately PLN 3333 (approximately 770 Euro) per measurement. There were additional costs. The total cost of the "EkoCyfrowi" project was PLN 151933,35 (approximately 35200 Euro). It consisted of a photography competition, the "Send a postcard instead of a text message" competition, a literary competition, an exhibition, a smartphone-free picnic, and other additional expenses. A total of 159 people participated in the competitions, and 500 participated in the picnic.[61]

Thanks to the institute's contacts, the International Conference Poland as a 5G-Free Zone, vital for the movement's development, took place in Krakow on February 28, 2019. There was an alliance between activist groups such as the Polish Association for Combating Electrocontamination "Right to Life," the STOPZET Association, and the James Clerk Maxwell Institute of Electromagnetic Research Foundation, POLISH NO TO 5G (led by Grzegorz Sacharuk), the National Institute of Social Affairs, and We Free People. The 5G Free Poland Coalition was also established, combining their activities. The conference featured lectures by Pawela, Wojtkowiak, Gremba, Gałdzińska–Calik, Gelzok, Rose, Łopata, Zięba and ufologist Janusz Zagórski.

---

61 Ibid.

CHAPTER 3

Turn Off the TV – Turn On Your Thinking

The Polish Anti-vaccination Movement: Activists, Significant Organizations, "Experts"

Older, and much more extensive, both numerically and organizationally, than the anti-5G movement, the anti-vaccination movement has developed a multi-node structure in Poland throughout its existence, in which, in addition to activists and their organizations, one can find festivals dedicated to the cause, magazines, a group of "experts" recruited from among doctors and scientists not directly connected to the health service, and numerous connections with the Polish political class. Relations with politicians and the participation of activists and experts in the activities of political parties are so frequent in the movement that they deserve separate treatment in subsequent chapters. The most visible figures of the anti-vaccination movement are not mainly recruited from niche groups of supporters of conspiracy narratives and the marginalized discursive group weakly connected to the mainstream but are often already part of Poland's professional and political elites. Despite this level of organizational and political development, in its essence, it remains very similar to the movement opposing the 5G telephone network, being a transmission belt between scientifically questionable theories, often of Russian origin or amplification, supported by media related to alternative medicine and the far right to mainstream Polish politics. However, it is possible to describe the stratification of the movement in much more detail and divide it into clear thematic segments, to which membership is determined by specific motivations, views, or ideological affiliations of activists and "experts" mentioned in public statements.

It is worth noting that among the people spreading conspiracy narratives about COVID-19 and, more broadly, vaccines, one can find numerous activists and "experts" previously active in accusations of the harmfulness of 5G networks to human health. Numerous statements challenging the scientific consensus and suggesting the conspiratorial nature of the pandemic

 | DOI:10.1163/9789004747852_005

("plandemic") were made by mentioned earlier Wojtkowiak,[1] Pawela,[2] Kazahari,[3] Zagórski,[4] and Zięba.[5]

However, Justyna Socha was the key figure in developing the Polish anti-vaccination movement. She represents the figure, typical also in other countries – a mother concerned about the health of her child who "discovers" that her offspring has been allegedly harmed by administering a vaccine.[6] Similar figures can be found as movement leaders in Australia,[7] the USA,[8] Ireland,[9] England,[10] and New Zealand.[11] Socha is a mother of four children, and, as she claims, the oldest of them had an adverse reaction to vaccination (psoriasis)

1 Diana Wojtkowiak, "The link between the SARS–CoV–2 pandemic and the 5G network", *Gazeta Warszawska,* 29.05.2020, accessed 06.09.2024, https://mail.gazetawarszawska.com/index.php/scientia–et–arte/5010–diana–wojtkowiak–the–link–between–the–sars–cov–2–pandemic–and–the–5g–network.

2 Anna Mierzyńska, "Epidemia dezinformacji w Polsce. Część fałszywek wskazuje na aktywność Rosji", *OKO.Press*, 14.03.2020, accessed 06.09.2024, https://oko.press/koronawirus–wywolal–dezinformacyjna–powodz.

3 CEPowiśle, "Hanna Kazahari (Tokio): Plandemia – zaplanowana przez globalistów i syjonistów fałszywa flaga wirusa", *BitChute*, 08.04.2020, accessed 06.09.2024, https://www.bitchute.com/video/ccEa9MRbElOQ.

4 Magdalena Kozioł, "Antycovidowcy mają własną telewizję. Wiele programów negujących koronawirusa bije rekordy popularności", *Wyborcza.pl,* 14.10.2020, accessed 06.09.2024, https://wroclaw.wyborcza.pl/wroclaw/7,35771,26393182,wroclawski–trzon–antycovidowego–protestu–ufolog–sygnalistka.html.

5 No author, "Witamina C na ciężki COVID-19? Fake newsy Jerzego Zięby!", *Demagog,* 17.02.2022, accessed 06.09.2024, https://demagog.org.pl/fake_news/witamina–c–na–ciezki–covid–19–fake–newsy–jerzego–zieby.

6 Stephanie Alice Baker and Michael James Walsh, "A mother's intuition: it's real and we have to believe in it': how the maternal is used to promote vaccine refusal on Instagram". *Information, Communication & Society* 26, no. 8 (2023):1675–1692.

7 Kirsty Johnston, "»It's like a cult«: How anti–vaccine »mumfluencers« are fuelling the Parliament occupation", *Stuff,* 22.02.2022, accessed 06.09.2024, https://www.stuff.co.nz/national/health/coronavirus/300520631/its–like–a–cult–how–antivaccine–mumfluencers–are–fuelling–the–parliament–occupation.

8 Rachel Martin, "Former Anti–Vaccine Mom Explains How Movement Pulled Her In, And How She Left", *NPR*, 18.01.2021, accessed 06.09.2024, https://www.npr.org/2021/01/18/957981974/how–anti–vaccine–movement–could–hurt–efforts–to–end–pandemic.

9 Aaron Rogan, "Anti–vaccine group to campaign in Ireland", *The Times*, 20.03.2017, accessed 06.09.2024, https://www.thetimes.co.uk/article/anti–vaccine–group–to–campaign–in–ireland–jnq5fpfkv.

10 Stephanie Alice Baker and Michael James Walsh, "COVID: how anti-vaccine influencers exploit mothers", *The Conversation*, 14.02.2022, accessed 06.09.2024, https://theconversation.com/covid–how–anti–vaccine–influencers–exploit–mothers–175670 .

11 Georgia Forrester, "Kiwi mother and healthy food advocate petitions health star rating", *Stuff,* 14.11.2016, accessed 06.09.2024, https://www.stuff.co.nz/life–style/well–good/86300986/kiwi–mother–and–healthy–food–advocate–petitions–health–star–rating.

after a series of mandatory vaccinations, which prompted her to become active in the movement opposing mandatory immunization of children. She has no medical education and has worked as an insurance agent for years. In 2011, she was one of the founders of the National Association of Vaccination Awareness STOP Adverse Post-Vaccination Reactions (STOP NOP), and in 2014, she was already the spokesperson for this organization.[12]

STOPNOP filed reports of alleged crimes committed by the Ministry of Health to the prosecutor's office for, in their opinion, the lack of proper response of the health service to adverse post-vaccination reactions (ADRS) in children. They also demanded the development of a new system for preventing ADRS, as well as the creation of a special compensation fund for people who suffered or who lost loved ones due to the ADRS, and demanded the creation of a list of experts independent of pharmaceutical companies, who would issue opinions in such cases in court.[13] Over time, the association began to organize street protests, particularly actively defending doctors who opposed the vaccination requirement and were punished by the Medical Chambers.

In December 2015, already as the president of the association, Socha published a petition to the then Prime Minister Beata Szydło and the Minister of Justice Zbigniew Ziobro, in which she accused nine Polish doctors – including Prof. Jacek Wysocki, then Rector of the Poznań University of Medical Sciences, and Dr. Paweł Grzesiowski, epidemiologist and expert in infection prevention – of collaborating with pharmaceutical companies producing vaccines.[14] The latter sued her twice for defamation – and won the court cases against Socha twice.

Socha, in 2018, ceased to be the president of STOP NOP. In the following four years, she was dismissed and re-elected to this position. The association began to fall into an increasing crisis resulting from accusations against Socha of embezzling the organization's funds.[15] Ultimately, the association's board

---

12 Izabela Smolińska, "Dzieci umierają w Polsce po szczepionkach? Sprawa w prokuraturze", *Wprost*, 27.06.2014, accessed 06.09.2024, https://www.wprost.pl/tylko-u-nas/454092/dzieci-umieraja-w-polsce-po-szczepionkach-sprawa-w-prokuraturze.html.

13 Ibid.

14 Sylwia Sałwacka,"»Żadnej skruchy«. Justyna Socha ze STOP NOP mimo wyroku sądu znów oskarża lekarzy", *Wyborcza.pl* 22.01.2019, accessed 06.09.2024, https://poznan.wyborcza.pl/poznan/7,36001,24387903,justyna-socha-ze-stop-nop-nie-sklada-broni-mimo-wyroku-sadu.html.

15 Zespół Pravdy, "Cała prawda o antyszczepionkowcach ze STOP NOP. Czy można z nimi wygrać?", *Pravda*, 12.06.2023, accessed 06.09.2024, https://pravda.org.pl/artykuly/antyszczepionkowcy-stop-nop.

filed a report against her to the prosecutor's office regarding alleged failure to account for money collected during fundraisers; an investigation is underway.[16] The prosecutor's office in Poznań is investigating whether Socha misappropriated PLN 470000 from public fundraisers.[17] Another trial is underway in the same city against the association because an appeal appeared on the STOP NOP Facebook profile, insinuating that doctors in intensive care units cause the death of patients by giving them lethal doses of morphine.[18] However, Socha remains active in protests, publishing in the nationalist and anti-Semitic "Warszawska Gazeta"[19] and the newspaper of alternative medicine supporters "Health Without Medicines" ("Zdrowie bez leków").

In addition to organizations associating activists, organizations associating doctors opposing mandatory vaccinations have also begun to emerge. The most visible in the media are the Polish Association of Independent Physicians and Scientists (PSNLiN),[20] the Ordo Medicus Institute, the Society of Integrated Medicine Physicians (TLMZ), and the Association Medics of Hope. PSNLiN can be considered a particularly active entity in this matter. Throughout the pandemic, members of this organization have opposed all government restrictions related to the coronavirus, wearing masks, performing tests, limiting interpersonal contacts, and were convincing public opinion about the effectiveness of treating COVID-19 with amantadine.[21] While PSNLiN tries to focus primarily on "independent" doctors, Ordo Medicus promotes its views using "independent" scientists. The foundation has published books disseminating non-scientific opinions on vaccines and the pandemic

16 MJM, MW, "Lekarka głosiła, że pandemia to »ściema«. Odebrano jej prawo wykonywania zawodu", *Puls Medycyny*, 02.11.2022,, za: https://www.pulsmedycyny.pl/lekarka-glosila-ze-pandemia-to-sciema-odebrano-jej-prawo-wykonywania-zawodu-1168346.

17 Tomasz Cylka, "Podejrzane 470 tys. zł Justyny Sochy. Prokuratura w Poznaniu zajęła się liderką antyszczepionkowców", *Wyborcza.pl*, 06.02.2022, accessed 06.09.2024, https://poznan.wyborcza.pl/poznan/7,36001,28078738,podejrzane-470-tys-zl-justyny-sochy-prokuratura-w-poznaniu.html#S.embed_article-K.C-B.1-L.2.zw.

18 Sylwia Sałwacka, "STOP NOP: »Lekarze zabijają na intensywnej terapii«. Wiceminister zdrowia składa zawiadomienie do prokuratury", *Wyborcza.pl*, 18.10.2019, accessed 06.09.2024, https://poznan.wyborcza.pl/poznan/7,36001,25321009,skandaliczny-wpis-stop-nop-wiceminister-zdrowia-sklada-zawiadomienie.html.

19 Sylwia Sałwacka, "»Żadnej skruchy«".

20 Miłada Jędrysik, "Ofensywa »niezależnych lekarzy i naukowców«: o pandemii postanowiono w Davos, amantadyna leczy", *OKO.Press*, 07.07.2021., za https://www.oko.press/niezalezni-lekarze-i-naukowcy-o-pandemii.

21 Dominika Wantuch, "Rekordowa liczba medyków przed sądem lekarskim: za negowanie pandemii i opowiadanie bzdur o szczepieniach", *Wyborcza.pl*, 01.06.2022, accessed 06.09.2024, https://www.krakow.wyborcza.pl/krakow/7,44425,28521312,rekordowa-liczba-medykow-przed-sadem-lekarskim-za-negowanie.html.

(like: *False Pandemic*;[22] Covidian dizziness;[23] White Paper on the Coronavirus Pandemic: Facts and Data Hidden from Public Opinion),[24] got involved in the creation of the report: *Restrictions on Rights and Freedoms Introduced in Poland in Connection with the COVID-19 Epidemic*,[25] and organized a conference devoted to the "effects of government restrictions on the functioning of the health service in 2020 and 2021."

Some magazines (such as *HARMONIA. Your Health. Your Responsibility* or *Health Without Medicine*") as well as the annual conference "What Your Doctor Won't Tell You" regularly publish materials created by the movement opposing mandatory vaccinations. Representatives of the movement are also regular guests and speakers during festivals of "alternative medicine," "new spirituality," or "esotericism," both the largest ones, such as: "Harmonia Kosmosu," "Wibracje," "Life Balance Congress," as well as smaller ones such as: "Festiwal Zdrowie," "Pełnia Zdrowia – Festiwal Kultury Zdrowia," "Festival Wolności i Zdrowia. Primum Non-Nocere," "Ekologiczne Dożynki," "Festival Topinamburu," "Kosmiczny Festiwal," festival "Żywie."

The number of doctors' opponents of mandatory vaccinations has increased to such an extent that it was possible to divide people practicing medical professions and professing anti-scientific concepts into several subgroups. The first category includes the most well-known physician, often involved in the movement for years, running their own video channels, publishing books and articles on "alternative" methods of treatment. They are considered authorities by others movement members, and, along with the prosecutions they face, such as suspensions from legal practicing medicine or court cases, also considered "martyrs" of the cause. Hubert Czerniak, Jerzy Jaśkowski,[26] Zbigniew

---

22 Mariusz Błochowiak, ed., *Fałszywa pandemia. Krytyka naukowców i lekarzy* (Warszawa: Fundacja Osuchowa, 2020).

23 Roman Zieliński and Kornelia Polok, *Kowidowy zawrót głowy* (Wiry: Wydawnictwo Indygobook, 2023).

24 Mariusz Błochowiak, "Biała księga pandemii koronawirusa – fakty i dane ukrywane przed opinią publiczną", *Polskie Stowarzyszenie Niezależnych Lekarzy i Naukowców*, 05.09.2022, accessed 06.09.2024, https://www.psnlin.pl/artykuly,biala-ksiega-pandemii-koronawirusa-fakty-i-dane-ukrywane-przed-opinia-publiczna,25,254.html.

25 Łukasz Bernaciński, Jerzy Kwaśniewski, ed., *Ograniczenia praw i wolności wprowadzone w Polsce w zw. z epidemią Covid-19. Aspekty prawne, medyczne i ekonomiczne* (Warszawa, Wydawnictwo Naukowe Instytutu na rzecz Kultury Prawnej Ordo Iuris, 2022).

26 No author, "Jerzy Jaśkowski kontra lekarze? W odpowiedzi podaje fake newsy!", *Demagog*, 09.12.2022, accessed 06.09.2024, https://www.demagog.org.pl/fake_news/jerzy-jaskowski-kontra-lekarze-w-odpowiedzi-podaje-fake-newsy.

Martyka,[27] Zbigniew Hałat,[28] and Dorota Sienkiewicz have become particularly well-known in this context.

They were eager to appear in video materials created by far-right and alternative medicine youtubers like: Janusz Zagórski, Aleksander Berdowicz, Rafał Mossakowski, Marcin Rola, or Eugeniusz Sendecki. They were also guests at many festivals of "alternative medicine," "new spirituality," or "esotericism." They often mixed their "medical" concepts with belief in other conspiracy narratives such as the secret plan to depopulate Europe and especially Poland (allegedly created by the WHO, the "world government," the EU, or the Jewish plutocracy), the "great replacement," the "genocide of the white race," the harmfulness of 5G and GMO, the manipulation of the government and the medical community by representatives of medical corporations (the so-called "Big Pharma"), the "Great Reset," the "New World Order," anti-Semitic, homophobic, pro-Russian narratives or stories about the "Great Lehia" or the so-called chemtrails. This group is currently the most important source of the spread of anti-scientific pseudo-medical narratives in Poland and is crucial for recruiting new supporters.

Hubert Czerniak is a doctor of internal medicine who has been with various right-wing groups for a dozen or so years. He ran for the Polish Parliament from the Kukiz'15 list in 2015 and the Confederation list in 2019 for the European Parliament. Czerniak believes that vaccines are not medically tested,[29] cause autism and brain damage, are a form of cannibalism, and are contrary to the Bible norms.[30] He links his opinion on vaccines with anti-Semitic conspiracy theories. He calls vaccinations the "Holocaust of Poles"[31] and "genocide and murder using syringes instead of gas chambers." He claims: "The origin of the expert council preparing the vaccination schedule should be checked.

27 No author, "Lockdowny nie działały, a szczepienia są nieskuteczne? Fałsz!", *Demagog*, 28.12.2022, accessed 06.09.2024, https://www.demagog.org.pl/fake_news/lockdowny-nie-dzialaly-a-szczepienia-sa-nieskuteczne-falsz.

28 No author, "Zbigniew Hałat o szczepionkach? Kolejny dezinformujący materiał!", *Demagog*, 11.01.2022, accessed 06.09.2024, https://www.demagog.org.pl/fake_news/zbigniew-halat-o-szczepionkach-kolejny-dezinformujacy-material.

29 Katarzyna Skrzydłowska-Kalukin, "Jak kłamią antyszczepionkowcy", *Wprost*, 16.07.2017, accessed 06.09.2024, https://www.wprost.pl/tygodnik/10065260/Jak-klamia-antyszczepionkowcy.html .

30 Tomasz Kowalewicz, "»Szczepienia obniżają IQ i powodują autyzm«. Skandaliczny wykład na KUL-u", *Wyborcza.pl*, 16.06.2017, accessed 06.09.2024, https://lublin.wyborcza.pl/lublin/7,48724,21969310,szczepienia-obnizaja-iq-i-powoduja-autyzm-skandaliczny-wyklad.html .

31 EMisja.tv, "dr Czerniak: »Zaczyna się holocaust Polaków«", *Youtube*, 25.09.2018, accessed 06.09.2024, https://www.youtube.com/watch?v=d3ykgqBbCk4.

What nationality are they, because the government may not know, they are preparing a place for some other nation in this country"[32] and demands the restoration of the category of nationality in the identity card, clearly suggesting that he wants to distinguish number of Jews in Polish elites. By the decision of the prosecutor's office, his license to legally practice medicine was suspended, and he was also charged with over 50 corruption offenses. The doctor was allegedly accepting bribes for issuing medical certificates.[33] Czerniak is a member of PSNLiN.

Jerzy Jaśkowski was employed for many years (1973–2000) at the Medical University of Gdańsk in the Department and Institute of Physics and Biophysics as its head. However, GUM points out that the management of the Ecophysical Laboratory of the Department of Biophysics was only a customary function.[34] Even thou, he has not worked at the university for twenty-four years, he still often refers to this employment in his bio. In 2017, the Supreme Medical Chamber suspended him from practicing medicine – he cannot practice it in any form.[35] He is the author of numerous publications, including the series "300 Years of Vaccine Fraud," devoted to the alleged falsification of vaccine effects data in Poland.[36] As he claims, "you have to be a real idiot to get vaccinated against the flu,"[37] and he calls vaccinations and their composition poison.[38] In his opinion, "the flu epidemic has always occurred only and exclusively in military units,"[39] and "where vaccinations were used, there were epidemics," arguing that in the last 200 years, he does not know of an example in which an epidemic broke out spontaneously.[40] He believes that the measles vaccine causes numerous neurological complications. "These complications do not go away and last for the rest of the person's life. The mechanism of brain damage may vary. Both the neurotoxins contained, such as mercury or

32 Tomasz Molga, "Powstał miesięcznik szarlatanów. Lekarze to »skończeni, sadystyczni bandyci«", *WP*, 17.12.2018, accessed 06.09.2024, https://wiadomosci.wp.pl/powstal-miesiecznik-szarlatanow-lekarze-to-skonczeni-sadystyczni-bandyci-6328518195103361a.

33 Magda Miesnik, "Wybory do PE 2019. »WiceZięba« dr Hubert Czerniak startuje do europarlamentu", *WP*, 16.04.2019, accessed 06.09.2024, https://wiadomosci.wp.pl/wybory-do-pe-2019-wicezieba-dr-hubert-czerniak-startuje-do-europarlamentu-6370935095813761a.

34 Ewa Palińska, "Szczepionkowy ping–pong doktora z GUMedem", *Trójmiasto.pl*, 23.11.2016, accessed 06.09.2024, https://www.trojmiasto.pl/zdrowie/Szczepionkowy-ping-pong-doktora-z-GUMedem-n107665.html?&vop=w&strona=4.

35 No author, "Jerzy Jaśkowski kontra lekarze".

36 Palińska, "Szczepionkowy ping–pong doktora z GUMedem".

37 No author, "Jerzy Jaśkowski kontra lekarze?".

38 Ibid.

39 Ibid.

40 Ibid.

aluminum, as well as antibiotics that destroy bacterial flora, can cause damage to the digestive tract. Through such a damaged digestive tract, i.e., its leakage, various types of toxins found in food can penetrate directly into the bloodstream,"[41] we read in one of Jaśkowski's articles, in which he repeats theses from Wakefield's texts rejected by the scientific community. He considers autism to be one of the side-effects of vaccination.[42] He also claims that sunbathing has a beneficial effect on the skin. Jaśkowski says, "UVB mainly produces vitamin D, which has a protective effect, eliminating cancerous changes." He also believes that women who want to get pregnant should pay attention to the sanitary pads they use. Jaśkowski explains that they are bleached with chlorine, and chlorine, one of the strongest poisons, is released, causing inflammation of the vulva, and that is why the mass use of those made of cotton should be restored.[43]

Another important figure in the movement opposing mandatory vaccinations is Zbigniew Martyka. He graduated from the Military Medical Academy in Łódź and worked as a doctor in military units (including the Military Hospital in Krynica). He is currently the head of the infectious disease ward in a small town of Dąbrowa Tarnowska.[44] He declares himself a deeply religious Catholic.[45] He gained popularity among opponents of mandatory vaccinations by expressing views questioning the application of sanitary restrictions during the COVID-19 pandemic, claiming that the lockdown did not help prevent the spread of the virus and caused more harm than good, and that isolation and limiting contact did not fulfil their task.[46] He also announced that vaccines were ineffective against the virus and that it did not pose a threat to children: "masks do not work. There is no inhibition of virus transmission and,

41 Ilona Godlewska, "Kontrowersyjne tezy lekarza. »Mimo 300 lat istnienia szczepień nie udowodnili, że one cokolwiek działają«", *Wyborcza.pl*, 29.11.2016, accessed 06.09.2024, https://trojmiasto.wyborcza.pl/trojmiasto/7,35612,21031892,kontrowersyjne-tezy-lekarza-mimo-300-lat-istnienia-szczepien.html.

42 Ibid.

43 Ibid.

44 Iwona Konarska, "Pierwszy zastępca przewodniczącego Okręgowego Sądu Lekarskiego w Tarnowie i jego poglądy", *Termedia*, 14.12.21, accessed 06.09.2024, https://www.termedia.pl/koronawirus/Pierwszy-zastepca-przewodniczacego-Okregowego-Sadu-Lekarskiego-w-Tarnowie-i-jego-poglady,44843.html.

45 Zbigniew Martyka, "Efekty wiary w pandemię", *Agencja Informacyjna*, 01.02.21, accessed 06.09.2024, https://www.agencja–informacyjna.com/dr–zbigniew–martyka–efekty–wiary–w–pandemie.

46 No author, "Lockdowny nie działały, a szczepienia są nieskuteczne? Fałsz!", *Demagog*, 28.12.2022, accessed 06.09.2024, https://www.demagog.org.pl/fake_news/lockdowny-nie–dzialaly–a–szczepienia–sa–nieskuteczne–falsz.

if anything, it is so negligible that it has absolutely no effect on the number of infections."[47] He declared that COVID-19 vaccinations are a "large-scale experiment"[48] and that there are no studies confirming that the COVID-19 vaccine can inhibit coronavirus transmission. He also claimed that there is simply no pandemic and falsely declared that information about the side effects of the vaccine is being increasingly widely reported. According to him, "only a trace amount is actually reported, about 1%, so what was officially reported can be multiplied by a hundred."[49] He also spread fake news that "we had proven forgery in the case of false deaths, when coffins of immigrants who sank as victims of the cutter disaster near Lampedusa were shown, and these were shown as coffins of victims who died of coronavirus."[50]

Until recently, an important figure in the Polish movement opposing mandatory vaccinations was Dr. Zbigniew Hałat, who died in 2022. In the early 1990s, he served as Deputy Minister of Health and Chief Sanitary Inspector (1991–1993). In the Lower Silesian branch of the National Health Fund, he was the director responsible for medical matters. During the COVID-19 pandemic, while remaining an active doctor, he criticized wearing masks and vaccinations. He stated in the media that "the World Health Organization is primarily concerned with promoting LGBT culture or so-called genderism." In turn, the doctor called the head of the WHO, Tedros Adhanom Ghebreyesus, "a well-known Ethiopian terrorist in the past, and a microbiologist, not a doctor."[51] Hałat described COVID-19 vaccines as a "medical experiment."[52] He claimed that this preparation is neither effective nor does it prevent serious conditions.[53] Those vaccinated against COVID-19 "spread the infection" and that administering COVID-19 vaccines is unethical.[54] He also published columns on the religious Radio Maryja website concerning, among other things,

47 Ibid.

48 Ibid.

49 Ibid.

50 Ibid.

51 No author, "Guru antyszczepionkowców »fotografuje młode dziewice«. »To był kiedyś niezły lekarz«". *Medonet*, 12.01.2022, accessed 06.09.2024, https://www.medonet.pl/zdrowie/wiadomosci,guru–antyszczepionkowcow–fotografuje–dziewice––to–byl–kiedys–niezly–lekarz,artykul,30960450.html.

52 No author, "Zbigniew Hałat o szczepionkach? Kolejny dezinformujący materiał!", *Demagog*, 11.01.2022, accessed 06.09.2024, https://www.demagog.org.pl/fake_news/zbigniew–halat–o–szczepionkach–kolejny–dezinformujacy–material.

53 No author, "Szczepionki to niepotrzebne i nieprzebadane zagrożenie? Fake news!", *Demagog*, 04.08.2021, accessed 06.09.2024, https://www.demagog.org.pl/fake_news/szczepionki–to–niepotrzebne–i–nieprzebadane–zagrozenie–fake–news.

54 Ibid.

alleged benefits of preserving virginity until marriage. He also posted similar content on his personal pages, where he stated that "virginity until marriage is the source of world harmony," "people who lose their virginity early are more likely to have multiple sexual partners, fuel epidemics and social pathologies," and "virginity until marriage" "saves lives, reduces illness, raises self-esteem, strengthens careers, opposes slavery, counteracts crime, strengthens families, strengthens communities, saves funds."[55] He also published photos of girls on his pages – some were captured as if they did not know they were being photographed – with captions such as "Ania, a God-loving student of the University of Wrocław, is an example of the beauty of the spirit and body of Slavic virgins." The photos of the "virgins," some of whom look like teenagers and some were wearing only swimsuits, were – if the captions are to be believed – taken by Hałat himself. One photo shows pre-teen children playing in a playground.[56]

One of the forerunners of the movement against mandatory vaccinations, active in this matter already in the years 2014–2015,[57] is the president of PSNLiN, Dorota Sienkiewicz, a specialist in children's diseases and medical rehabilitation, a retired employee of the Medical University of Białystok (UMB) and a former assistant at the UMB Children's Disease Clinic, the UMB Children's Neurology and Rehabilitation Clinic, and then the UMB Children's Rehabilitation Clinic with the Early Care Center for Disabled Children "to Give a Chance." She is the author of publications in which she suggests that vaccines can cause autism, ADHD, and allergies and proves their alleged ineffectiveness.[58] In the autumn of 2018, she spoke during a joint session of the Parliaments' Health and Social Policy and Family Committees together with activists of the movement fighting against mandatory vaccinations, including Justyna Socha.[59] She argued that the vaccines used for immunization are not

---

55 No author, "Guru antyszczepionkowców »fotografuje młode dziewice«".

56 Jakub Maciejewski, "Guru antyszczepionkowców fotografuje młode "dziewice", promuje rosyjskie treści, twierdzi, że szczepionka przyciąga magnes", *wPolityce.pl*, 28.12.2021, accessed 06.09.2024, https://wpolityce.pl/polityka/579752-antyszczepionkowe-fake newsy-zbigniewa-halata.

57 Beata Falkowska, "Czy szczepić dzieci?", *Nasz Dziennik*, 10.10.2014, accessed 06.09.2024, https://naszdziennik.pl/mysl/102489,czy–szczepic–dzieci.html.

58 Krystian Lurka, "Sienkiewicz przestrzega przed szczepieniami. UMB i białostocki OIL wyciągną konsekwencje", *Termedia*, 15.11.2018, accessed 06.09.2024, https://www.termedia.pl/mz/Sienkiewicz–przestrzega–przed–szczepieniami–UMB–i–bialostocki–OIL–wyciagna–konsekwencje,32090.html.

59 Maciej Chołodowski, "Białystok. Pediatra przeciwna szczepieniom dzieci stanęła przed sądem lekarskim. Bronią jej antyszczepionkowcy", *Wyborcza.pl*, 09.02.2022, accessed

safe because, in her opinion, they contain aluminum.[60] She is also a co-author of the *White Paper on the Coronavirus Pandemic: Facts and Data Hidden from the Public Opinion*, published by the Ordo Medicus Institute.[61] She has repeatedly downplayed the risk of severe COVID-19 in children and adolescents, denied the risk of SARS-CoV-2 transmission from children and adolescents to the elderly, questioned the indications and effectiveness of vaccinations against COVID-19 in children and adolescents, and questioned the possibility of obtaining medical assistance in the event of adverse reactions following vaccination. She also made a direct appeal to children and adolescents, but also adults, not to get vaccinated due to the alleged high harmfulness of the COVID-19 vaccine, trivialized COVID-19. She said that no one should be afraid of it, and compared this disease to cold, argued that having COVID-19 will allow one to gain better immunity,[62] and taking the vaccine against this disease, lowers immunity,[63] is very dangerous and not only carries numerous complications, e.g., heart disorders, but can also lead to death.[64] Additionally, she repeated claims that are inconsistent with medical knowledge that coronavirus vaccines are experimental and that no one tested them in Poland before they were approved,[65] and therefore – they are dangerous, while the PCR tests themselves, which are used to detect the virus, are unreliable.[66] Due to statements contrary to scientific knowledge denying the pandemic and the effectiveness of vaccines, in 2022, she was deprived of the right to practice medicine for one year. As a result, activists of the movement organized a

06.09.2024, https://bialystok.wyborcza.pl/bialystok/7,35241,28090535,bialystok-pediatra-przeciwna-szczepieniom-dzieci-przed-sadem.html.

60 AM, "Uniwersytet Medyczny odcina się od opinii »ekspertki« antyszczepionkowców", *Polityka Zdrowotna*, 14.11.2018, accessed 06.09.2024, https://www.politykazdrowotna.com/artykul/uniwersytet-medyczny-odcina-sie-od-opinii-ekspertki-antyszczepionkowcow/835455.

61 Błochowiak, "Biała księga pandemii koronawirusa".

62 Małgorzata Solecka, "Kolejny lekarz bez prawa wykonywania zawodu za działania antynaukowe", *MP.pl*, 23.11.2022, accessed 06.09.2024, https://www.mp.pl/szczepienia/aktualnosci/311437,kolejny-lekarz-bez-prawa-wykonywania-zawodu-za-dzialania-antynaukowe.

63 No author, "Dezinformacja na temat szczepień w wywiadzie z dr Sienkiewicz", *Demagog*, 14.02.2022, accessed 06.09.2024, https://www.demagog.org.pl/fake_news/dezinformacja-na-temat-szczepien-w-wywiadzie-z-dr-sienkiewicz.

64 Solecka, "Kolejny lekarz bez prawa wykonywania zawodu za działania antynaukowe".

65 No author, "Dezinformacja na temat szczepień w wywiadzie z dr Sienkiewicz".

66 Ibid.

march through the city, praying the rosary and a demonstration in front of the District Medical Chamber in Białystok.[67]

Another category of healthcare and medical professionals who hold anti-scientific views can be described as "propagators-activists." Most of them are less well-known than those mentioned above, do not have their own YouTube channels, and only some run businesses based on pseudo-medical services. They are sometimes invited to "alternative medicine" festivals and conferences as speakers. Still, in most cases, this is not their primary source of income, and they have not yet earned a "reputation" in the community. They "work" for it by organizing demonstrations, speaking critically in the media about vaccines, lockdowns, or other anti-pandemic restrictions, and operating in organizations they have established.

A specific group of doctors who oppose mandatory vaccinations are in the same time deeply religious opponents of abortion. For them, the essential issue that discourages them from vaccinations is the connection between the preparations and cells obtained in the past through the procedure of terminating a pregnancy. An example of such a person is family medicine doctor Dorota Łucja Jarczewska. In her private life, the mother of six children is a regular contributor to religious media and anti-abortion associations and foundations. Jarczewska was a lecturer at the anti-abortion Pro-Life Summer Academy organized by the Polish Association of Defenders of Human Life in 2023.[68] She also collaborates with the anti-abortion Life and Family Foundation and the Pro-Right to Life Foundation, as well as the Voice of the Faithfull Association[69] and the Institute for Legal Culture Foundation Ordo Iuris, associated with the Brazilian religious group TFP. In the media related to the Redemptorist monk Tadeusz Rydzyk (Radio Maryja and TV Trwam),[70] as

---

67 Ewelina Sadowska-Dubicka, "Marsz i protest w obronie lekarki, która odradza szczepionki", *Białystok Online*, 08.02.2022, accessed 06.09.2024, https://www.bialystokonline.pl/marsz-i-protest-w-obronie-lekarki-ktora-odradza-szczepionki-zdjecia,artykul,128534,1,1.html.

68 Redakcja Portalu, "»Potrzebujemy spotkań, które dodają odwagi«. Rusza Letnia Akademia Pro-life", *Stacja7.pl*, 31.05.2023, accessed 06.09.2024, https://www.stacja7.pl/z-kraju/potrzebujemy-spotkan-ktore-dodaja-odwagi-rusza-letnia-akademia-pro-life.

69 No author, "15.X.2022 Stowarzyszenie Głos Wiernych zaprasza na wykład dr. n. med. Doroty Jarczewskiej", *Katolickie Stowarzyszenie Lekarzy Polskich. Oddział Mazowiecki bł. Ewy Noiszewskiej*, 15.10.2022, accessed 06.09.2024, https://www.kslpmazowsze.pl/2022/10/15-x-2022-stowarzyszenie-glos-wiernych-zaprasza-na-wyklad-dr-n-med-doroty-jarczewskiej.

70 Fundacja Życie i Rodzina, "O liniach płodowych z aborcji – dr n. med. Dorota Jarczewska", *YouTube*, 21.10.2022, accessed 06.09.2024, https://www.youtube.com/watch?v=e4gOX4vSAPs.

well as during events organized by the aforementioned non-governmental organizations, she appeared speaking out on the issue of the conscience clause in medical professions"[71] and the harmfulness of hormonal contraception. She even edited a monograph entitled *Harmfulness of oral hormonal contraception,* – in which she stated that contraception could cause an early abortion, increases the risk of thrombotic disease (and one of its consequences may be death due to pulmonary embolism), and is an oncological burden, and the most aggressive and difficult to treat subtype of breast cancer is causally related to birth control pills.[72]

The critical topic raised by Jarczewska in the context of vaccinations was for her the connection between this process and abortion. She gave a lecture on this subject for the Life and Family Foundation,[73] spoke on this subject in the media of the Redemptorist monk Tadeusz Rydzyk,[74] and in an interview conducted with her for the Pro-Right to Life Foundation. She also devoted one of the chapters to this issue (*Fetal lines and their relationship with COVID-19 vaccinations – ethical aspects*) in the Ordo Iuris publication *Restrictions on Rights and Freedoms introduced in Poland in connection with the COVID-19 Epidemic Part Two – Health Aspects.*[75] She spoke in the debate "Medical aspects of restrictions in connection with the COVID-19 epidemic," organized by the same organization.[76] Her participation in the video shared on the YouTube channel by the Education Board in Krakow also had an openly anti-vaccination dimension. "The COVID-19 pandemic ended at the turn of April and May 2021"; "Vaccines are experimental therapy"; "There is no justification for conducting a medical experiment on children"; "The flu is much more

71 Nasz Dziennik, "lek. med. Dorota Łucja Jarczewska – Wykład – 14 maja 2016 r.", *YouTube*, 31.10.2016, accessed 06.09.2024, https://www.youtube.com/watch?v=lkSJu8wMqSI.

72 Magdalena Guziak-Nowak, "Antykoncepcja – druga strona medalu", *Przewodnik Katolicki*, no data, accessed 06.09.2024, https://www.przewodnik-katolicki.pl/Archiwum/2016/Przewodnik-Katolicki-16-2016/Spoleczenstwo/Antykoncepcja-druga-strona-medalu.

73 Faustyna Rewucka, "Covid ujawnił problem wykorzystywania linii płodowych z aborcji", *Fundacja Życie i Rodzina*, 31.05.2022, accessed 06.09.2024, https://ratujzycie.pl/covid-ujawnil-problem-wykorzystywania-linii-plodowych-z-aborcji.

74 Małgorzata Solecka, "Małopolska kurator jak antyszczepionkowy harcownik", *MP.pl*, 07.02.2022, accessed 06.09.2024, https://www.mp.pl/szczepienia/aktualnosci/291069,malopolska–kurator–jak–antyszczepionkowy–harcownik.

75 Dorota Łucja Jarczewska, "*Linie płodowe i ich związek ze szczepieniami przeciwko Covid–19 – aspekty etyczne*", in: "*Ograniczenia praw i wolności wprowadzone w Polsce w związku z epidemią Covid–19. Część druga – aspekty zdrowotne*", ed. Bernaciński and Kwaśniewski.

76 Wantuch, "Rekordowa liczba medyków przed sądem lekarskim".

common than COVID-19, and it is a disease that no one cares about"; "The risk of COVID symptoms in people after taking the so-called vaccine is 27 times greater, and the risk of hospitalization is 8 times greater compared to people who have natural immunity" – these are just a few examples of statements that were included in this film.[77] Jarczewska herself stated that vaccination means administering an "experimental therapy" to a young organism that will bear the effects of this therapy for several decades. "Autoimmune diseases may appear" and "cascades of reactions, e.g., autoaggressive diseases, may develop,"[78] she said. She also referred to a statement by the British Joint Committee on Vaccination and Immunisation (JCVI) from July 2021, which recommended that only adolescents aged 12–15 from risk groups be vaccinated against COVID-19. It is worth adding that the JCVI recommendations later changed, and the entire adolescent population was vaccinated.[79]

Another religious orthodox is the head of the Marianna Orańska clinic in Stronie Śląskie, cardiologist Anna Martynowska.[80] She has gained tremendous popularity among opponents of mandatory vaccinations. Even before the pandemic she was politically involved in right-wing parties. In 2015, she ran for the Parliament in district no. 2 (former Wałbrzych province) on the KORWiN party list, and in 2019, for Confederation. With the outbreak of the COVID-19 pandemic, she openly denied the existence of both the global epidemic and the existence of the virus itself, criticizing, among others, doctors and prosecutors. She accused the former of abandoning professional ethics and morality for money. She often appeared in right-wing media. On the YouTube nationalist channel, "Media Narodowe," when asked whether the virus exists, she answered: "I don't think so."[81] She compared information about the pandemic to Orson Wells' radio broadcast about the Martian invasion of Earth.[82] When asked, "What are the media, the Minister of Health, and other doctors who

---

77 RN, MP, "Antyszczepionkowe wykłady na stronie Małopolskiego Kuratorium Oświaty", *Life in Kraków*, 14.02.2022, accessed 06.09.2024, https://www.lifeinkrakow.pl/w–miescie/4140,antyszczepionkowe–wyklady–na–stronie–malopolskiego–kuratorium–oswiaty.

78 Wantuch, "Rekordowa liczba medyków przed sądem lekarskim".

79 Solecka, "Małopolska kurator jak antyszczepionkowy harcownik".

80 Who also appears, under two names, as Anna Prześlica–Martynowska.

81 Ewa Wilczyńska, "Lekarka przekonuje, że żadnego COVID–u nie ma. »Biedaki w kombinezonach bez sensu się trudzili«", *Wyborcza.pl*, 11.07.20, accessed 06.09.2024, https://wroclaw.wyborcza.pl/wroclaw/7,35771,26109828,lekarka-przekonuje-ze-zadnego-covidu-nie-ma-sa-ludzie-starsi.html.

82 Barbara Kwiatkowska, "Dolny Śląsk. »Rzekoma epidemia«. Lekarka nie wierzy w COVID–19. Odsunięta od zawodu", *Wp.pl*, 04.08.2020, accessed 06.09.2024, https://wiadomosci.wp.pl/wroclaw/dolny-slask-rzekoma-epidemia-lekarka-nie-wierzy-w-covid-19-odsunieta-od-zawodu-6539494913538880a.

endorse this selling us?" she replied: "Paranoia."[83] She argued that the coronavirus is a common seasonal infection.[84] Martynowska claims that COVID-19 has never been isolated and cannot be diagnosed. PCR tests are unsuitable for diagnosing infected people because they give completely incorrect results. Vaccines are useless because they are not supposed to protect us in any way, as the disease does not exist. Additionally, in her opinion, the vaccine itself was developed too quickly.[85] She also believes that the coronavirus can be diagnosed in anyone and that the Sanitary Inspectorate is "a sword in the hands of the Minister of Health" who "has prepared a health, economic and civilizational collapse." Martynowska declares: "With the help of the Sanitary Inspectorate, my compatriots have been humiliated and degraded. What are these face masks supposed to be used for? Precisely the degradation of the nation."[86] She believes covering the mouth and nose with them does not fulfil its protective function. She calls masks "muzzles" and considers them a manifestation of the loss of civil liberties.[87] Additionally, as a deeply religious person,[88] she mixes the scientific dimension with the eschatological one. She claims that the fight against the pandemic is an "attack by Satan" on human civilization, which aim is "intensifying the crisis."[89] As she publicly stated: "The fear of COVID was used to carry out psychopathic plans. It was prepared for years. It can be best described by saying that if someone does not believe in God, they will believe in anything.[90] "I read the Apocalypse of St. John, and I encourage you all to read it. None of us was sure how it going to happen. Now

83 Wilczyńska, "Lekarka przekonuje, że żadnego COVID-u nie ma".

84 Kak, "Lekarka głośno powiedziała, co myśli o pandemii. Zakazano jej wykonywania zawodu", *Fronda.pl*, 07.09.2020, accessed 06.09.2024, https://www.fronda.pl/a/lekarka-glosno-powiedziala-co-mysli-o-pandemii-zakazano-jej-wykonywania-zawodu,149823.html.

85 No author, "Teoria Anny Martynowskiej o pandemicznych oszustwach jest fałszywa", *Demagog*, 23.02.2021, accessed 06.09.2024, www.demagog.org.pl/fake_news/teoria-dr-anny-martynowskiej-o-pandemicznych-oszustwach-jest-falszywa.

86 E. Wilczyńska, "Lekarka przekonuje, że żadnego COVID-u nie ma".

87 Kak, "Lekarka głośno powiedziała, co myśli o pandemii".

88 Kontrrewolucja.net, "Dr Anna Martynowska oficjalnie wraca do wykonywania zawodu. Wygrała w sądzie", *Kontrrewolucja.net*, 26.08.2021., accessed 06.09.2024, https://www.kontrrewolucja.net/wiadomosci/dr-anna-martynowska-oficjalnie-wraca-do-wykonywania-zawodu-wygrala-w-sadzie.

89 No author, "Teoria Anny Martynowskiej o pandemicznych oszustwach jest fałszywa".

90 Rafał Zieliński, "Lekarka z Dolnego Śląska o pandemii: Psychopatyczne plany, to było przygotowywane latami. Na szczęście mamy swoich księży", *Wyborcza.pl*, 17.05.2021, accessed 06.09.2024, https://wroclaw.wyborcza.pl/wroclaw/7,35771,27095329,lekarka-z-dolnego-slaska-o-pandemii-psychopatyczne-plany.html.

everyone who wants to know, knows,"[91] she added. She also propagates a conspiratorial narrative about a planned, international conspiracy of the pharmaceutical lobby, which invented COVID-19[92] for its own needs. Martynowska herself, however, notices a recurring feature of many doctors who oppose mandatory vaccinations, which may be a clue as to the group that is particularly susceptible to the anti-vaccination narrative among doctors: "I have noticed a pattern that most of those doctors who dare to tell it like it is are people of deep faith. We are not afraid because we are free people, no matter how it ends. […] We must settle accounts elsewhere, not here, on Earth. And we try to do what our conscience dictates,"[93] declares Martynowska. This thesis of the doctor is confirmed by her intensive contacts with the Confederation party, with its leader, MEP Grzegorz Braun, she has appeared publicly.[94] Ordo Iuris[95] also supports Martynowska. Even after the medical chamber suspended her license to practice the profession, Martynowska continued her activity in the movement of opponents of mandatory vaccinations, appearing, among others, at the protest of this group in Kłodzko.[96]

Another doctor who opposes mandatory vaccinations and is deeply religious, psychiatrist Katarzyna Ratkowska, was also visible in the national media. She is also a member of PSNLiN.[97] Her appearance on Polsat News, was widely commented on in the media. Her case is critical because it allows us to precisely trace the train of thought of opponents of mandatory vaccinations, who do not consider themselves to be contributing to the spread of COVID-19 but, on the contrary, to be saving others from the extraordinary scale of adverse reactions to vaccination, which in their opinion are the real

91 Ibid.

92 Nadia Szagdaj, "Lekarka głosiła, że Covid–19 nie istnieje i straciła prawo do wykonywania zawodu. Teraz znów przyjmuje pacjentów", *I.pl*, 27.08.2021., accessed 06.09.2024, https://www.i.pl/lekarka–glosila–ze–covid19–nie–istnieje–i–stracila–prawo–do–wykonywania–zawodu–teraz–znow–przyjmuje–pacjentow/ar/c1–15773736.

93 Kontrrewolucja.net, "Dr Anna Martynowska oficjalnie wraca do wykonywania zawodu".

94 Posoborowe absurdy, "Wykład »Fałszywa pandemia« – Grzegorz Braun, dr Anna Martynowska, dr Tomasz Wołoszyn i Roman Szostek.", *Gloria.tv*, 17.11.2021, accessed 06.09.2024, https://www.gloria.tv/post/ypQg84EEbg2b1xMesFgDfzTZA#2080.

95 No author, "Dr Anna Martynowska skazana na karę pieniężną przez Naczelny Sąd Lekarski".

96 Bp, "Manifestowano przeciwko przymusowi szczepień", *DKL24*.pl, 15.01.2022, accessed 06.09.2024, https://www.dkl24.pl/pl/a/12046/manifestowano–przeciwko–przymusowi–szczepien.html.

97 Suwerenni–TV, "Dr Katarzyna Ratkowska z PSNLIN. EKSPERYMENTY na dzieciach w Polsce? AMANTADYNA pomaga w leczeniu?", *Youtube*, 06.09.21, accessed 06.09.2024, https://www.youtube.com/watch?v=Bs4YLZpVO8Q.

cause of deaths during the pandemic. According to Ratkowska, vaccines pose a "much more significant threat to health and life than the disease itself."[98] "The first reports of deaths of minors after so-called "vaccination" are coming from around the world. Pharmaceutical companies that have disclaimed responsibility and are making huge profits from these "vaccines" are allowing themselves more and more," she said in an interview with the "DoRzeczy.pl" portal.[99] In her opinion, in this "medical experiment," "we observe complications, including deaths, the frequency of which is higher than the officially reported statistics." She claims that "millions of people around the world die after getting injected with these liquids"[100] and there are numerous "reports of deaths of minors after so-called »vaccination«,"[101] and in Great Britain in 2021, there were "six times more deaths among vaccinated people," of which 50% of infections were people vaccinated with two doses."[102] According to Ratkowska, this is due to the way the vaccine works. In her opinion, the implanted protein "is toxic and causes severe post-vaccination complications, including death. The key one would be in her opinion, the new VITT disease, or "post-vaccination immune thrombosis with thrombocytopenia",[103] which allegedly causes strokes and heart attacks, as well as autoimmune complications caused by our tissues producing the spike protein, a foreign antigen, a toxin. In her opinion, as a result of this process, patients would suffer from inflammation of various organs: the heart muscle, liver, brain, or spinal cord with paralysis. According to Ratkowska the implanted proteins are later produced by various human body cells, e.g., by endothelial cells (lining the arteries) or brain cells. As a result, in her opinion, our immune system can attack any organ that

---

98 Jacek Sądej, "Eksperymenty na polskich dzieciach", *Nasz Dziennik*, 26.06.2021 accessed 06.09.2024, https://www.naszdziennik.pl/polska-kraj/239403,eksperymenty-na-polskich-dzieciach.html.

99 Paweł Zdziarski, "Psychiatra: Szczepienia dzieci i nastolatków są niedopuszczalne", *Do Rzeczy*, 11.06.2021, accessed 06.09.2024, https://www.dorzeczy.pl/opinie/187785/psychiatra-szczepienia-dzieci-i-nastolatkow-sa-niedopuszczalne.html.

100 Faustyna Rewucka, "Dr Ratkowska: Nie można rozwijać nauki kosztem życia ludzi", *Fundacje Życie i Rodzina*, 01.03.2022, accessed 06.09.2024, https://www.ratujzycie.pl/dr-ratkowska-nie-mozna-rozwijac-nauki-kosztem-zycia-ludzi.

101 Tw, "Polsat News krytykowany po wypowiedzi psychiatry o wyższej śmiertelności zaszczepionych na covid", *Wirtualne Media*, 30.06.2021, accessed 06.09.2024, www.wirtualnemedia.pl/artykul/polsat-news-krytykowany-po-wypowiedzi-psychiatry-o-wyzszej-smiertelnosci-zaszczepionych-na-covid.

102 Red, "Prof. Fal: szczepienia są największą zdobyczą nauki", *Polsat News*, 29.06.2021, accessed 06.09.2024, https://www.polsatnews.pl/wiadomosc/2021-06-29/prof-andrzej-fal-i-dr-katarzyna-ratkowska-w-programie-gosc-wydarzen-transmisja-od-1920.

103 Tw, "Polsat News krytykowany po wypowiedzi psychiatry o wyższej śmiertelności".

produces the spike protein. So, when the attack occurs on the heart muscle cells, it causes inflammation of the heart muscle. When on the cells lining the arterioles from the inside, it causes the formation of clots, and when the clot flows with the bloodstream to the brain, it creates an embolism there and, as a result, a stroke.[104] The third group of complications, according to Ratkowska, is neurological, such as dementia, comparable to the old Creutzfeldt–Jakob disease (the so-called "mad cow disease"), which causes loss of memory, consciousness, and death within three months.[105] According to Ratkowska, these diseases do not end the list. The vaccine is also supposed to cause kidney failure and galloping cancers (within three weeks, cancer with metastases develops and death occurs), haemophilia, paresis and epilepsy,[106] and even loss of vision.[107] She also believes that vaccines contain graphene.[108] None of these claims have been scientifically confirmed. Ratkowska also believes that other methods of treating COVID-19 are blocked by provisions included in agreements on vaccine preparations between pharmaceutical companies and governments[109], which is the reason for the "false indoctrination that «vaccines» are the only solution.[110] According to her, COVID-19 may be treated with amantadine, ivermectin, and hydroxychloroquine combined with zinc, fluvoxamine, and mirtazapine.[111] She also claims that with those medications she cured 1/3 of her patients.[112] Thanks to her gained media visibility. She became recognizable and publicly active. She spoke at demonstrations of opponents of mandatory vaccinations in Katowice,[113] Warsaw,[114]

104 Rewucka, "Dr Ratkowska: Nie można rozwijać nauki kosztem życia ludzi".

105 Ibid.

106 Ibid.

107 Sądej, "Eksperymenty na polskich dzieciach".

108 No author, "Katarzyna Ratkowska ujawnia tajne dokumenty Pfizera? Fake news!", *Razem Przeciw Dezinformacji*, 23.06.2022, accessed 06.09.2024, www.razemprzeciwdezinformacji.pl/katarzyna-ratkowska-ujawnia-tajne-dokumenty-pfizera-fake-news-2.

109 Rewucka, "Dr Ratkowska: Nie można rozwijać nauki kosztem życia ludzi".

110 Zdziarski, "Psychiatra: Szczepienia dzieci i nastolatków są niedopuszczalne".

111 Ibid.

112 Red, "Prof. Fal: szczepienia są największą zdobyczą nauki".

113 Suwerenni-TV, "Dr Katarzyna Ratkowska z PSNLIN. EKSPERYMENTY na dzieciach".

114 PSNLiN, "Ratujmy nasze dzieci przed eksperymentalną szczepionką inżynierii genetycznej – Warszawa", *Polskie Stowarzyszenie Niezależnych Lekarzy i Naukowców*, 12.07.2021, accessed 06.09.2024, https://www.psnlin.pl/artykuly,ratujmy-nasze-dzieci-przed-eksperymentalna-szczepionka-inzynierii-genetycznej-warszawa,34,88.html and No author, "Trumna dla Ministra Zdrowia – VI Międzynarodowy Protest Przeciw Przymusowi Szczepień – cz.2", *Polskie Stowarzyszenie Niezależnych Lekarzy i Naukowców*, 08.06.2022, accessed 06.09.2024, https://www.psnlin.pl/video,trumna-dla-ministra-zdrowia-vi-miedzynarodowy-protest-przeciw-przymusowi-szczepien-cz-2,99.html.

and Wrocław[115] and at meetings in Opole Lubelskie and on YouTube far-right channels: wRealu24.pl and Pl1.tv.[116] She also with Dorota Sienkiewicz and MEP Braun,[117] represented opponents of mandatory vaccinations during the 135th session of the Polish Parliament Health Committee. She also wrote an open letter to the president on the subject of vaccinations for children against COVID-19, in which she described vaccines as a "medical experiment."[118] The letter was published on June 15, 2021, on the Ordo Medicus Institute website. Like Martynowska and many other doctors who oppose mandatory vaccinations, she is an opponent of abortion and is a deeply religious Catholic.[119] She believes that tissues obtained from foetuses killed by abortion were used for scientific research, including the production of vaccines against COVID-19. In her opinion, "tissues and organs for such experiments are taken from aborted children when they are still alive, aged 2 to 5 months. For example, organs such as kidneys, liver, or eyes are taken. The removal of organs is the direct cause of the death of these children. These children die in agony."[120] She also claims that there are countries where abortion can be performed even after birth.[121] In reality, such cases do to exist.

In a small northern Poland town, Chojnice, the treasurer of PSNLiN, practices medicine another religious Orthodox, Ewelina Gierszewska. She spoke at the founding conference of PSNLiN, describing the process of vaccination

115 WRealu24.pl, "MARSZ PAMIĘCI DR ZBIGNIEWA HAŁATA – Wrocław 3 maja 2022 r.", *Polskie Stowarzyszenie Niezależnych Lekarzy i Naukowców*, 04.05.2022, accessed 06.09.2024, https://www.psnlin.pl/news,marsz-pamieci-dr-zbigniewa-halata-wroclaw-3-maja-2022-r,149.html.

116 Redakcja, "Dr Katarzyna Ratkowska: Kto wydał zgodę, żeby Pfizer robił eksperymenty na naszych dzieciach?", *CAI24*.pl, 16.06.2021, accessed 06.09.2024, https://www.cai24.pl/polska/33200/dr-katarzyna-ratkowska-kto-wydal-zgode-zeby-pfizer-robil-eksperymenty-na-naszych-dzieciach.

117 No author, "Polscy lekarze bronią swoich praw – Konferencja prasowa w Sejmie RP", *Polskie Stowarzyszenie Niezależnych Lekarzy i Naukowców*, 12.04.2022, accessed 06.09.2024, https://www.psnlin.pl/video,polscy-lekarze-bronia-swoich-praw-konferencja-prasowa-w-sejmie-rp,80.html.

118 No author, "Katarzyna Ratkowska w Polsat News. Gdzie minęła się z prawdą?", *Demagog*, 02.07.2021, accessed 06.09.2024, https://www.demagog.org.pl/fake_news/katarzyna-ratkowska-w-polsat-news-gdzie-minela-sie-z-prawda.

119 Katolickie Stowarzyszenie Lekarzy Polskich, "Listy otwarte do Prezydenta i Premiera RP Katolickiego Stowarzyszenia Lekarzy Polskich ws. publikacji orzeczenia TK", *Radio WNET*, 10.12.2020, accessed 06.09.2024, https://www.wnet.fm/2020/12/10/listy-otwarte-do-prezydenta-i-premiera-rp-katolickiego-stowarzyszenia-lekarzy-polskich-ws-publikacji-orzeczenia-tk.

120 Rewucka, "Dr Ratkowska: Nie można rozwijać nauki kosztem życia ludzi".

121 Ibid.

against COVID-19 as a medical experiment on a global scale and defining the vaccine itself as an "incompletely tested, experimental liquid," the taking of which can harm health or even cause death.[122] She stated, also during public appearances, that the public is not informed about post-vaccination reactions in the form of cerebral venous sinus thrombosis, visceral vein thrombosis, arterial thrombosis coexisting with thrombocytopenia, Guillain-Barré syndrome, myocarditis, and deaths that, in her opinion, are in direct temporal coincidence with taking the vaccine.[123] She also encouraged people not to get vaccinated against COVID-19. She also spread anti-scientific conspiracy narratives about graphene contamination of vaccines and offered lectures and ways to cleanse the body of its compounds.[124] She is also an advisor to natural medicine and the president of the Christian Foundation of the Center for Good News. On the organization's website, apart from religious issues, readers will find encouragement not to use sunscreens – contrary to the recommendations of dermatologists and oncologists. The foundation also organizes retreats with so-called "Daniel fasts," which, in reality, are six-week fasts, criticized by dietitians as incomplete and ruining the metabolism.[125] Her activities and views resulted in five disciplinary proceedings against her. The first hearing against the doctor before the District Medical Court at the District Medical Chamber in Gdańsk gathered a group of almost 100 opponents of mandatory vaccinations in front of the institution's headquarters, with flags, banners, and a megaphone.[126]

122 Anna Zajkowska, "W Okręgowym Sądzie Lekarskim w Gdańsku ruszyła sprawa chojnickiej lekarki Eweliny Gierszewskiej", *Weekendfm.pl*, 24.06.2022, accessed 06.09.2024, https://www.weekendfm.pl/?n=101590&chojnice_-w_okregowym_sadzie_lekarskim_w_gdansku_ruszyla_sprawa_chojnickiej_lekarki_eweliny_gierszewskiej .

123 Redakcja Płońsk24, "Stowarzyszenie o pandemii, szczepionkach i szczepieniu dzieci", *Płońsk24*.pl, 10.06.2021, accessed 06.09.2024, https://www.plonsk24.pl/aktualnosci/news/id/3918.html.

124 No author, "Zdrowie to wolność – dr Ewelina Gierszewska PSNLiN", *Polskie Stowarzyszenie Niezależnych Lekarzy i Naukowców*, 18.03.2022, accessed 06.09.2024, https://www.psnlin.pl/video,zdrowie-to-wolnosc-dr-ewelina-gierszewska-psnlin,72.html.

125 Anna Brzeska, "Medycy – antyszczepionkowcy. Co z nimi zrobić?", *Onet*, 01.08.2021, accessed 06.09.2024, https://wiadomosci.onet.pl/kraj/koronawirus-i-medycy-antyszczepionkowcy-co-z-nimi-zrobic/p25bnjx .

126 Jacek Wierciński, "Lekarka z Chojnic przed sądem i protest antyszczepionkowców", *Zawsze Pomorze. Niezależny Portal Regionalny*, 22.06.2022, accessed 06.09.2024, https://www.zawszepomorze.pl/lekarka-z-chojnic-przed-sadem-i-protest-antyszczepionkowcow.

Another frequent participant in the anti-vaccination movement, Paweł Basiukiewicz, also has ties to religious orthodoxy. He is a specialist in internal medicine, cardiologist, electrocardiologist, and head of the COVID ward from October 2020 to June 2021. In addition to being a co-author of the Ordo Iuris report, he is also the author of the report *Not a Single Tear – Health Care in a Pandemic*. In the media, he opposed the lockdown, which he believed was ineffective.[127] He also claimed that "the media have bewitched us," vaccines did not prevent the transmission of the virus, and "COVID-19 is clinically indistinguishable from other flu diseases."[128] Basiukiewicz also participated in numerous events of the anti-vaccination movement (especially those related to religious circles), such as the Toruń symposium "Faces of the Pandemic" (organized by Tadeusz Rydzyk), the Week of Life and Freedom (organized by Ordo Iuris[129]), meetings of the Ordo Medicus Investigative Committee on the Pandemic[130] and wrote part of the *White Paper on the Coronavirus Pandemic: Facts and Data Hidden from the Public Opinion* (prepared by Ordo Medicus)[131] and was a co-author of the Ordo Iuris report entitled *The Effects of Government Restrictions on the Functioning of the Health Service in 2020 and 2021*.[132] He also participated in the debate organized by Ordo Iuris entitled "Medical Aspects of Restrictions in Connection with the COVID-19 Epidemic."[133] He also signed letters to the

127 Miłada Jędrysik, "Dr Basiukiewicz, lockdowny i odporność zbiorowiskowa, czyli dlaczego Polska nie jest Szwecją", *OKO.Press*, 13.04.2021, za: https://www.oko.press/basiukiewicz-lockdown-nie-dzial.

128 WMa, "Dr Paweł Basiukiewicz o Covid–19: "zostaliśmy zaczarowani przez media". Szczepionki nie znosiły transmisji wirusa", *PCh24*.pl, 24.02.2024, accessed 06.09.2024, https://www.pch24.pl/dr-pawel-basiukiewicz-o-covid-19-zostalismy-zaczarowani-przez-media-szczepionki-nie-znosily-transmisji-wirusa.

129 No author, "Zdrowie i Pro Life to tematy dwóch ostatnich dni Tygodnia Życia i Wolności", *Stowarzyszenie im. Piotra Skargi*, 11.06.2021, accessed 06.09.2024, https://www.piotrskarga.pl/zdrowie-i-pro-life-to-tematy-dwoch-ostatnich-dni-tygodnia-zycia-i-wolnosci,15568,i.html.

130 mariusz błochowiak, "xvi POSIEDZENIE KOMISJI ŚLEDCZEJ – LEK. PAWEŁ BASIUKIEWICZ", *Fundacja Ordo Medicus*, 09.05.2022, accessed 06.09.2024, https://www.ordomedicus.org/xvi-posiedzenie-komisji-sledczej-lek-pawel-basiukiewicz.

131 Błochowiak, "Biała księga pandemii koronawirusa".

132 Mariusz Błochowiak, "SKUTKI RESTRYKCJI RZĄDOWYCH DLA FUNKCJONOWANIA SŁUŻBY ZDROWIA W 2020 I 2021 ROKU", *Fundacja Ordo Medicus*, 01.03.2022, accessed 06.09.2024, https://www.ordomedicus.org/skutki-restrykcji-rzadowych-dla-funkcjonowania-sluzby-zdrowia-w-2020-i-2021-roku.

133 No author, "Mocny głos ekspertów Ordo Iuris w sprawie zarządzania kryzysem COVID-19. Wystąpili: Chazan, Rieske, Basiukiewicz, Witczak", *Stowarzyszenie Polonia Christiana*, 13.01.2022, accessed 06.09.2024, https://www.poloniachristiana.org/mocny-glos-eksper

Polish authorities calling for lifting the lockdown and the non-introduction of mandatory vaccinations.[134]

A large group of medical professionals who oppose mandatory vaccinations are supporters of so-called "alternative medicine." Despite using different treatment methods, they are united by a keen exploration of non-standard and often scientifically unproven treatment methods.

A representative of this group is Andrzej Więckowski, who graduated from medical school in 1975 at the Faculty of Medicine of the Medical Academy in Łódź. Initially, he specialized in internal diseases. Over time, he became interested in "natural," "holistic," and "integrated" medicine. He began educating himself in non-medical "therapy" types such as acupuncture, homeopathy, herbal medicine, iridology, and chiropractic. He began providing services in their scope as part of the so-called "Holistic Vital Integrated Medicine Center." There, based on these aforementioned pseudo-medical practices, he offers patients paid treatment for chronic diseases (including many resulting, according to him, from autoimmunity), including cancer, as well as "treatment" for autism (through "detoxification of toxins") and "treatment" for food intolerances. He expresses his views not only on his YouTube channel but also on the "Holistic Vital" YouTube channel. He also performs at festivals of opponents of mandatory vaccinations, such as "Pełnia Zdrowia – Festiwal Kultury Zdrowia" and "conferences" of alternative medicine, such as "Czego Ci lekarz nie powie." He is a member of PSNLiN and a board member of the Society of Physicians of Integrated Medicine.

Another physician from this group is Andrzej Frydrychowski, a surgeon and employee of the Department and Institute of Physiology at the Medical University of Gdańsk. Today, the university officially distances itself from his views, and he presents himself as a surgeon specialist in holistic medicine, acupuncture, and clinical physiology.[135] With linguist Michał Lange he is the author of a multi-volume series of guidebooks entitled *Effectively Heal* (*Skutecznie Leczyć*). They claim that diseases commonly referred to as incurable are compatible. They write about "unconventional" "treatments" for COVID-19,

---

tow-ordo-iuris-w-sprawie-zarzadzania-kryzysem-covid-19-wystapili-chazan-rieske-basiukiewicz-witczak.

134 Media, "Lekarze sceptyczni wobec szczepień. Naczelna Izba Lekarska wzywa na przesłuchania", *Onet,* 09.04.2021, accessed 06.09.2024, https://www.wiadomosci.onet.pl/bialystok/szczepienia-naczelna-izba-lekarska-wzywa-na-przesluchania-sceptycznych-lekarzy/mzdx51s.

135 No author, "Prof. Andrzej Frydrychowski", *Zdrowie bez Leków,* no data, accessed 06.09.2024, www.zdrowiebezlekow.pl/blog/andrzej-au96.

cancer, heart failure, hypertension, diabetes, psoriasis, multiple sclerosis, Alzheimer's disease, and degenerative joint disease. Frydrychowski claims that COVID-19, like cancer, can be defeated by intravenous infusions of "vitamin C, DMSO, and appropriately diluted hydrogen peroxide." DMSO is dimethyl sulfoxide – an organic chemical compound used as a solvent. It is believed to have anti-inflammatory properties, but its use in medicine is limited. It is toxic in large doses. A 36-year-old woman died in Poznań in 2019 due to an infusion of vitamins and DMSO.[136]

Additionally, Frydrychowski is sure that conventional medicine only treats symptoms, not causes. Doctors want the patient to be sick for as long as possible, and one-third of the number of deaths in the world is the result of complications from drugs. He believes that no organism needs drugs, only to not be hindered in its self-healing. At the same time, he supports the conspiracy narrative about the existence of "electromagnetic smog" emitted by telecommunications devices,[137] which is part of the concept of the anti-5G movement. For administering DMSO intravenously to a patient between November 21, 2018, and January 14, 2019, without their consent, at the "100 Integrative Medicine Clinic," his license to practice medicine was revoked for one year.[138] As part of a protest against such a verdict, Frydrychowski decided to publicly burn the document *The Law on the Practice of the Medical Profession* and called medical chambers criminal organizations.[139] He was also a speaker at "conference" of opponents of mandatory vaccinations "What Your Doctor Won't Tell You,"[140] and performed at the most prominent "alternative medi-

136 Sylwia Sałwacka, "»Czego Ci lekarz nie powie«. Pseudomedyczny kongres na targach w Poznaniu. Gwiazdami są autorzy fake newsów", *Wyborcza.pl*, 17.06.2021, accessed 06.09.2024, https://www.poznan.wyborcza.pl/poznan/7,36001,27212801,czego-nie-powie-ci-lekarz-pseudomedyczny-kongres-na-targach.html.

137 No author, "Każda choroba przewlekła jest do wyleczenia? To nie jest prawda", *Demagog*, 23.08.2023, accessed 06.09.2024, https://www.demagog.org.pl/fake_news/kazda-choroba-przewlekla-jest-do-wyleczenia-to-nie-jest-prawda.

138 Anna Jackowska, "Ilu lekarzy-antyszczepionkowców ukarał samorząd lekarski?", *Co w zdrowiu*, 15.12.21, accessed 06.09.2024, https://www.cowzdrowiu.pl/aktualnosci/post/ilu-lekarzy-antyszczepionkowcow-ukaral-samorzad-lekarski-dane.

139 No author, "Prof. Andrzej Frydrychowski: izby lekarskie to organizacje przestępcze", *Pl1.tv*, no data, accessed 06.09.2024, https://www.pl1.tv/title/prof-andrzej-frydrychowski-izby-to-organizacja-przestepcza.

140 Nicole Młodziejewska, "Pseudonaukowa konferencja medyczna tuż przy szpitalu tymczasowym i punkcie szczepień? Stowarzyszenie »Otwarte Ramiona« jest przeciwne", *Głos Wielkopolski*, 25.06.2021, accessed 06.09.2024, https://www.gloswielkopolski.pl/pseudonaukowa-konferencja-medyczna-tuz-przy-szpitalu-tymczasowym-i-punkcie-szczepien-stowarzyszenie-otwarte-ramiona-jest/ar/c1-15682027.

cine" festival, Harmonia Kosmosu.[141] He is a member of PSNLiN and president of the Society of Integrated Medicine Physicians.

Another member of PSNLiN, Anna Furmaniuk, claims that natural medicine is deeply rooted in her family. She declares that from an early age, she learned medicines from "God's pharmacy," "understanding natural cycles," "trusts intuition," and had "deep contact with nature, with the animal world."[142] That is why she started to "treat" people with homeopathy and Bert Hellinger's constellations. As research shows, the former does not have a healing effect,[143] while the latter, according to psychologists and scientists, is a mixture of pseudo-psychology and claims based on magical thinking.[144] Furmaniuk, as she declares, also uses other activities from the field of alternative medicine in her "natural therapy," such as Lowen's bioenergetics, Levine's work with trauma, process psychology, and total biology,[145] including the consumption of psychedelics in the form of toadstools of the Amanita Muscaria mashroom[146] and regularly publishes in the pseudomedical journal "Zdrowie bez Leków." She also promoted the "alternative" therapies mentioned above on social media. At the same time, on the same profiles, she denied the existence of the COVID-19 pandemic and encouraged people not to get vaccinated against the disease. She wrote that the "plandemic" was a "hoax" and discouraged vaccinations, claiming that the health service would

141 Medyczne Grono, "Harmonia Kosmosu 2022: prof. Andrzej Frydrychowski o leczeniu chorób przewlekłych", *Medyczne Grono*, 28.06.2022, accessed 06.09.2024, https://www.medycznegrono.pl/wydarzenia/harmonia-kosmosu-2022-prof-andrzej-frydrychowski-o-leczeniu-chorob-przewleklych.

142 Anna Furmaniuk, "O mnie", *Furmaniuk.com*, no data, accessed 06.09.2024, https://www.furmaniuk.com.pl/id/6/o-mnie.

143 Aleksandra Smolińska, "Hamankiewicz: pacjenci powinni wiedzieć, że homeopatia nie działa", *Polityka Zdrowotna*, 22.11.2016, accessed 06.09.2024, https://www.politykazdrowotna.com/artykul/hamankiewicz-pacjenci-powinni-wiedziec-ze-homeopatia-nie-dziala/841647.

144 Bartosz Zalewski, "Ustawienia rodzinne Hellingera – szkodliwa szarlataneria?", *Strefa Psyche Uniwersytetu SWPS*, no data, accessed 06.09.2024, https://web.swps.pl/strefa-psyche/blog/relacje/18620-ustawienia-rodzinne-berta-hellingera-szkodliwa-szarlataneria.

145 Anna Jackowska, "Lekarka straciła PWZ na rok. Jak często zapadają takie wyroki?", *Cowzdrowiu.pl*, 03.11.2022, accessed 06.09.2024, https://cowzdrowiu.pl/aktualnosci/post/lekarka-stracila-pwz-na-rok-jak-czesto-zapadaja-takie-wyroki.

146 Tomasz Wypych, "Lekarka ukarana przez sąd za antyszczepionkowe wpisy. Teraz chce leczyć homeopatią", *Salon24*, 05.11.2022, accessed 06.09.2024, https://www.salon24.pl/newsroom/1262337,lekarka-ukarana-przez-sad-za-antyszczepionkowe-wpisy-teraz-chce-leczyc-homeopatia.

not be able to treat its side effects for the next three decades.[147] "This is a medical experiment, and you are all taking part in it," she wrote in her social media. "Then you have these minimal percentages of adverse reactions to vaccination on the leaflet. No. This is a hundred times more and a hundred times more dangerous," Furmaniuk argued.[148] Due to the trial before the medical court, she lost her license to practice for a year. On the day the verdict was announced, she appeared on wReal24 together with Justyna Socha,[149] where she said that she "will continue to write what I wrote" and "I will not rest until the truth comes out.[150] She also intends to continue to "accept patients" as a "homeopath" and "Hellinger systemic settings therapist."[151] The cost of the first such visit to her office in 2023 is PLN 800[152] (about 200 Euro).

PSNLiN Vice-President Katarzyna Bross von Walderdorff graduated from medical school in Munich in 1989 and works in Kozy near Bielsko-Biała. She has no specialization. She is a propagator of pseudoscientific homeopathy and a "holistic" approach to medicine. She considers homeopathic preparations equivalent to antibiotics,[153] while in her opinion, diseases result from energy disorders of the whole.[154] She has become involved in activities against compulsory vaccinations, closely cooperating with the STOP NOP association. In September 2015, she appeared in the Parliament during a vaccination debate and began appearing in programs on national television. In 2016, she participated in the work of the Parliamentary Team for the Safety of the Child and Adult Vaccination Program, founded by members of the right-wing populist

147 Paulina Wójtowicz, "»Plandemia to ściema«. Lekarka, która zniechęcała do szczepień straciła prawo wykonywania zawodu", *Medonet*, 03.11.2022, accessed 06.09.2024, https://www.medonet.pl/zdrowie/wiadomosci,plandemia–to–sciema––lekarka––ktora–zniechecala–do–szczepien–stracila–prawo–wykonywania–zawodu,artykul,16113385.html.

148 Katarzyna Górniak, "Lekarka rozpowszechniała teorie spiskowe. Została ukarana, ale zapewnia, że nie spocznie", *TVN24*, 05.11.2022, accessed 06.09.2024, https://www.fakty.tvn24.pl/zobacz–fakty/lekarka–anna–furmaniuk–rozpowszechniala–teorie–spiskowe–zostala–ukarana–ale–zapewnia–ze–nie–spocznie–ra1123698–6324178.

149 Jackowska, "Lekarka straciła PWZ na rok. Jak często zapadają takie wyroki?".

150 Górniak, "Lekarka rozpowszechniała teorie spiskowe".

151 MP, "»Plandemia to ściema«. Naczelny Sąd Lekarski surowo ukarał lekarkę", *Salon24*, 02.11.2022, accessed 06.09.2024, https://www.salon24.pl/newsroom/1261719,plandemia–to–sciema–naczelny–sad–lekarski–surowo–ukarl–lekarke.

152 Ibid.

153 NATULI dzieci są ważne, "Antybiotyki są nadużywane! Rozmowa z Katarzyną Bross–Walderdorff", *NATULI. Dzieci są ważne,* no data, accessed 06.09.2024, https://www.dziecisawazne.pl/antybiotyki–sa–naduzywane–rozmowa–katarzyna–bross–walderdorff.

154 Indi, "Kulkami w chorobę", *ox.pl*, 13.05.2014, accessed 06.09.2024, https://www.wiadomosci.ox.pl/kulkami–w–chorobe,26578.

Kukiz'15 parliamentary club, where she questioned the validity of vaccination. "Parents who do not want to vaccinate their children are not parents who neglect them. These parents attach particular importance to their children's health and have slightly different views. These are those who want to have a natural pregnancy, who do not necessarily want the child to have an ultrasound twice a month or every month, and who want to «give birth humanly,» Bross von Walderdorff said during this meeting.[155] Also, in 2018, she argued in Parliament for abolishing mandatory vaccinations.[156] She said then that "the increase in the frequency of children with allergies observed in recent years is a consequence of the large number of vaccinations in the first year of life. (…) In my experience, it seems to be particularly important not to vaccinate newborns in their first 24 hours of life."[157] She referred to COVID-19 vaccines as "an experimental genetic engineering liquid" and a "biocidal poison" that poses a threat to the health and life of pregnant women and their children.[158] Bross von Walderdorff also declared that her goal was to "inform about the threats resulting from the celebrity event known as the plandemic."[159] She publicly questioned the effectiveness of vaccinations and vaccines against COVID-19, downplayed the risks posed by the disease in the age group of children and people between 19 and 50, and compared vaccinations against to medical experiments conducted during World War II in extermination camps. In 2019, the media reported that Bross von Walderdorff treated patients without a registered practice and did not keep medical records. She appeared before the Supreme Medical Court for the theories she propagated, which punished her with a reprimand. In November 2021, her license to practice medicine was suspended for a year.[160] During the Supreme Medical Court hearing, a dem-

---

155 Ewa Furtak, "Protestowała w Poznaniu, wspiera ruch antyszczepionkowy. Kim jest lekarka Katarzyna Bross–Walderdorff?", *Wyborcza.pl*, 02.08.2021, accessed 06.09.2024, https://www.bielskobiala.wyborcza.pl/bielskobiala/7,88025,27402965,leczy–homeopatia–i–wspiera–ruch–antyszczepionkowy–kim–jest.html.

156 Ewa Furtak, "Lekarka niezłomna. Rodzice polecają ją sobie, gdy chcą zaświadczenie o przeciwwskazaniach do szczepienia", *Wysokie obcasy*, 20.11.2021, accessed 06.09.2024, https://www.wysokieobcasy.pl/wysokie–obcasy/7,173910,27825864,katarzyna–bross–walderdorff–lekarka–niezlomna–rodzice.html.

157 Iwona Dudzik, "Mobilizacja wiedzy kontra demonstracja", *Podyplomie.pl*, 08.07.2017, accessed 06.09.2024, https://podyplomie.pl/medical-tribune/27896,mobilizacja-wiedzy-kontra-demonstracja.

158 No author, "»Plandemia« i szczepionki groźne dla kobiet w ciąży? Fake news!", *Demagog*, 26.07.2022, accessed 06.09.2024, https://www.demagog.org.pl/fake_news/plandemia–i–szczepionki–grozne–dla–kobiet–w–ciazy–fake–news.

159 Ibid.

160 Ibid.

onstration was held in front of the Chamber building in Warsaw, attended by, among others, Justyna Socha. After the verdict was announced, the doctor declared that she would continue to treat patients but would not be able to write prescriptions or issue certificates, postponing the implementation of mandatory vaccinations for children.[161]

Alternative to the scientific consensus methods of treatment are also used by another member of the PSNLiN, an internist from Gomunice (Kamieńsk commune, near Radomsko, Łódź province) Izabela Kaczmarek–Matyszczyk. In 2015, she became known in the media as recommending very unusual "therapies." In addition to classical family medicine, she specializes in Chinese medicine and "total biology"[162] and "treats" with acupuncture and homeopathy.[163] Instead of medication, she suggested converting to Hinduism or Buddhism, hugging a birch tree, "taking a shower," and "having more sex."[164] "A human is a spiritual-physical-psychological being. You can't be healthy if you ignore any of these elements," the doctor emphasized, and she called those mentioned above "hugging a birch tree" silvotherapy and replied that there is a branch of science that deals with stimulating the body to heal itself by spending time in the presence of trees and shrubs.[165] She accepted and charged money for this advice.[166] When she signed an appeal by scientists and doctors who were opponents of mandatory vaccinations, Medical Court started a case against her. STOP NOP organized a March for Freedom in Poznań in her defence when this happened. The president of the nationalistic party National Movement (then Confederation MP Robert Winnicki), and a

161 Agata Szczerbiak, "Stowarzyszenie wątpiących medyków", *Polityka*, 02.06.2021, accessed 06.09.2024, https://www.polityka.pl/tygodnikpolityka/kraj/2120787,1,stowarzyszenie-watpiacych-medykow.read.

162 No author, "Pomoc specjalistów. lekarz Izabela Kaczmarek", *Polskie Stowarzyszenie Niezależnych Lekarzy i Naukowców*, no data, accessed 06.09.2024, https://www.psnlin.pl/pomoc-specjalistow,45.html.

163 Mk, "Na zdrowie ... brzoza i seks? Czyli na co skarżą się pacjenci ośrodka w Gomunicach", *Radomsko. Nasze Miasto,* 09.11.2015, accessed 06.09.2024, https://www.radomsko.naszemiasto.pl/na-zdrowie-brzoza-i-seks-czyli-na-co-skarza-sie-pacjenci/ar/c1-3565602.

164 Ibid.

165 Bż, "Piszą skargi na lekarkę, bo »kazała się przytulać do drzew«. »Nawet nie wiem, jak się bronić«", *TVN24*, 19.11.2015, accessed 06.09.2024, https://www.tvn24.pl/lodz/wiejska-lekarka-jak-szamanka-kazala-nam-przytulac-sie-do-drzew-ra595889-3318448.

166 No author, "Wiejska lekarka jak szamanka. Kazała pacjentom przytulać się do drzew", *WP.pl*, 20.11.2015, accessed 06.09.2024, https://www.kobieta.wp.pl/wiejska-lekarka-jak-szamanka-kazala-pacjentom-przytulac-sie-do-drzew-5982777517777537a.

journalist and opponent of mandatory vaccinations, religious orthodox Jan Pospieszalski,[167] came to attend.

There is no shortage of opponents of mandatory vaccinations among dentists. The most famous case is Krystyna Lisiecka-Opalko, a retired employee of the Department of Dental Propedeutics and Physical Diagnostics of the Pomeranian Medical University in Szczecin and a member of the board and treasurer of PSNLiN. She, like many doctors who oppose mandatory vaccinations, advocates treating COVID-19 with amantadine or ivermectin and also considers the Board of the Polish Society of Epidemiologists and Infectious Disease Physicians to be people associated with "Big Pharma." In her opinion, COVID-19 vaccines are not vaccines but untested genetic engineering liquids, from which the human body becomes a toxin producer, where the protein molecule introduced into the body is similar to the protein in the muscle fibres of the heart and attacks them, resulting in the formation of an adverse reaction after vaccination (ADR).[168] Similarly to Jarczewska and Bross von Walderdorff, in the film "Detailed information on the vaccination of students against COVID-19" on the official YouTube channel of the Małopolska Board of Education, another member of PSNLiN[169] and like Krystyna Lisiecka-Opalko, a dentist, Daniel Mazur, spoke out. He claimed that to avoid one hospitalization, 223 children must experience a severe adverse reaction after vaccination.[170] He also took part in Jaworzno, with Piotr Szlachtowicz, a publicist sympathetic to the Confederation, and Grzegorz Płaczek, an activist of the anti-vaccination movement and Confederation MP, in a pseudoscientific conference entitled "Human health in an unhealthy world."[171]

---

167 Sylwia Sałwacka, "114 lekarzy antyszczepionkowców w sądzie. Protest przed izbą lekarską, przyjechali Pospieszalski i Winnicki", *Wyborcza.pl*, 17.11.2023, accessed 06.09.2024, https://www.poznan.wyborcza.pl/poznan/7,36001,29569297,sprawa-114-lekarzy-antyszczepionkowcow-protest-przed-izba-lekarska.html.

168 No author, "KTO NAM ZMIENIŁ MEDYCYNĘ? Prof. Krystyna Lisiecka-Opalko PSNLiN", *Polskie Stowarzyszenie Niezależnych Lekarzy i Naukowców*, 17.05.2022, accessed 06.09.2024, https://www.psnlin.pl/video,kto-nam-zmienil-medycyne-prof-krystyna-lisiecka-opalko-psnlin,93.html.

169 No author, "Pomoc specjalistów. lek. stom. Daniel Mazur", *Polskie Stowarzyszenie Niezależnych Lekarzy i Naukowców*, no data, accessed 06.09.2024,https://www.psnlin.pl/pomoc-specjalistow,17.html.

170 Olga Szpunar, "Antyszczepionkowy film zamówiony przez Barbarę Nowak wciąż na kanale kuratorium. Katarzyna Lubnauer z KO: »Będziemy interweniować«", *Wyborcza.pl*, 20.07.22, accessed 06.09.2024, https://www.krakow.wyborcza.pl/krakow/7,44425,28705950,antyszczepionkowy-film-zamowiony-przez-barbare-nowak-wciaz-na.html.

171 No author, "Jaworzno. Konferencja niezależnych naukowców i lekarzy", *Tarnogorski.info*, 07.10.2021, accessed 06.09.2024, https://www.tarnogorski.info/jaworzno-w-niedziele-17-pa.

A specific group among the members of PSNLiN are dieticians, such as Jolanta Nalewaj-Nowak and Emilia Sykut Domańska. The first one became widely known due to the tragedy in Poznań, in the clinic she managed then. In mid-January 2019, an ambulance brought a 36-year-old patient in critical condition to the emergency department of the hospital on Szwajcarska Street in Poznań. Despite medical help, the woman died, leaving behind eight-month-old twins.[172] The probable cause of cardiac arrest and death was the DMSO preparation. This substance is an organic chemical compound used as a solvent. In large doses, it is very toxic. In Poland, it is not registered as a medicinal product. Proponents of alternative medicine consider it an effective remedy for cancer and muscle atrophy.[173] DMSO is administered through intravenous infusions, usually twice a week. The therapy lasts from 6 to 10 weeks. The cost of one infusion is PLN 400–800 (about 100–200 Euro).[174] It was used in the clinic and was managed by Nalewaj-Nowak at the time. The dietitian is also a supporter of other pseudoscientific therapies and an opponent of vaccinations. She also lectured at a conference on opponents of mandatory vaccinations: "What won't your doctor tell you?"[175] Another dietitian active in PSNLiN is Emilia Sykut Domańska. Her main public activity is intensive criticism of the WHO.[176] She also cooperates with the far-right party Confederation.[177] She participates in conferences of opponents of mandatory vaccinations, such as "The time of the pandemic and prospects for the future," which took place in Opole Lubelskie.[178]

172 Sylwia Sałwacka, "36-latka zmarła po pseudoterapii w ośrodku medycyny naturalnej. Odpowie za to pielęgniarka", *Wyborcza.pl*, 04.09.2021, accessed 06.09.2024, https://www.poznan.wyborcza.pl/poznan/7,36001,27533062,smierc-w-osrodku-medycyny-naturalnej-pielegniarka-odpowie-za.html.

173 Ibid.

174 Sylwia Sałwacka, "Co się wydarzyło w Klinice 100? Pacjentka nie żyje, szefowie ośrodka milczą, strona internetowa znika", *Wyborcza.pl*, 23.01.2019, accessed 06.09.2024, https://www.poznan.wyborcza.pl/poznan/7,36001,24392005,co-sie-wydarzylo-w-klinice-100-pacjentka-nie-zyje-szefowie.html.

175 Ibid.

176 No author, "Traktat antypandemiczny WHO – Wysłuchanie publiczne – dr Emilia Sykut-Domańska z PSNLiN",, *Polskie Stowarzyszenie Niezależnych Lekarzy i Naukowców*, 12.04.2022, accessed 06.09.2024, https://www.psnlin.pl/news-psnlin,traktat-antypandemiczny-who-wysluchanie-publiczne-dr-emilia-sykut-domanska-z-psnlin,56.html.

177 Paweł Januszek, "[KORONA LUBLIN / WYKŁAD] Dr n. rol. Emilia Sykut-Domańska: »Żyć zdrowo w niezdrowych czasach«", *Postaw na wolność*, 15.12.2021, accessed 06.09.2024, https://www.postawnawolnosc.pl/2021/12/15/korona-lublin-wyklad-dr-n-rol-emilia-sykut-domanska-zyc-zdrowo-w-niezdrowych-czasach.

178 Region Lubelski – media obywatelskie. "Konferencja dla wolności, bezpieczeństwa i zdrowia. Opole Lubelskie, 20.02.2022", *Rumble*, no data, accessed 06.09.2024,https://

Among the people listed as "experts" in the movement opposing mandatory vaccinations, in addition to doctors, there are also scientists and academic lecturers, including those associated with medical sciences. An important propagator of non-scientific vaccine narratives is the physicist by education, Mariusz Błochowiak, and the Ordo Medicus Institute, which he manages and is affiliated with the Ordo Iuris Institute for Legal Culture, mentioned earlier. Although he does not have a medical education, Błochowiak has repeatedly spoken out critically about the methods of treating COVID-19 recognized by science and the restrictions related to the pandemic. Błochowiak is the editor of the *White Paper on the Coronavirus Pandemic. Facts and Data Hidden from the Public Opinion*[179] and four books from the *False Pandemic* series, in which he claims that there is no epidemic or coronavirus pandemic, neither in Poland nor in any other country[180] and that we are witnessing a "repeat of history," (like in the case of the "false swine flu pandemic"), but this time on a much larger, global scale. As Błochowiak states: "Today we already know that the pandemic announced by WHO in 2009 was fake news, and the vaccines that were purchased by various countries at the time were not only medically unnecessary but also harmful.[181] He also believes that every vaccine carries a risk of complications and death,[182] and masks do not protect against infection.[183]

Additionally, in his opinion, the mortality rate of COVID-19 in adults is at the level of the mortality rate of seasonal flu – 0.15%.[184] Błochowiak believes that neither the CDC nor the WHO has presented scientific data on the positive effects of using masks in public places, the tests do not detect whether someone is infected or not,[185] and the excess deaths during the pandemic were

---

www.rumble.com/vvh1d4–konferencja–dla–wolnoci–bezpieczestwa–i–zdrowia.–opole–lubelskie–20.02.2022.html.

179 Błochowiak, "Biała księga pandemii koronawirusa".

180 See further: Mariusz Błochowiak, ed., *Fałszywa pandemia. Krytyka naukowców i lekarzy* (Warszawa: Fundacja Osuchowa, 2020).

181 See further: Mariusz Błochowiak, ed., *Fałszywa pandemia. Krytyka naukowców i lekarzy 4* (Warszawa: Fundacja Osuchowa, 2021).

182 Ibid.

183 See further: Mariusz Błochowiak, ed., *Fałszywa pandemia. Krytyka naukowców i lekarzy 3* (Warszawa: Fundacja Osuchowa, 2021).

184 No author, "Ordo Medicus o skuteczności szczepionki przeciw Covid–19 i szczepieniach dzieci", *Stowarzyszenie Polonia Christiana*, 20.08.2021, accessed 06.09.2024, https://www.poloniachristiana.org/ordo–medicus–o–skutecznosci–szczepionki–przeciw–covid–19–i–szczepieniach–dzieci.

185 PZ, "Fizyk w poznańskiej księgarni neguje covid–19. »Maski nieskuteczne, pandemia niegroźna«", *Epoznan.pl*, 18.09.2022, accessed 06.09.2024, https://www.epoznan.pl/

caused not by COVID-19, but by government restrictions and the paralysis of the health service.[186] The "social commission of inquiry on the pandemic" created by opponents of mandatory vaccinations operates in Warsaw under his leadership.[187]

The PSNLiN audit committee member, Dr Piotr Witczak, a medical biologist, is very active in public. He studied at the Medical University of Lodz in the Department of Experimental Immunology. He also worked at the government Agency for Health Technology Assessment and Tariff System. Witczak often appears in the media and, in addition to PSNLiN, also cooperates with the Ordo Medicus Institute. He also spoke at the "What Your Doctor Won't Tell You" conference. Witczak spoke very critically about COVID-19 vaccination programs. He demanded an immediate suspension of vaccinations among children because, in his opinion, the mortality rate of SARS-CoV-2 in them is close to zero, similar to seasonal flu.[188] Witczak said in the media that the most popular mRNA vaccines can cause more harm than good, and "the fight against the virus has killed more people than the virus itself."[189] He falsely increased the number of serious post-vaccination complications in his statements, saying that there were 1 in 1,000 vaccinations when, in reality, it was 1 in 100,000.[190] Witczak also suggested that during the testing phase, there was a deliberate plan to conceal the true frequency of adverse events of the vaccine being tested.[191] He placed his theses in a direct conspiratorial narrative. As he stated: "The recipe was simple. Choose an infectious disease that people commonly fall ill with, caused by various pathogens, develop a test with the previously

---

news–news–132558–fizyk_w_poznanskiej_ksiegarni_neguje_covid_19_maski_nieskutec zne_pandemia_niegrozna.

186 No author, "O nas", *OrdoMedicus.org*, no data, accessed 06.09.2024,https://www.ordome dicus.org/o–nas.

187 Its members include Piotr Witczak, Dorota Sienkiewicz, Kornelia Polok, Roman Zieliński, Ryszard Rutkowski, Marek Błażejak, Jerzy Milewski, Paweł Basiukiewicz, Marek Sobolewski and Paweł Klimczewski (Confederation party candidate in the 2023 Polish parliamentary elections).

188 No author, "Szczepionki nie powstrzymują koronawirusa? Fake news!", *Demagog*, 12.10.2021, accessed 06.09.2024, https://www.demagog.org.pl/fake_news/szczepionki–nie–powstrzymuja–koronawirusa–fake–news.

189 Kontrrewolucja.net, "Dr Witczak przypomina, że na temat walki z tzw. pandemią »nie było żadnej debaty eksperckiej«. »Walka z wirusem zabiła więcej ludzi niż on sam«", *Kontrrewolucja.net*, 01.10.2021, accessed 06.09.2024, https://www.kontrrewolucja.net/wia domosci/dr–witczak–przypomina–ze–na–temat–walki–z–tzw–pandemia–nie–bylo–z adnej–debaty–eksperckiej–walka–z–wirusem–zabila–wiecej–ludzi–niz–on–sam.

190 No author, "Szczepionki nie powstrzymują koronawirusa? Fake news!".

191 Dorota Tuńska, "Dr Witczak: Szczepienie dzieci na COVID–19 to skandal".

mentioned flaws, and test it on a mass scale. In this way, we have a situation in which we can practically turn every infectious season into an epidemic of 'one virus.'"[192] Witczak declared that the deaths were caused by an unhealthy lifestyle, which caused many elderly people to die every year as a result of respiratory infections and weakened immunity. He stated: "This was very well exploited by the «designers» of this phenomenon because we have to talk about the purposefulness of what happened, about specific political goals,"[193] and "the studies on childhood vaccines were deliberately designed to hide the true number of adverse events from the public."[194] At the same time, Witczak provides pseudo-medical services. He offers remotely "consultations." A package of four "consultations" by phone or via Messenger costs PLN 850 (about 200 Euro). They concern dozens of diseases, including depression, bulimia, panic attacks, and even multiple sclerosis.[195] He advises "complementary therapy for autism" and "treats holistically." He claims that practically every person with autism spectrum disorder (ASD) can be effectively helped with a chance of complete recovery. However, he has only completed a dietetics course.[196] The list of reservations raised by doctors and scientists about Piotr Witczak's activity is long. This list includes Cutler's chelation method – therapy, which involves removing heavy metals from the body, is neither a tested nor medically approved treatment method. However, it was supposed to be used by Witczak on children with quadriplegia, autism, and metabolic disorders. The court ordered the prosecutor's office to investigate Witczak's case thoroughly.

The lists of "independent" scientists and doctors and the movement's demonstrations and protests often feature the names of Roman Zieliński and Kornelia Polok, who work closely together. Together, they signed all the letters to the authorities questioning the widespread vaccination against COVID-19 and the existence of the pandemic, and they cooperated with both PSNLiN and

192 Rafał Górski, "Pandemia COVID. Po czyjej stronie są nauka i fakty?", *Tygodnik Spraw Obywatelskich,* 07.03.2023, accessed 06.09.2024, https://instytutsprawobywatelskich.pl/pandemia-covid-po-czyjej-stronie-sa-nauka-i-fakty.

193 Ibid.

194 Piotr Witczak, "Trybunał Konstytucyjny – oświadczenie dr Piotra Witczaka z PSNLiN nt. obowiązku szczepienia dzieci", *Polskie Stowarzyszenie Niezależnych Lekarzy i Naukowców,* 08.05.2023, accessed 06.09.2024, https://www.psnlin.pl/video,trybunal-konstytucyjny-oswiadczenie-dr-piotra-witczaka-z-psnlin-nt-obowiazku-szczepienia-dzieci,135.html.

195 Jacek Wykowski, »Nie każdy doktor nauk medycznych może leczyć, ale każdy może doradzać«. »Skala zjawiska poraża«", *Puls Medycyny,* 18.01.2023, accessed 06.09.2024, https://www.pulsmedycyny.pl/nie-kazdy-doktor-nauk-medycznych-moze-leczyc-ale-kazdy-moze-doradzac-skala-zjawiska-poraza-1175123.

196 Ibid.

Ordo Medicus. They also participate in the "Week of Life and Freedom!" organized by the Father Piotr Skarga Association of Christian Culture (a sister organization of Ordo Iuris). Until 2014, Zieliński worked as an assistant professor at the departments of genetics of various, but mainly small, universities. However, Zieliński is not an expert in medical, but plants and small animals, genetics. None of Zieliński's scientific works concern human genetics, but only, for example, microorganisms or mussels.[197] Despite this, he speaks very often and widely about COVID-19. Zieliński questions the validity of PCR tests for coronavirus and the use of the COVID-19 vaccine. He is also one of the signatories of the "Appeal of scientists and doctors to stop vaccinations against SARS-CoV-2 and return to normality." During a conference organized by STOP NOP, he warned that COVID-19 vaccines are dangerous because they can cause, for example, infertility. "It is a product of genetic engineering," he emphasized.[198] As he declared in one of his interviews: "The preparation cannot be called a vaccine. It is not a vaccine in its essence. I call it an mRNA construct, recombinant RNA, a preparation that is a work of genetic engineering."[199] To sum up, when asked "what can PCR tests say about actual infection," he said, "They cannot say anything of the sort."[200] So why were the tests introduced? "They are intended to lend credibility to the coronavirus pandemic,"[201] believes Zieliński. In his opinion, "recommendations have been introduced that contradict ancient knowledge and the essence of man as a biological species. A ban on leaving the house, masks, excessive sterilization of rooms and the body – these are recommendations that lead to increased susceptibility to the disease, and not those that protect against it."[202] He sees this as a conspiracy, thanks to which the government wants to weaken the

197 Sylwia Sałwacka, "Genetyk z Radomia został guru antyszczepionkowców. Naukowcy z PAN: »To ekspert od małży«", *Wyborcza.pl*, 12.01.2020, accessed 06.09.2024, https://www.poznan.wyborcza.pl/poznan/7,36001,26677114,genetyk-z-radomia-zostal-guru-antyszczepionkowcow-naukowcy.html.

198 Ibid.

199 Agnieszka Kisielewska, "PROF. ZW. DR HAB. ROMAN ZIELIŃSKI "O SZCZEPIONCE GENETYCZNEJ PFIZERA I TESTACH PCR", za: *Biuletyn Informacji Publicznej Starostwa powiatowego w Siedlcach*, no data, accessed 06.09.2024, https://www.bip.powiatsiedlecki.pl/plik,8953,zalacznik-nr-10-prof-zw-dr-hab-roman-zielinski-szczepienia-inzynieria-genetyczna-sars-cov-2.pdf.

200 See further: Mariusz Błochowiak, ed., *Fałszywa pandemia. Krytyka naukowców i lekarzy* 2 (Warszawa: Fundacja Osuchowa, 2020).

201 No author, "Wywiad z Romanem Zielińskim zawiera nieprawdziwe informacje", *Demagog*, 15.01.2021, accessed 06.09.2024, https://www.demagog.org.pl/fake_news/wywiad-z-romanem-zielinskim-zawiera-nieprawdziwe-informacje.

202 Ibid.

population and prepare fertile ground for other infections, e.g., by "creating" more viruses.[203]

Kornelia Polok, a close associate of Zieliński and co-author of his book *Covidian dizzyness*[204] (as well as texts published in the *False Pandemia* book series), has a similar opinion. She, in turn, is a plant geneticist. Her opinion on COVID-19 and vaccines is generally analogous to that of Zieliński. She is convinced that the mortality rate of the disease is comparable to that of the flu ("The flu or cold has been changed to COVID"[205]) and has the same percentage of a more severe course and mortality in the event of the disease as in the case of receiving the vaccine. She believes that having the coronavirus ("a mild cold virus"[206]) has fewer side effects than the vaccine itself, and it will be profitable for pharmaceutical companies to hide data on complications so that more and more people get vaccinated, which is why "something like this cannot be trusted"[207]. She adds that, in her opinion, the incidence of the disease at a level of about 1% of the population does not justify mass vaccinations, and "it cannot be ruled out that the interference of virus fragments in our cells may activate some of our previously inactive genes, and deactivate some active genes – which will result in cancers.[208] In her opinion, PCR tests are ineffective because they involve multiplying viral DNA, while the SARS-CoV-2 coronavirus is an RNA virus. "The enzyme that carries out this reaction will not multiply RNA," she explains, and to examine the genetic material of the RNA virus, "the RNA must be converted to DNA." In addition, in her opinion, this reaction creates "a fairly unstable hybrid, a molecule that is partly composed of RNA and DNA," which "is not suitable for direct analysis".[209] Meanwhile, RNA isolation has been routine for years. An RNA-DNA hybrid is not created;

203 Ibid.

204 Zieliński and Polok, *Kowidowy zawrót głowy.*

205 Krzysztof Jabłonowski, "Wywiad o nieskuteczności testów na COVID–19: wyjaśniamy manipulacyjne tezy", *TVN24*, 03.09.2020, accessed 06.09.2024, https://konkret24.tvn24.pl/zdrowie/wywiad-o-nieskutecznosci-testow-na-covid-19-wyjasniamy-manipulacyjne-tezy-ra1028628-ls5789901.

206 Jan Bodakowski, "Za duża jest liczba zgonów po wstrzyknięciu preparatu zwanego szczepionką na covid [OPINIA]", *Media Narodowe*, 19.08.2021, accessed 06.09.2024, https://www.medianarodowe.com/2021/08/19/za–duza–jest–liczba–zgonow–po–wstrzyknieciu–preparatu–zwanego–szczepionka–na–covid–opinia.

207 Redakcja *Plonsk24.pl*, "»Gdy pójdziemy drogą szczepień, to jaka czeka nas przyszłość?«", *Plonsk24.pl*, 02.01.2021, accessed 06.09.2024, https://www.plonsk24.pl/aktualnosc–14–3777–gdy_pojdziemy_droga_szczepien_to_jaka.html.

208 Jan Bodakowski, "Za duża jest liczba zgonów".

209 Jabłonowski, "Wywiad o nieskuteczności testów na COVID–19: wyjaśniamy manipulacyjne tezy".

it is only so-called complementary DNA, and the resulting material must not be subjected to additional processing.[210]

Another scientist associated with Ordo Medicus is Marek Sobolewski, a statistician from the University of Rzeszów. He is a very active participant in "symposia," "conferences," and publications of opponents of mandatory vaccinations, especially those related to religious activity. He contributed to the *White Paper on the Coronavirus Pandemic. Facts and Data Hidden from the Public Opinion* published by Ordo Medicus,[211] and the Ordo Iuris report *Restrictions on Rights and Freedoms Introduced in Poland in Connection with the Covid-19 Epidemic.*[212] He spoke during the meeting of the "pandemic commission" convened by Ordo Medicus and at a conference convened by this organization devoted to "the effects of government restrictions on the functioning of the health service in 2020 and 2021," as well as at the "Faces of the Pandemic" symposium organized in Toruń by monk Tadeusz Rydzyk and many times in the media he manages, Radio Maryja and TV Trwam.[213] There, Sobolewski claimed that "sanitary segregation" has no scientific basis,[214] there is no evidence of the effectiveness of lockdowns, masks, and other non-pharmaceutical measures,[215] and that there is no basis for treating vaccinated and unvaccinated people differently. He also suggested that infection rates are higher in "voivodeships with a higher level of vaccination." He said that although he did not conclude that vaccines stimulate the number of infections or weakened immunity, there is no basis for treating vaccinated and unvaccinated people differently because the latter are not the cause of virus transmission.[216]

---

210 Ibid.

211 Błochowiak, "Biała księga pandemii koronawirusa".

212 Bernaciński and Kwaśniewski, ed., Ograniczenia praw i wolności wprowadzone w Polsce".

213 Radio Maryja, "»Oblicza pandemii«: dr Marek Sobolewski", *YouTube,* 29.01.2022, accessed 06.09.2024, https://www.youtube.com/watch?v=NbxX-yD4oRc.

214 Jakub Czermiński and Jakub Mikulski, "Ojciec Rydzyk bierze się za covid. W Toruniu zorganizuje sympozjum", *Rzeczpospolita,* 24.01.2022, accessed 06.09.2024, https://www.rp.pl/ochrona-zdrowia/art19320451-ojciec-rydzyk-bierze-sie-za-covid-w-toruniu-zorganizuje-sympozjum.

215 DoRzeczy.pl, "Dr Sobolewski: Nie ma dowodów skuteczności środków niefarmaceutycznych", *DoRzeczy,* 12.02.2022, accessed 06.09.2024, https://www.dorzeczy.pl/opinie/262321/dr-sobolewski-nie-ma-dowodow-skutecznosci-srodkow-niefarmaceutycznych.html.

216 Paulina Błaszkiewicz, "Koronasceptycy w szkole Rydzyka: Lockdown to katastrofa, maseczki to absurd, a pandemia generuje hajs w mediach", *Wyborcza.pl,* 29.01.2021, accessed 06.09.2024, https://torun.wyborcza.pl/torun/7,48723,28052653,koronasceptycy-w-szkole-rydzyka-lockdown-to-katastrofa-maseczki.html.

Among the scientists associated with the movement against mandatory vaccination (participants in debates, signatories of open letters, speaking publicly on the issue of vaccines contrary to scientific knowledge), there were more names of academics in the field of medical sciences. It is worth noting that almost half of this list are scientists from the Medical University of Białystok. Among them are Prof. Maria Sobaniec-Łotowska and Prof. Stanisław Sulkowski, who, commissioned by the Białystok Metropolitan Curia, were investigating the so-called "miracle in Sokółka," which allegedly took place in October 2018. They claimed that fragments of heart muscle tissue were found on the *hostia* (sacramental bread) in the Sokółka church. Then, in an official statement, they expressed outrage at suggestions that they had conducted their research in a pseudoscientific manner. The Medical University of Białystok distanced itself from the results of these studies in its official position.[217] The *hostia* was most likely dyed red by a bacteria called *serratia marcescens*. It produces a red pigment and can grow on bread.[218]

217 Maciej Chołodowski, "Koronawirus. Prawicowi lekarze i naukowcy przeciwni szczepieniom. Prof. Flisiak: Ten apel to nieodpowiedzialność", *Wyborcza.pl*, 02.12.2020, accessed 06.09.2024, https://www.bialystok.wyborcza.pl/bialystok/7,35241,26567278,koronawirus-prawicowi-lekarze-i-naukowcy-sprzeciwprzeciwni.html.

218 No author, "»Cud w Sokółce« to bakteria?", *TVN24*, 29.10.2009, accessed 06.09.2024, https://www.tvn24.pl/polska/cud-w-sokolce-to-bakteria-ra113674-3698513.

CHAPTER 4

Conspiracy Influencers

Alternative Media and Non-governmental Organizations Promoting the Postulates of Movements Fighting against Mandatory Vaccinations and 5G

The pattern of spreading anti-scientific conspiracy narratives outside the milieu of people interested in them looked similar in Poland to other European Union countries.[1] At first, the prominent mouthpieces were internet televisions run by ufologists, supporters of alternative medicine and new spirituality. Over time, the topic reached far-right YouTube channels, which could count on a much larger audience. As a result, non-governmental organizations became interested in it, adding the "anti-scientific narratives" segment to their general activities. All these activities moved the topic from a niche and margin to politicians of parties considering themselves "anti-system" and, thanks to them, to the mainstream of Polish public discourse.

Internet television (primarily YouTube channels) aimed at supporters of alternative medicine, ufologists, and people associated with groups of new spirituality were the main transmission belts in Poland at the beginning of existence of the movements opposing fifth-generation telephones and mandatory vaccinations, publicizing their postulates and activity. These media outlets broadcast live videos at demonstrations, rallies, festivals, and lectures for these groups, published interviews with activists and "experts" of the movement, disseminated numerous conspiracy narratives regarding the COVID-19 virus, vaccines, pandemics, and 5G, and repeatedly combined them into various hybrid narratives. The key media channels for these ideas in Poland were "Niezależna Telewizja" by Janusz Zagórski, "Porozmawiajmy.tv" by Aleksander Berdowicz and "Ukryte Terapie" by Jerzy Zięba.

Created in 2008 by Janusz Zagórski, "Niezależna Telewizja" is primarily a YouTube channel with 14 thousand subscribers. Until recently, however, it had 330 thousand and was the largest of this type of media (approx. 8000 video recordings) in Poland and one of the oldest. However, it was removed from the portal by YouTube, and the currently existing version is a new version slowly rebuilding its reach. Zagórski is, as he calls himself, an ufologist and "researcher of unexplained phenomena." He is a hotelier by profession. Since

1 Insa Backhaus et al., "Far-right political ideology and COVID-19 vaccine hesitancy: Multilevel analysis of 21 European countries", *Social Science & Medicine* 335 (2023):116–227.

 | DOI:10.1163/9789004747852_006

the late 80s, he has been actively researching the "UFO phenomenon." Since 1997, he has organized the largest cyclical event in Poland devoted to ufological topics – "UFO Forum." From 1998 to 2007, he was one of the experts on "unexplained phenomena" in the programme "Nie do wiary," broadcast on biggest private Polish television, TVN.[2] Zagórski also organizes festivals for supporters of alternative medicine and ufology, such as the "Harmonia Kosmosu," the "Festiwal Wolności, Świadomości i Zdrowia" i "Festiwal Topinamburu."

In his video materials and on the stage of the festivals he organizes, "experts' and activists of the movement against mandatory vaccinations, such as Jerzy Zięba,[3] Justyna Socha,[4] and Hubert Czerniak,[5] have repeatedly appeared, as well as "experts" (Jacek Nowak,[6] Diana Wojtkowiak[7]) and activists (Ewa Pawela,[8] Jadwiga Łopata, and Julian Rose[9]) fighting the fifth-generation telephone network and "electrosmog." Zagórski has organized demonstrations combining the topics of the pandemic, 5G, and vaccinations.[10] He has

2 No author, "Ufolog u Prokopa. Janusz Zagórski wierzy, że obcy są wśród nas", *WP*, no data, accessed 06.09.2024, https://web.archive.org/web/20191120185150/https://ztymicosieznaja.wp.pl/artykul/ufolog-u-prokopa-janusz-zagorski-wierzy-ze-obcy-sa-wsrod-nas.

3 NTV, "Debata – Punkt zwrotny dla Polski i Świata – prof. M. Matyja, J. Zięba, W. Sumliński, J. Zagórski", *YouTube*, 22.06.2022, accessed 06.09.2024, https://www.youtube.com/watch?v=L9EM47tkY44.

4 NTV, "Obrona suwerenności zdrowotnej – Justyna Socha", *YouTube*, 24.06.2022, accessed 06.09.2024, https://www.youtube.com/watch?v=K7sFiTdupvA.

5 NTV, "Wystąpienie dr Huberta Czerniaka", *YouTube*, 25.06.2022, accessed 06.09.2024, https://www.youtube.com/watch?v=acvfCTSHSEI.

6 Koalicja Polska wolna od 5G, "5G – Stop 5G – Dlaczego protestujemy – Janusz Zagórski w rozmowie z prof. Jacek Nowak", *YouTube*, 20.01.2020, accessed 06.09.2024, https://www.youtube.com/watch?v=tOAhMLRcU5Y.

7 No author, "5G – Pola torsyjne a dewastacja ludzkiego zdrowia – Janusz Zagórski w rozmowie z Dianą Wojtkowiak", *Niezależna Telewizja*, 01.01.2020, accessed 06.09.2024, https://niezaleznatelewizja.pl/index.php/aiovg_videos/5g-pola-torsyjne-a-dewastacja-ludzkiego-zdrowia-janusz-zagorski-w-rozmowie-z-diana-wojtkowiak.

8 Niezależna Telewizja, "5G powstrzymać zbrodniczą falę _ Janusz Zagórski i Ewa Pawela", *BitChute*, 03.06.2020, accessed 06.09.2024, https://www.bitchute.com/video/yzEMZyrvug2v.

9 No author, "GLOBALNY PROTEST STOP 5G – Jadwiga Łopata i Julian Rose", *Niezależna Telewizja*, 07.04.2020, accessed 06.09.2024, https://niezaleznatelewizja.pl/index.php/aiovg_videos/globalny-protest-stop-5g-jadwiga-lopata-i-julian-rose.

10 Magdalena Kozioł, "Antycovidowcy mają własną telewizję. Wiele programów negujących koronawirusa bije rekordy popularności", *Wyborcza.pl*, 14.10.2020, accessed 06.09.2024, https://wroclaw.wyborcza.pl/wroclaw/7,35771,26393182,wroclawski-trzon-antycovidowego-protestu-ufolog-sygnalistka.html.

also given public lectures[11] or posted recordings on his Niezależna Telewizja website, spreading anti-scientific conspiracy narratives. To a large extent, he mixes his activities, which can be described as activism, with the function of a medium for movements postulates. At the same time, Zagórski seeks contacts with mainstream politicians and cooperates with Kukiz'15 Jarosław Sachajko.

The second important channel for propagating conspiratorial and anti-scientific narratives from the circles of supporters of "alternative medicine" is the YouTube channel "Porozmawiajmy.tv" run by Aleksander Berdowicz. He is a retired soldier and a Higher Officers' Radiotechnical School and English Philology graduate. After leaving service, he became one of the leaders of the Polish new spirituality circles. His YouTube channel is a crucial hub for the exchange of information, advertising, and promotion for followers of the occult, ufology, and spiritualism, where this content is mixed with conspiratorial narratives, including those of a political nature. Porozmawiajmy.tv has 133 thousand subscribers on YouTube (2024). The topic of vaccines has appeared there many times. This channel has featured activists and "experts" of the movement, such as Justyna Socha,[12] Jerzy Jaśkowski,[13] Hubert Czerniak,[14] and Włodzimierz Julian Korab-Karpowicz.[15] Similarly, 5G in a conspiratorial and non-scientific approach, has appeared many times in Porozmawiajmy.tv video materials, in which Diana Wojtkowiak,[16] Jadwiga

11 No author, "Konferencja w Krakowie organizowana przez J. Rose i J. Łopata Wystąpienie Janusza Zagorskiego", *Niezależna Telewizja*, 01.04.23, accessed 06.09.2024, https://niezaleznatelewizja.pl/index.php/aiovg_videos/konferencja-w-krakowie-organizowana-przez-j-rose-i-j-lopata-wystapienie-janusza-zagorskiego.

12 Porozmawiajmy TV, "Co powinieneś wiedzieć o szczepieniach – Justyna Socha", *YouTube*, 23.11.2017, accessed 06.09.2024, https://www.youtube.com/watch?v=bpv835EHk30.

13 Porozmawiajmy TV, "Patologie współczesnej medycyny i inne sprawy do naprawy, część 1 – dr Jerzy Jaśkowski", *YouTube*, 06.02.2022, accessed 06.09.2024, https://www.youtube.com/watch?v=E3HzEDxPFnE.

14 Porozmawiajmy TV, "Czy szczepić dziecko przeciwko wzw typu B – dr Hubert Czerniak", *YouTube*, 30.04.2018, accessed 06.09.2024, https://www.youtube.com/watch?v=1BFrkQiFoCQ.

15 Porozmawiajmy TV, "Wielokulturowość degraduje i niszczy ciągłość pokoleń – prof. Włodzimierz Julian Korab-Karpowicz", *Youtube*, 31.01.2018, accessed 06.09.2024, https://www.youtube.com/watch?v=LvkP_QLdW30&t=57s.

16 No author, "Czym jest pole torsyjne i jak z niego korzystać – dr Diana Wojtkowiak", *Porozmawiajmy TV*, 01.09.2020, accessed 06.09.2024, https://porozmawiajmy.tv/czym-jest-pole-torsyjne-i-jak-z-niego-korzystac-dr-diana-wojtkowiak.

Łopata,[17] and Ewa Pawela[18] appeared. Jerzy Zięba[19] and Hanna Kazahari,[20] who combine both narratives, were also highlighted there. What is also worth mentioning Berdowicz had previously collaborated closely with Łopata and Julian Rose in the GMO-Free Poland Association.

The Porozmawiajmy.tv channel is also an essential space for the interpenetration of conspiracy narratives related to vaccines, 5G, and the pandemic with the postulates of the Polish far-right. Berdowicz's recordings featured, among others, Confederation MEP Grzegorz Braun,[21] former Confederation MPs Robert Majka[22] and Janusz Korwin-Mikke,[23] Confederation candidate for MP and MEP Hubert Czerniak,[24] as well as Kukiz'15 MP Jarosław Sachajko.[25]

Another important channel for the spread of anti-scientific narratives about vaccines, the COVID-19 pandemic, and 5G are the publications, video materials, and websites of Jerzy Zięba. This certified mining engineer does not

---

17 No author, "Uczyńmy Polskę wolną od 5G – Jadwiga Łopata", *Porozmawiajmy TV*, 07.03.19, accessed 06.09.2024, https://porozmawiajmy.tv/uczynmy-polske-wolna-od-5g-jadwiga-lopata.

18 No author, "Technologia 5G jako broń technotroniczna – Ewa Pawela", *Porozmawiajmy TV*, 28.01.2019, accessed 06.09.2024 https://porozmawiajmy.tv/technologia-5g-jako-bron-technotroniczna-ewa-pawela.

19 No author, "Wszystko co musisz wiedzieć o tej epidemii – Jerzy Zięba", *Porozmawiajmy TV*, 19.04.2020, accessed 06.09.2024, https://porozmawiajmy.tv/wszystko-co-musisz-wiedziec-o-tej-epidemii-jerzy-zieba.

20 No author, "W Polsce stopniowo wprowadza się faszyzm – Benjamin Fulford, Fabiola Tsugami-Shaba i Hanna Kazahari", *Porozmawiajmy TV*, 29.04.2020, accessed 06.09.2024, https://porozmawiajmy.tv/w-polsce-stopniowo-wprowadza-sie-faszyzm-benjamin-fulford-fabiola-tsugami-shaba-i-hanna-kazahari.

21 No author, "Współczesne niewolnictwo, plany Żydów przeciw Żydom i palenie zielarek – Grzegorz Braun", *Porozmawiajmy TV*, 18.08.2016, accessed 06.09.2024, https://porozmawiajmy.tv/wspolczesne-niewolnictwo-plany-zydow-przeciw-zydom-i-palenie-zielarek-grzegorz-braun.

22 No author, "Ludobójstwo Polaków w XXI-wieku – poseł Majka i grupa anty5G", *Porozmawiajmy TV* 09.06.2019, accessed 06.09.2024, https://porozmawiajmy.tv/ludobojstwo-polakow-w-xxi-wieku-posel-majka-i-grupa-anty5g.

23 No author, "Handel dziećmi, gwiazdy Dawida i znaczenie sądownictwa – Janusz Korwin-Mikke, z komentarzem Leszka Bubla", *Porozmawiajmy TV*, 17.03.2020, accessed 06.09.2024, https://porozmawiajmy.tv/handel-dzieci-gwiazdy-dawida-i-znaczenie-sadownictwa-janusz-korwin-mikke-z-komentarzem-leszka-bubla.

24 No author, "Dr Hubert Czerniak w Gdańsku, 17.02.2018, część 1", *Porozmawiajmy TV*, 17.03.2018. accessed 06.09.2024, https://porozmawiajmy.tv/dr-hubert-czerniak-w-gdansku-17-02-2018-czesc-1.

25 No author, "Zakazane krojenie marchewki, roundup na deser i referendum bez mocy – dr Jarosław Sachajko", *Porozmawiajmy TV*, 09.06.2016, accessed 06.09.2024, https://porozmawiajmy.tv/zakazane-krojenie-marchewki-roundup-na-deser-i-referendum-bez-mocy-dr-jaroslaw-sachajko.

have a medical education. He studied at the Faculty of Mining and Metallurgy Machines at the University of Science and Technology in Kraków, later cleaned restaurants in Germany, and worked as a cook, diver, and salesman in Australia. He describes himself as an independent journalist and a certified hypnotherapist and naturopath.[26] In his publications, he promotes therapies inconsistent with current medical knowledge. Jerzy Zięba questions the validity of vaccinations,[27] believes that cancer can be treated with vitamin C[28] and amygdalin[29] and schizophrenia with vitamin B3 supplementation,[30] claims that hypnosis can enlarge a woman's breasts, and declares that oesophageal reflux, hyperthyroidism, and even Crohn's disease can be cured in a few days. He has spoken about vaccines many times, each time critically and repeating numerous conspiracy narratives related to vaccination.[31] He also linked them to 5G technology, accusing it of causing to the pandemic.[32]

The contacts mentioned above between Zagórski and Berdowicz and far-right politicians are not the only channel through which conspiracy narratives about 5G and vaccines from alternative medicine circles penetrate political organizations. However, participation in the shows of the channels mentioned above allowed the anti-5G and anti-vaccine movements to go beyond

26 DSZ, "»A piersi rosną i rosną«. Eksperci ostrzegają przed jego teoriami, on nadal chce »leczyć« ludzi", *Onet*, 13.12.2021, accessed 06.09.2024, https://wiadomosci.onet.pl/kraj/kim–jest–jerzy–zieba–i–czemu–lekarze–go–nie–uznaja–tresci–zagrazajace–zyciu/8q6nb4b.

27 Katarzyna Burda, "Szczepionkowy i nowotworowy szarlatan oraz jego wyznawcy", *Newsweek*, 16.04.2016, www.archive.is/20170928113638/http://www.newsweek.pl/wiedza/nauka/szczepionkowy–szarlatan–i–jego–wyznawcy,artykuly,384030,1.html.

28 Luq, "»Jestem listonoszem, który przyniósł niektórym lekarzom złą wiadomość«. Zięba w *Skandalistach*", *Polsat News*, 29.04.2017, accessed 06.09.2024, https://www.polsatnews.pl/wiadomosc/2017–04–29/jerzy–zieba–w–skandalistach.

29 Paulina Łopatniuk, "Wyleczy raka i powiększy biust, czyli Zięba na uniwersytecie", *Patolodzy na klatce* (blog), no data, accessed 06.09.2024, https://www.web.archive.org/web/20170607024459/https://patolodzynaklatce.wordpress.com/2015/11/24/wyleczy–raka–i–powiekszy–biust–czyli–zieba–na–uniwersytecie.

30 Anna Kaczmarek, "Czego naucza Jerzy Zięba w Polskim Radiu? Przesłuchałam audycję i boję się, że ktoś uwierzy w to, co on tam wygaduje", *naTemat.pl*, 31.10.2016, accessed 06.09.2024, https://natemat.pl/zdrowie/193489,czego-naucza-jerzy-zieba-w-polskich-radiu-tym-razem-wzial-sie-za-psychiatrie.

31 Jerzy Zięba, "PRZERAŻAJĄCE, NIGDZIE NIENAGŁAŚNIANE FAKTY O SZCZEPIONCE MRNA", *Jerzy Zięba TV*, 03.01.2021, accessed 06.09.2024, https://jerzyzieba.com/szczepienia.

32 Paweł Walewski, "Zięba i Socha znów w natarciu. Kto zyskuje, a kto traci na zdrowotnej dezinformacji?", *Polityka*, 19.04.2020, accessed 06.09.2024, https://www.polityka.pl/tygodnikpolityka/nauka/1953210,1,zieba–i–socha–znow–w–natarciu–kto–zyskuje–a–kto–traci–na–zdrowotnej–dezinformacji.read.

the bubble of conspiracy theorists. More well-known far-right "internet journalists" and YouTubers such as Rafał Mossakowski, Eugeniusz Sendecki, and Piotr Korczarowski also began to contact them. Additionally, the conspiratorial nature of anti-scientific narratives means that their followers most often engage in political movements that describe themselves as "anti-systemic." The most active political actor in Poland in this segment is the far right.

A crucial element of this network are the YouTube channels controlled by Rafał Mossakowski.[33] In the last decade, they were one of the main hubs in Poland for the dissemination of conspiracy narratives and fake news, as well as the far-right vision of the world. Mossakowski's recordings included[34] conspiratorial narratives about mind control through 5G telephony,[35] secret Illuminati organizations, the planned nature of the COVID-19 pandemic, or Hillary Clinton operating a paedophile network in the name of Satan and consuming children's blood,[36] as well as reports of the direct intervention in the history of Poland by the Virgin Mary, who stopped the alleged incoming world war in 1877. However, despite the dissemination of often downright conspiratorial and unrealistic narratives, it was one of the most important contact points for the Polish far right. Same channels also featured some participants of the mainstream of Polish political life: publicist Rafał Ziemkiewicz, former PiS senator and ex-director of the Institute of National Thought Heritage Jan Żaryn, historian Ewa Kurek, and nationalist priest Roman Kneblewski.[37]

Ewa Pawela, presented as a "mind control whistleblower," also appeared on his programmes on equal terms with these well-known figures in Polish public life. She called for breaking the sanitary rules introduced due to the pandemic, such as wearing masks or maintaining social distance. Citing doctors and professors, she does not mention them by name, but she claims that according to them "all these rules related to the epidemic actually harm our immunity." At the same time, Mossakowski calls the COVID-19 pandemic a "plandemic" in his

---

33 Such as the Powiśle Educational Center (Centrum Edukacyjne Powiśle), also known as the Educational Center Poland (Centrum Edukacyjne Polska), the Educational Center Brynica (Centrum Edukacyjne Brynica).

34 Until the channel was deleted by YouTube.

35 Polskie Sprawy, "Ewa Pawela: Jak się przeciwstawić depopulacji Polaków?", *CDA*, 08.01.2019, accessed 06.09.2024, https://www.cda.pl/video/1321708011.

36 Przemysław Witkowski, "Służby, mafie i loże vs >>szczęść boże<<, czyli magiczny świat Grzegorza Brauna", *Krytyka Polityczna*, 07.10.2019, accessed 06.09.2024, https://krytykapolityczna.pl/kraj/grzegorz–braun–witkowski.

37 Ibid.

speeches,[38] declares that Zionists have taken over the WHO and denies the number of coronavirus victims,[39] and his video materials features such activists and advocates of the anti-vaccination movement as Socha, Płaczek, and Szlachtowicz.[40] The topic of 5G also appeared in Mossakowski's recordings many times,[41] whether mentioned by Pawela, Kazahari,[42] or the host himself.

One can also notice a significant component of sympathy for Russian narratives in the context of the content propagated on his channels. This approach, in addition to the roots of the theory about the harmfulness of fifth-generation telephones, is another element indicating Russian influence in anti-scientific conspiracy narrative circles. Mossakowski's programmes feature many activists sympathetic to the Russian point of view on crucial geopolitical and ideological issues, such as Bartosz Bekier, Henryk Pająk,[43] Adam Wielomski,[44] Stanisław Michalkiewicz, Iwan Komarenko,[45] Maciej Poręba,[46] Janusz Korwin-Mikke, Grzegorz Braun, Magdalena Ziętek-Wielomska and Kamil Klimczak.[47] Mossakowski himself also took part in the so-called "anti-war debates" organized by "Myśl Polska," weekly where he appeared alongside the central figures of the camp sympathizing with Russia in Poland, namely Tomasz Jankowski, Mateusz Piskorski, Ronald Lasecki and Jan Engelgard. When Piskorski, accused of collaborating with the intelligence of the Russian

---

38 Chrobry Szlak, "Polska w cieniu p(l)andemii – Rafał Mossakowski", *YouTube*, 20.04.2020, accessed 06.09.2024, https://www.youtube.com/watch?v=-RiXUVcM1DY.

39 Rafał Pankowski et al., *Wirus nienawiści: »Brunatna Księga« czasu epidemii* (Warsaw: Nigdy Więcej, 2020), 18.

40 CEPowiśle, "Etyka informacyjna: Justyna Socha, Grzegorz Płaczek, Rafał Mossakowski, Piotr Szlachtowicz – panel", *CDA*, 23.09.21, accessed 06.09.2024, https://www.cda.pl/video/84786661a/vfilm.

41 Once a channel is deleted from YouTube, these recordings are no longer available.

42 Pankowski, *Wirus nienawiści*, 18.

43 Bomba Prawdy Kontynuacja 3, "Rosja na Ukrainie to gwarant normalizacji zgniłej i upadającej tkanki tego kraju Henryk Pająk", *YouTube*, 20.02.2022, accessed 06.09.2024, https://www.youtube.com/watch?v=WpACcYDGgEA.

44 Adam Wielomski, Magdalena Ziętek-Wielomska, "Rozmowy małżeńskie Wielomskich: Bolesław Piasecki", *Konserwatyzm.pl*, 16.01.2021, accessed 06.09.2024, https://konserwatyzm.pl/rozmowy-malzenskie-wielomskich-boleslaw-piasecki.

45 Bomba Prawdy Kontynuacja 3, "Obecne Państwo ukraińskie które nie zasługuje na pomoc Polski Iwan Komarenko", *YouTube*, 24.02.2022, accessed 06.09.2024, https://www.youtube.com/watch?v-prQQahOOUdM.

46 Maciej Poręba Kanał Oficjalny, "Maciej Poręba w CEP – co dalej z Ukrainą, plany Putina, czy będzie wojna w Donbasie, Новороссия", *YouTube*, 25.02.2022, accessed 06.09.2024, https://youtu.be/LpwMH3tkonU.

47 Chrobry Szlak, "Wojna na Ukrainie i obłęd »patriotów« – Kamil Klimczak, Rafał Mossakowski", *YouTube*, 24.02.2022, accessed 06.09.2024, https://youtu.be/JqPx791CoTM.

Federation, was released from custody in 2019, he gave several interviews to Mossakowski.[48] The latter was also Braun's electoral proxy in the 2015 presidential election and has been one of his closest associates for years. It was with Mossakowski and Ewa Stankiewicz that Braun occupied the State Electoral Commission in 2015.[49] Mossakowski also runs the YouTube channel "Magna Polonia," which is a YouTube representation of the magazine (of the same name) of the former Neo-fascist National-Radical Camp leader Przemysław Holocher, in which Braun regularly publishes his texts. Mossakowski also joined the pro-Russian party 1polska.pl, along with former influential activist of the most well-known pro-Russian Zmiana party, Nabil al-Malazi, and activists of the anti-5G movement Hanna Kazahari, Tobiasz Żuchiewicz, Jarosław Bogusławski, and Robert Majka (former Confederation MP).

He is also very active in the Ukrainian issue. Since 2017, Mossakowski has organized "Volhynian monthly manifestation." These demonstrations are an initiative of the so-called "eastern borderlands[50] organisations." Officially, the manifestations, which often gather in front of the Ukrainian embassy, are to remind us of the murders of Poles in Volhynia[51] and to demand that the Polish Parliament passed a law on the National Day of Remembrance of the Victims of Ukrainian Genocide against Poles from Volhynia and Eastern Lesser Poland.[52] In practice, they are one of the platforms for presenting a radically anti-Ukrainian narrative, presenting this nation as "Banderites," "rezuns,"[53] "Nazis," and "murderers."

Mossakowski is supported here by the organization he belongs, the Patriotic Union of Borderland and Veteran Organizations, where he was even a member of the audit committee. This organization is a continuation of the Borderland Patriotic Movement of the former leader of the national commu-

48 Konrad Szczygieł, "ABW ściga, Sejm wpuszcza", *FrontStory*, 23.08.2022, accessed 06.09.2024, https://frontstory.pl/mateusz-piskorski-szpiegostwo-sejm.

49 Jacek Gądek, "Ewa Stankiewicz i Grzegorz Braun. Kim są", *Onet*, 21.11.2014, https://wiadomosci.onet.pl/tylko-w-onecie/ewa-stankiewicz-i-grzegorz-braun-kim-sa/ykkfd.

50 Eastern Borderlands was a term coined for the eastern part of the Second Polish Republic during the interwar period (1918–1939). As a result of the post–World War II border changes, all of the territory was ceded to the USSR, and none of it is in modern Poland.

51 The massacres of Poles in Volhynia and Eastern Galicia were carried out in German-occupied Poland by the Ukrainian Insurgent Army (UPA) against the Polish minority in Volhynia, Eastern Galicia, parts of Polesia, and the Lublin region from 1943 to 1945.

52 Biuro Prasowe OWP, "Warszawa: Miesięcznica Wołyńska", *Obóz Wielkiej Polski*, no data, accessed 06.09.2024, https://www.owp.org.pl/index.php/703-warszawa-miesiecznica-wolynska-2.

53 Rezun – in Old Polish meaning a murderer, assassin, slaughterer, borrowing from the Ukrainian language from the word *rizun* (Ukrainian різун), meaning a butcher.

nists from the period of the People's Republic of Poland, Bohdan Poręba. The union is headquartered in the Rembelinsky Palace et al. Ujazdowskich 6a, the former headquarters of the pro-soviet Union of Fighters for Freedom and Democracy (ZBOWiD),[54] additionally sharing it with the General Zygmunt Berling Association of Traditions of the Polish People's Army, which still sympathizes with Russia. It was there, in the Rembelinsky Palace, that Mossakowski held many of his public meetings, including with some of the most active Kremlin supporters among the Western far right and participants in many of the propaganda activities of the Russian Federation – namely, former Neofascists Nick Griffin of the British National Party and Roberto Fiore[55] of Forza Nuova. Braun led the latter event.

Thanks to such extensive contacts with far-right politicians and their frequent presence in Mossakowski's recordings and activities, the themes of antiscientific conspiracy narratives can freely permeate a wider audience and often gain the status of an element of the programmes or views propagated by said politicians, additionally mixing with narratives sympathetic to Russia.

The following three popular far-right channels (eMisja.tv, Telewizja Narodowa, and NNTV) did not devote much attention to 5G. However, vaccines and the pandemic were one of their main themes. And again, pro-Russian themes are noticeable in the content they promote.

The eMisja.tv channel on YouTube is run by Piotr Korczarowski and his partner Sara Kostecka. It has 140,000 subscribers on the portal, and 14,000 people follow its fan page on Facebook. Korczarowski promotes anti-vaccination attitudes,[56] demands that Poland stop supporting Ukraine,[57] and has participated in numerous anti-Ukrainian demonstrations, such as the one on November 11, 2022, organized in Wrocław by Jacek Międlar, a former priest and current nationalist activist.[58] Korczarowski believes that a Jewish conspir-

---

54 Now existing under the name: "Związek Kombatantów i Byłych Więźniów Politycznych RP"

55 Przemysław Witkowski, "Faszyzm? My jesteśmy gorsi!", *Krytyka Polityczna*, 02.02.2019, accessed 06.09.2024, https://krytykapolityczna.pl/kraj/faszyzm-nop.

56 Karolina Kozakiewicz, "Putinowska propaganda na marszu narodowców we Wrocławiu. Kim są ludzie, którzy zastąpili Międlara?", *Wyborcza.pl*, 12.11.2022, accessed 06.09.2024, https://www.wroclaw.wyborcza.pl/wroclaw/7,35771,29133285,rosyjska-propaganda-na-marszu-narodowcow-we-wroclawiu-kim-sa.html.

57 Piotr Korczarowski, "Ponownie publikuję mój wczorajszy komentarz, w którym ośmieliłem się postawić znak równości wobec śmierci żyda i goja", *Facebook*, 23.02.2022, accessed 06.09.2024, https://www.facebook.com/piotr.korczarowski/posts/4489834687787128.

58 Kozakiewicz, "Putinowska propaganda na marszu narodowców".

acy caused the war in Ukraine. "This is the price of this war – to cleanse this area of Slavs,"[59] he declares.

Another popular far-right YouTube channel that posts anti-vaccination content is Telewizja Narodowa, created by a doctor and nationalist activist Eugeniusz Sendecki.[60] He is a former activist of the Christian National Union, All-Polish Youth, and the League of Polish Families, initially associated with PiS MP Artur Zawisza and later with the anti-Semitic and nationalist Polish diaspora activist Jan Kobylański and the former leader of the national-communist Grunwald Patriotic Union, film director Bohdan Poręba. Sendecki's channel promotes all sorts of conspiracy theories, from the existence of a secret world government through the concept of a "plandemic" to content intended to justify Poland's exit from the European Union. Sendecki's recordings also feature leaders of the anti-vaccination movement (Justyna Socha,[61] Paweł Skutecki[62]) and anti-vaccination activists sympathetic to Russia (Iwan Komarenko,[63] Wojciech Olszański[64]).

From these last two YouTube channels (eMisja.tv and Telewizja Narodowa), another loud community supporting the fight against mandatory vaccinations has also emerged. These are Compatriots Comrades, a network of informal autonomous local nationalist, paramilitary, and anti-vaccination groups (whose members became politically active primarily during the COVID-19 pandemic) connected by the person of their leader – the actor sympathetic to Russia, Wojciech Olszański.[65] In all three of these entities, one can

---

59 No author, "Czy grozi nam wojna? Antyukraińska konferencja w Kielcach", *Demagog*, 03.04.2023, accessed 06.09.2024, https://www.demagog.org.pl/fake_news/czy-grozi-nam-wojna-antyukrainska-konferencja-w-kielcach.

60 Anna Mierzyńska, "Rosyjski ślad. Prokremlowscy działacze wśród uczestników napaści z udziałem antyszczepionkowców", *OKO.Press*, 06.08.2021, accessed 06.09.2024, https://www.oko.press/prokremlowscy-dzialacze-wsrod-antyszczepionkowcow.

61 EugeniuszSendecki71, "PROTEST PRZECIWKO NADUŻYCIOM W SYSTEMIE SZCZEPIEŃ W POLSCE NOP (6)", *YouTube* 04.06.2017, accessed 06.09.2024, https://www.youtube.com/watch?v=24fBHI2sZ90.

62 EugeniuszSendecki71, "OPÓR POLSKI PRZECIWKO PRZYMUSOWI SZCZEPIEŃ I UBECKI ZARZĄD NAD POLSKĄ SŁUŻBĄ ZDROWIA", *YouTube*, 03.06.2017, accessed 06.09.2024, https://www.youtube.com/watch?v=dJ3ZOb_JPZs&t=91s.

63 EugeniuszSendecki71, "IWAN KOMARENKO I MACIEJ PORĘBA ROZMAWIAJĄ W TELEWIZJI NARODOWEJ", *YouTube*, 06.11.2021, accessed 06.09.2024, https://www.youtube.com/watch?v=4FKydSpDLtA.

64 EugeniuszSendecki71, "ALEKSANDER JABŁONOWSKI I EUGENIUSZ SENDECKI NA PROTEŚCIE PRZECIWKO PRZYMUSOWI SZCZEPIEŃ 03-06-2017", *YouTube*, 03.06.2017, accessed 06.09.2024, https://www.youtube.com/watch?v=MsAEu3SjO64.

65 Piotr Głuchowski, "»Tworzymy listy śmierci!«. Wojciech Olszański i Kamractwo Rodaków na drodze do Sejmu", *Wyborcza.pl*, 17.10.2022., www.wyborcza.pl/duzyformat/

notice a mix of pro-Russian content with anti-scientific narratives on the COVID-19 pandemic and vaccinations.

Far-right YouTubers who are more technically proficient than those discussed above have also begun to make contact with opponents of mandatory vaccinations and 5G telephony. Their programmes no longer look like amateur recordings, as is often the case with programmes on alternative medicine. The image is recorded from several cameras, and information bars are displayed at the bottom of the screen. The hosts and guests are elegantly dressed. All this is supposed to make them credible as serious media that can be trusted. Some new right-wing Internet television stations already have more regular viewers than traditional media of this type. They reach people who no longer watch traditional television but obtain information via smartphones. The materials posted on these channels promote a clearly defined world vision. They propagate numerous conspiracy narratives and undermine scientific authorities. As a result, it was an accessible milieu for activists and organizations fighting against 5G telephony and mandatory vaccinations to infiltrate.

Until recently, the YouTube channel wRealu24.pl was an essential channel for promoting the anti-vaccinations and anti-5G movements and their demands. This channel had been blocked several times due to controversial content and was eventually permanently banned. When the portal finally removed it, it had 436,000 subscribers. Today, it broadcasts in a limited form on the BanBye portal. Marcin Rola, a former politician of the Kukiz'15 movement, runs it. His media is the most crucial news channel for the Polish far right. The entire leadership of the Confederation party (Krzysztof Bosak, Robert Winnicki, Janusz Korwin-Mikke, Konrad Berkowicz, and others) and politicians from the right wing of PiS (Dariusz Matecki, Janusz Kowalski, Dominik Tarczyński, and others) have appeared on it, as well as many other politicians from this ideological circle (Marek Jakubiak, Jacek Międlar, Paweł Kukiz, Robert Bąkiewicz). Rola himself is closely associated with the Confederation party circles, from whose list he was supposed to run for the Sejm in 2023, according to media reports.[66]

On Rola's programmes, far-right politicians are often met by celebrities who believe in conspiracy narratives (such as singers Edyta Górniak and Ivan Komarenko, and private investigator Krzysztof Rutkowski). Their role in spread-

---

7,127290,29022949,kamractwo-rodakow-na-drodze-do-sejmu-pis-to-dla-nich-zdrajcy.html.

66 Ultimately, despite media announcements, he was not included on the party's electoral list.

ing conspiracy narratives related to the COVID-19 or 5G pandemics is worth noting. Their popularity, which goes far beyond the circles of those interested in alternative medicine or activists and supporters of far-right parties and organizations, means that the anti-vaccination message can reach much wider circles. As a result, the meeting places of these groups, including the media managed by Marcin Rola, play a crucial role in strengthening the power of the message of anti-scientific conspiracy narratives. In his materials, messages of this content are also mixed with anti-Ukrainian and pro-Russian ones, and his programmes often feature people who mix these issues, such as Ivan Konarenko[67] and Sebastian Pitoń.[68]

Regarding COVID-19, Rola constantly declares that there was no coronavirus pandemic. In his opinion, it was invented by the World Health Organization and "Western bankers" associated with this organization, who want to weaken the economies of independent countries so that they can then take them over for next to nothing.[69] He is also against mandatory vaccinations against coronavirus. The topic of 5G has also appeared on wRealu24.pl, many times. MPs Jarosław Sachajko[70] and Robert Majka,[71] as well as Zbigniew Gelzok[72] and Rafik Ennaoui[73] from the Polish National Association for Combating Electrocontamination "Right to Life" spoke about it in the programmes of Rola.

---

67 Michał Jelonek, "Niedawny gość wRealu24 – Rosjanin Ivan Komarenko", *Facebook*, 14.06.2022, accessed 06.09.2024, https://www.facebook.com/watch/?v=3316023765345327.

68 Piotr Głuchowski, Katarzyna Bielecka, "Mroczni siewcy Putina. Sprawdzamy, komu w Polsce mieszają w głowie", *Wyborcza.pl*, 14.08.2023, accessed 06.09.2024, https://wyborcza.pl/duzyformat/7,127290,30065814,mroczni–siewcy–putina–sprawdzamy–komu–w–polsce–mieszaja–w.html.

69 Krystian Rusiniak, "Znany dziennikarz Marcin Rola opublikował książkę o manipulacjach przy epidemii koronawirusa. Jeden z redaktorów TV Media Narodowe stał się bohaterem. Uchwycił on ciekawą wpadkę autora [FOTO]", *Media Narodowe*, 09.06.23, accessed 06.09.2024, https://tvmn.pl/Znany-dziennikarz-Marcin-Rola-opublikowal-ksiazke-o-manipulacjach-przy-epidemii-koronawirusa-Jeden-z-redaktorow-TV-Media-Narodowe-stal-sie-bohaterem-Uchwycil-on-ciekawa-wpadke-autora-2112a.

70 KMJ Archiwum Polityczne, "Co z tym 5G? »Zostaliśmy tyle razy oszukani, że dmuchamy na zimne«. Jarosław Sachajko", *YouTube*, 11.03.2022, accessed 06.09.2024, https://www.youtube.com/watch?v=uNsbJfiu7wg.

71 KMJ Archiwum Polityczne, "Kto w polskim parlamencie ukrywa prawdę o 5G Poseł Robert Majka wRealu24 cz 1", *YouTube*, 25.08.2019, accessed 06.09.2024, https://www.youtube.com/watch?v=19CaR6LXm8o.

72 Barlinek 5G, "Eksperci ostrzegają przed zgubnymi skutkami 5G Zbigniew Gelzok i Rafik Ennaoui!", *YouTube*, 24.11.2019, accessed 06.09.2024, https://www.youtube.com/watch?v=zmei_MF_FUQ.

73 Ibid.

Another essential medium for the far right, serving as a channel for disseminating the content of an anti-scientific conspiracy narrative nature, is "Media Narodowe." This is a media project launched in 2017 under the patronage of the Independence March Association, the organizer of the largest Polish nationalist demonstration held on November 11 each year in Warsaw. The initiative combines several different nationalist media entities, such as: "Dziennik Narodowy," "Myśl.pl," "Prosto z Mostu," "Kierunki.info," "Polityka Narodowa," "wMeritum.pl"; "Prawapolityka.pl"; "3Droga.pl." Until 2023, 180,000 users subscribed to it on YouTube.[74] Presenters host it in a professional-looking studio dressed in suits and is eagerly visited by politicians from the Confederation party and PiS. It is maintained and developed thanks to financial support from the National Institute of Freedom and the Institute of National Thought Heritage.[75]

"Media Narodowe" is a channel presenting content of an extreme right-wing nature and is another element in the chain along which conspiracy narratives are being moved into the mainstream. Materials about vaccines and the pandemic appeared there regularly. Very often, "experts," politicians and activists associated with the movement opposing the vaccination requirement, such as Zbigniew Hałat,[76] Mirosław Piotrowski,[77] Daniel Galus,[78] Marcin Rola[79] and Anna Siarkowska,[80] spoke out or were invited to the studio. The topic of 5G also appeared on the National Media channel many times. MPs Sachajko and Tomasz Rzymkowski,[81] as well as Olaf Swol-

74 In 2023 the channel was blocked by YouTube.

75 Mateusz Bałuka, "Narodowcy z rządowym wsparciem. Ich media promują antyszczepionkowe treści", *Onet*, 21.09.2021, accessed 06.09.2024, https://www.wiadomosci.onet.pl/kraj/koronawirus-narodowcy-z-rzadowym-wsparciem-promuja-antyszczepionkowe-tresci/wold46f.

76 Media Narodowe, "Dr Zbigniew Hałat gościem Tematu Dnia | MEDIA NARODOWE", *CDA*, no data, accessed 06.09.2024, https://www.cda.pl/video/7769911e6.

77 Media Narodowe, "Media Narodowe – Wojsko WYMUSZA szczepienia? Piotrowski OSTRO do Błaszczaka: TAK NIE MOŻE BYĆ! – 27.07.2021", *CDA*, no data, accessed 06.09.2024, https://www.cda.pl/video/80448802d.

78 Bałuka, "Narodowcy z rządowym wsparciem".

79 Rusiniak, "Znany dziennikarz Marcin Rola".

80 Piotr Motyka, "Anna Maria Siarkowska wyraźnie o segregacji sanitarnej: Stop sanitaryzmowi! [NASZ WYWIAD]", *Media Narodowe*, 26.11.21, accessed 06.09.2024, https://medianarodowe.com/2021/11/26/anna-maria-siarkowska-wyraznie-o-segregacji-sanitarnej-stop-sanitaryzmowi-nasz-wywiad.

81 Media Narodowe, "5G – realne czy kreowane ZAGROŻENIE? Dostawca 5G z zarzutami korupcyjnymi | dr T. Rzymkowski", *CDA*, no data, accessed 06.09.2024, https://www.cda.pl/video/48391401c.

kień[82] from the Institute of Civil Affairs Foundation[83] and Mateusz Jarosiewicz[84] from the National Institute of Strategic Studies Foundation,[85] spoke critically about this generation of telephony. The channel's recordings were also reported on 5G telephony in a very alarming tone.[86]

Another entity important for the spread of anti-scientific conspiracy narratives in Poland is the Institute of Civil Affairs Foundation (INSPRO).[87] It is a non-governmental organization whose president is Rafał Górski, a former columnist for the *nouvelle droite* "Obywatel" magazine and a former candidate for the position of mayor of Łódź on behalf of the right-wing populist Kukiz'15 movement.[88] The foundation's deputy director is a sociologist and historian Olaf Swolkień, a right-wing ecological activist, currently a supporter of the Confederation party and a columnist for the weekly "Myśl Polska," which sympathizes with Russian narratives.[89] The leaders of INSPRO operate in the milieu of libertarians, former anarchists, nationalists, supporters of conspiracy narratives, ecologists, and small entrepreneurs.

In the context of anti-scientific conspiracy narratives, the organization is involved in promoting movements opposing mandatory vaccinations and the introduction of fifth-generation telephones, and demands a ban on the use of genetically modified food.[90] A regular columnist for the INSPRO-published weekly *The Civil Affairs Weekly* ("Tygodnik Spraw Obywatelskich")

---

82 No author, "Ekolog mówi prawdę o 5G! | Olaf Swolkień", *CDA*, no data, accessed 06.09.2024, https://ebd.cda.pl/620x368/5263092ac.

83 Patrz dalej.

84 Media Narodowe, "SZOKUJĄCE, ale czy PRAWDZIWE? Jak wprowadza się 5G w Polsce? | M. Jarosiewicz", CDA, no data, accessed 06.09.2024, https://www.cda.pl/video/3673192d0.

85 See further.

86 For example: "Shocking information about electromagnetic radiation! Does it cause serious illnesses" or "Spaniards will be controlled at every step! With the help of 5G, cameras, smartphone applications".

87 Weronika Kursa, "Biała księga pandemii koronawirusa – podsumowanie chorych czasów", *Tygodnik Spraw Obywatelskich,* 27.12.2023, accessed: 06.09.2023, https://instytutsprawobywatelskich.pl/biala-ksiega-pandemii-koronawirusa-podsumowanie-chorych-czasow.

88 Piotr Wesołowski, "Wybory 2018 Łódź. Rafał Górski kandydatem na prezydenta ruchu Kukiz'15?", *Wyborcza.pl,* 29.03.2018, accessed 06.09.2024, https://lodz.wyborcza.pl/lodz/7,35136,23200938,rafal–gorski–kandydatem–na–prezydenta–lodzi–ruchu–kukiz–15.html.

89 Olaf Swolkień, "Brzydka gra Katyniem", *Myśl Polska,* no data, accessed 06.09.2024, https://myslpolska.info/2023/04/13/brzydka–gra–katyniem.

90 Ewa Rembiałkowska et al., "Organizmy genetycznie zmodyfikowane. W kierunku ogólnopolskiej debaty o GMO/NGT", *Instytut Spraw Obywatelskich,* August 2023, accessed

is Paweł Wypychowski, a member of the board of the James Clerk Maxwell Institute of Electromagnetic Research foundation, owner of EMF3 – Electromagnetic Safety, which aims to supply devices that allegedly protect against electromagnetic fields (EMF). At the same time, INSPRO published a large brochure authored by him ("Mega-Law 5G – is this Book really White?"), in which the author repeated some of the myths and conspiracy narratives associated with 5G in the form of questions that requested a nationwide debate on the opportunities and threats posed by the new technology and demanded that the proceedings of the draft amendment to the Act on Supporting the Development of Telecommunications Services and Networks concerning 5G be suspended until a debate with the participation of supporters and opponents is held.[91] INSPRO also published another brochure, this time edited by Górski, entitled "Save the Children! – Technological Education: e-addictions and electrosmog"[92] and organize meetings during which Wypychowski talks about the harmfulness of "electrosmog".[93] In another magazine published by INSPRO ("Aktywność Obywatelska"), Wypychowski's text guide "Zrób to sam! A guide on measuring and assessing electromagnetic radiation",[94] in which he recommended meters available in his shop.[95]

06.09.2024, https://instytutsprawobywatelskich.pl/publikacje/organizmy-genetycznie-zmodyfikowane-w-kierunku-ogolnopolskiej-debaty-o-gmo-ngt.

91 Paweł Wypychowski, "Megaustawa 5G – czy ta Księga rzeczywiście jest Biała?", *Instytut Spraw Obywatelskich*, 2019, accessed 06.09.2024, https://instytutsprawobywatelskich.pl/publikacje/megaustawa-5g-czy-ta-ksiega-rzeczywiscie-jest-biala.

92 Rafał Górski, ed., "Ratuj Dzieci! – Edukacja technologiczna: e-uzależnienia i elektrosmog" *Instytut Spraw Obywatelskich*, 2020, accessed 06.09.2024, https://instytutsprawoby watelskich.pl/publikacje/ratuj-dzieci-edukacja-technologiczna-e-uzaleznienia-i-elektrosmog.

93 Zbigniew Podjadek, "Anatomia polskiego ruchu anty5G. Kim są ludzie, którzy walczą z budową sieci nowej generacji?", *Wyborcza.pl*, 16.09.2020, accessed 06.09.2024, https://wyborcza.biz/biznes/7,177150,26290907,anatomia-polskiego-ruchu-anty5g-kim-sa-ludzie-ktorzy-walcza.html.

94 No author, "Zrób to sam! Przewodnik jak mierzyć i oceniać promieniowanie elektromagnetyczne", *Tygodnik Spraw Obywatelskich*, 21.07.2020, accessed 06.09.2024, https://insty tutsprawobywatelskich.pl/zrob-to-sam-przewodnik-swiadomego-obywatela-pomia r i ocena-promieniowania-elektromagnetycznego.

95 The publication of both brochures ("5G Mega-Law – is this Book really White?" and "Save the Children! – Technological Education: e-addictions and electrosmog") was supported by the National Institute of Freedom – Centre for the Development of Civil Society – an executive agency established in 2017 with the task of supporting the development of civil society, as well as public benefit and volunteering activities.

INSPRO also published on its website and in its magazine's conversations with Zbigniew Gelzok[96] (president of the Polish Association for Combating Electrocontamination "Right to Life"), with Barbara Gałdzińska-Calik[97] (from the board of the James Clerk Maxwell Institute for Electromagnetic Research Foundation). Wypychowski is, in practice, an expert in technological solutions for INSPRO, and conversations with him on this topic have already been published on the INSPRO website a dozen or so times.[98] The organization also provided information on how to stop the construction of telecommunications masts in your area,[99] published petitions of the movement opposing 5G,[100] and reported on its conferences.[101] The movement opposing mandatory vaccinations received similar support from INSPRO. The organization's website and the magazines it publishes have published interviews with "experts" from the movement, such as Piotr Witczak,[102] Zbigniew Hałat,[103] and Zbigniew Martyka,[104] and have encouraged people to read the *White Paper on the Coronavirus Pandemic* published by the Ordo Medicus Foundation,[105] and books important to the movement opposing mandatory vaccina-

96 Rafał Górski, "Elektrosmog – wyrok sądu, rak mózgu i telefon komórkowy", *Tygodnik Spraw Obywatelskich*, 21.02.2023, accessed 06.09.2024, https://instytutsprawobywatelskich.pl/elektrosmog-wyrok-sadu-rak-mozgu-i-telefon-komorkowy.

97 Małgorzata Jankowska, "Blaski i cienie 5G, czyli jak ucywilizować cyfrową cywilizację (1)", *Instytut Spraw Obywatelskich*, 20.12.2019, accessed 06.09.2024, https://instytutsprawobywatelskich.pl/blaski-i-cienie-5g-czyli-jak-ucywilizowac-cyfrowa-cywilizacje-1.

98 Rafał Górski, "O nauce i Nauce", *Tygodnik Spraw Obywatelskich*, 12.05.2020, accessed 06.09.2024, https://instytutsprawobywatelskich.pl/o-nauce-i-nauce.

99 Jolanta Wolska, "Czy Dawid może pokonać Goliata?", *Tygodnik Spraw Obywatelskich*, 01.09.20, accessed 06.09.2024, https://instytutsprawobywatelskich.pl/czy-dawid-moze-pokonac-goliata.

100 No author, "Stop smog 5g", *Obywatele Decydują*, no data, accessed 06.09.2024, https://www.obywateledecyduja.pl/kampania/stop-smog-5g.

101 Ilona Sobul, "Elektrosmog – o co tyle hałasu? Relacja z debaty", *Tygodnik Spraw Obywatelskich*, 20.07.2021, accessed 06.09.2024, https://instytutsprawobywatelskich.pl/elektrosmog-o-co-tyle-halasu-relacja-z-debaty.

102 Rafał Górski, "Piotr Witczak: System szuka na mnie »haka«", *Tygodnik Spraw Obywatelskich*, 05.12.2023, accessed 06.09.2024, https://instytutsprawobywatelskich.pl/piotr-witczak-system-szuka-na-mnie-haka.

103 Rafał Górski, "Szczepienia. Trzy pytania do Ministra Zdrowia", *Tygodnik Spraw Obywatelskich*, 18.05.2021, accessed 06.09.2024, https://instytutsprawobywatelskich.pl/szczepienia-trzy-pytania-do-ministra-zdrowia.

104 Rafał Górski, "Polowania na lekarzy. Dr Martyka, COVID i utajnione rozprawy sądowe", *Tygodnik Spraw Obywatelskich*, 11.04.2023, accessed 06.09.2024, https://instytutsprawobywatelskich.pl/polowania-na-lekarzy-dr-martyka-covid-i-utajnione-rozprawy-sadowe.

105 Kursa, "Biała księga pandemii koronawirusa – podsumowanie".

tions and spreading conspiracy narratives about the pandemic, such as *Pandemic of Lies* by Judy Mikovits and Kent Heckenlively and Damian Garlicki's *The Deadly System? A Paramedic Breaks The Conspiracy of Silence*. The Civil Affairs Weekly has also published two open letters to the Minister of Health written by one of the leaders of the movement opposing mandatory vaccinations, Włodzimierz Julian Korab-Karpowicz,[106] and the movement's "expert" Ryszard Rutkowski.[107] Both INSPRO leaders have also actively supported the postulates of opposition to compulsory vaccinations, and both Górski[108] and Swolkień[109] have encouraged in their texts resistance to COVID-19 vaccinations. Both were open to controversial milieus from the beginning of their public activities. In September 1999, the later editor-in-chief of "Nowy Obywatel" Remigiusz Okraska, as well as Górski and Swolkień, attempted to create a common platform for anti-globalization groups, wanting to combine ecological and anarchist groups with neo-fascist ones. Representatives of neo-pagan and national-socialist groups (such as the Association for Tradition and Culture "Niklot," the Polish National Community of Bolesław Tejkowski, the Świaszczyca Association, the Native Faith association and the Polish Branch of the Slavic Council), neo-fascist magazines ("Trygław," "Odala" and "Securius"), as well as monarchist groups (the Organization of Polish Monarchists) also took part in this deliberations.[110] At that time, an "alliance of extremists" (called the Confederation for Our Land) was established. It also included Mateusz Piskorski, later officially accused by the Polish counterintelligence of espionage against Russia and China. When the alliance failed to create permanent structures, Swolkień, Górski, and Okraska worked closely together for the

---

106 Włodzimierz Julian Korab-Karpowicz et al., "Trzy pytania do Ministra Zdrowia w sprawie bezpieczeństwa i skuteczności szczepień na SARS-CoV-2", *Tygodnik Spraw Obywatelskich*, 29.04.21, accessed 06.09.2024, https://instytutsprawobywatelskich.pl/trzy-pytania-do-ministra-zdrowia-w-sprawie-bezpieczenstwa-i-skutecznosci-szczepien-na-sars-cov-2 and Włodzimierz Julian Korab-Karpowicz et al., "Apel naukowców i lekarzy w sprawie szczepień na koronawirusa SARS-CoV-2", *Instytut Spraw Obywatelskich*, 30.11.20, accessed 06.09.2024, https://instytutsprawobywatelskich.pl/apel-naukowcow-i-lekarzy-w-sprawie-szczepien-na-koronawirusa-sars-cov-2.

107 Ryszard Rutkowski, "COVID-19. List prof. Rutkowskiego do ministra zdrowia", *Tygodnik Spraw Obywatelskich*, 03.11.2020, accessed 06.09.2024, https://instytutsprawobywatelskich.pl/covid-19-list-prof-rutkowskiego-do-ministra-zdrowia.

108 Rafał Górski, "Pandemia udawana, pieniądze prawdziwe", *Tygodnik Spraw Obywatelskich*, 05.06.2020, accessed 06.09.2024, https://instytutsprawobywatelskich.pl/gorski-prowokuje-pandemia-udawana-pieniadze-prawdziwe.

109 Olaf Swolkień, "Rodacy do pracy", *Tygodnik Spraw Obywatelskich*, 13.10.2020, accessed 06.09.2024, https://instytutsprawobywatelskich.pl/rodacy-do-pracy.

110 Przemysław Witkowski, *Partia rosyjska* (Warsaw: Arbitror, 2023),90–93.

next ten years, creating the "Obywatel" magazine (later "Nowy Obywatel"). The funds for its establishment were obtained by forcing the Victus company to pay a fee for withdrawing creators of the magazine from the blockade of its investment on the site of the former "Norbelan" factory in Łódź.[111] The magazine included discussions and advertisements for writings of various types of neo-fascism (national-radical, third position, new-right), monarchist, neo-pagan, and orthodox catholic,[112] as well as texts by the pro-Russian Holocaust denier Roger Garaudy, former member of the terrorist organization Red Army Faction (also pro-Russian) Horst Mahler, and the first translations in Poland of articles by the neo-fascist Russian philosopher Alexander Dugin.[113]

"Obywatel" also published articles defending the activities of terrorists: Timothy McVeigh (the perpetrator of the Oklahoma City attack, in which 168 people died, including 19 children) and Theodore "Unabomber" Kaczynski (his bombs killed three people and injured twenty-nine). After a conflict based on personal and financial issues in the editorial office of "Obywatel"[114] and the resulting division, the magazine "Nowy Obywatel" and the Institute of Civil Affairs were established.

In the case of the Institute of Civil Affairs case, pro-Russian contexts and references also appear in addition to propagating anti-scientific conspiracy narratives. The organization's Tygodnik Spraw Obywatelskich publishes texts by journalists sympathetic to the Russian narrative, such as Leszek Sykulski,[115] Tomasz Gabiś,[116] Magdalena Ziętek–Wielomska,[117] Gracjan

---

111 Mon, "Łódzcy ekolodzy: dostali pieniądze, zakończyli protest", *WP*, 06.09.2001, accessed 06.09.2024, https://wiadomosci.wp.pl/lodzcy-ekolodzy-dostali-pieniadze-zakonczyli-protest-6109071767045249a.

112 Zbigniew Marcin Kowalewski et al., "LIST DO KIEROWNICTW ATTAC–POLSKA I ATTAC –FRANCJA", *Archiwum LBC*, 06.12.01, accessed 06.09.2024, http://www.geocities.ws/lbc_1917_2/128attaclist.htm.

113 Aleksander Dugin, "Post modernizm?", *Obywatel* 1, no. 9 (2003).

114 Rafał Górski et al., "Do Członków Stowarzyszenia Obywatele Obywatelom", *Silo Tips*, 23.11.2017, accessed 06.09.2024, https://silo.tips/download/od-do-czonkow-stowarzyszenia-obywatele-obywatelom.

115 Rafał Górski, "Dlaczego Rosja zaatakowała Ukrainę? Jak bronić Polski?", *Tygodnik Spraw Obywatelskich*, 24.05.2022, accessed 06.09.2024, https://instytutsprawobywatelskich.pl/dlaczego-rosja-zaatakowala-ukraine-jak-bronic-polski.

116 Mateusz Rolik, Tomasz Gabiś, "O psychiatrii. Rozmowa z Robertem Whitakerem", *Tygodnik Spraw Obywatelskich*, 04.04.2023, accessed 06.09.2024, https://instytutsprawobywatelskich.pl/o-psychiatrii-rozmowa-z-robertem-whitakerem.

117 Rafał Górski, "O Klausie Schwabie, Światowym Forum Ekonomicznym i globalistach z Davos", *Tygodnik Spraw Obywatelskich*, 21.06.2022, accessed 06.09.2024, https://instytutsprawobywatelskich.pl/o-klausie-schwabie-swiatowym-forum-ekonomicznym-i-globalistach-z-davos.

Cimek[118] and Ronald Lasecki.[119] Swolkień also debuted as a journalist in "Myśl Polska".[120]

Another non-governmental organization supporting anti-scientific conspiracy narratives is the National Institute of Strategic Studies Foundation (NISS). Similarly to the Institute of Civil Affairs, here, the scientific-sounding name conceals the nature of this organization associated with conspiracy narratives. NISS is headed by Wojciech Edward Dobrzyński and Marek Sas–Kulczycki. Both co-founded the far-right movement 1Polska.pl, which also included the aforementioned Mossakowski, Nabil al Malazi, Kazahari, Żuchiewicz, Bogusławski, and former Confederation MP Majka, known for their protests against 5G. Dobrzyński also ran for the Mazovian Voivodeship Assembly on the list of the Jedność Narodu party, which sympathized with Russian narratives, while Sas–Kulczycki co-created the party's programme. Both also frequently collaborate with Mossakowski, appearing as experts on his programmes.

On the organization's official YouTube channel, NISS TV, the organization publishes technophobic videos, including those spreading conspiracy narratives about 5G, vaccines, and the COVID-19 pandemic. It suggests that the pandemic was planned that vaccines are harmful to health,[121] and mixes them with anti-Semitic,[122] and pro-Russian conspiracy narratives related to the concept of a secret world government (New World Order),[123] the promotion of alternative medicine,[124] and opposition to anti-pandemic

---

118 Gracjan Cimek, "Czas Globalczyka (1 i 2)", *Tygodnik Spraw Obywatelskich*, 07.03.2023, accessed 06.09.2024, https://instytutsprawobywatelskich.pl/czas-globalczyka-1-i-2.

119 Ronald Lasecki, "Światopogląd biocentryczny", *Tygodnik Spraw Obywatelskich*, 06.09.2022, accessed 06.09.2024, https://instytutsprawobywatelskich.pl/swiatopoglad-biocentryczny.

120 Swolkień, "Brzydka gra Katyniem".

121 NISS TV (Narodowy Instytut Studiów Strategicznych), "Stop NOP – demonstracja w Warszawie (Justyna Socha, Grzegorz Braun, Paweł Nogal, Piotr Jankowski)", *YouTube*, 08.06.2020, accessed 06.09.2024, https://www.youtube.com/watch?v=A1NQwFjK5Sk.

122 NISS TV (Narodowy Instytut Studiów Strategicznych), "Jedwabne – 81 lat po ... Nadal czekamy na prawdę. Domagamy się ekshumacji!", *YouTube*, 12.07.2022, accessed 06.09.2024, https://www.youtube.com/watch?v=C--4wIDsM4I.

123 NISS TV (Narodowy Instytut Studiów Strategicznych), "Zatrzymujemy NWO – tworzymy alternatywne społeczeństwo! Zjazd Wolnych Ludzi w Warszawie. 14.09.2020", *YouTube*, 10.09.2020, accessed 06.09.2024, https://www.youtube.com/watch?v=SuHi2DkkQSg.

124 NISS TV (Narodowy Instytut Studiów Strategicznych), "Dorota Gudaniec – Przestańcie się bać. Medycyna naturalna – największym zagrożeniem dla big farmy!", *YouTube*, 14.09.2020, accessed 06.09.2024, https://www.youtube.com/watch?v=oCOiWI6edlc.

restrictions.[125] The NISS channel recordings featured activists opposing mandatory vaccinations, such as Agata Osiniak,[126] Anna Martynowska,[127] Grzegorz Płaczek,[128] Justyna Socha,[129] Piotr Szlachtowicz[130] and Hubert Czerniak,[131] as well as leaders of groups fighting against fifth-generation telephony, such as Jadwiga Łopata,[132] Janusz Zagórski,[133] Ewa Pawela,[134] Zbigniew Gelzok[135] and Hanna Kazahari.[136]

And once again, anti-scientific narratives are mixed in the message of the NGO involved in their propagation with narratives sympathetic to the Russian

125 NISS TV (Narodowy Instytut Studiów Strategicznych), "Precz z maseczkami, żadnej drugiej fali »pandemii«!", *YouTube*, 28.07.2020, accessed 06.09.2024, https://www.youtube.com/watch?v=UgWyyHB_Efo.

126 NISS TV (Narodowy Instytut Studiów Strategicznych), "Uwolnienie zawodu lekarza i naukowca od zniewolenia. Dr Agata Osiniak. Kraków 21 sierpnia 2022 roku.", *YouTube*, 09.09.22, accessed 06.09.2024, https://www.youtube.com/watch?v=i5Qbrr–5Oyo.

127 NISS TV (Narodowy Instytut Studiów Strategicznych), "Dr Anna Martynowska oskarża posła RP. Również o: Ludobójcach, złodzieju, psychopatach i TW SANEPID–u", *YouTube*, 08.10.2022, accessed 06.09.2024, https://www.youtube.com/watch?v=mH2A7fDtmVU.

128 NISS TV (Narodowy Instytut Studiów Strategicznych), "»Nie planuję samobójstwa« – P. Szlachtowicz, o czym mówi A. Wolska, to się samo komentuje – G. Płaczek", *YouTube*, 17.01.2022, accessed 06.09.2024, https://www.youtube.com/watch?v=7UY5WeHIscQ.

129 NISS TV (Narodowy Instytut Studiów Strategicznych), "Ważny apel Justyny Sochy. Sobota – demonstracje w Polsce przeciw segregacji sanitarnej. Godzina W19.", *YouTube*, 12.01.2022, accessed 06.09.2024, https://www.youtube.com/watch?v=SIyT7RhBZgw.

130 NISS TV (Narodowy Instytut Studiów Strategicznych), "Piotr Szlachtowicz o zniewolonym umyśle. Marcin Flach o pieniądzu, który zniewala.", *YouTube*, 20.10.2022, accessed 06.09.2024, https://www.youtube.com/watch?v=Kx1yw5id9Sc.

131 NISS TV (Narodowy Instytut Studiów Strategicznych), "Grzyby cz.1. Mózg jest plastyczny! Soplówka jeżowata regeneruje i oczyszczać mózg. Synestezja." *YouTube*, 02.06.2023, accessed 06.09.2024, https://www.youtube.com/watch?v=Dpfhj3kpIzI.

132 NISS TV (Narodowy Instytut Studiów Strategicznych), "Bitwa o ludzką duszę–Jadwiga Łopata. Konferencja –"Od Fałszu i Faszyzmu do Prawdy i Wolności, Razem!", *YouTube*, 01.03.2023, accessed 06.09.2024, https://www.youtube.com/watch?v=XamEzKa49Ao.

133 NISS TV (Narodowy Instytut Studiów Strategicznych), "W. Dobrzyński – Świadomość dwuwalutowości. J. Zagórski – NOWA ZIEMIA, początek kolejnego etapu drogi", *YouTube*, 27.09.2022, accessed 06.09.2024, https://www.youtube.com/watch?v=vqhRDTuwUcA.

134 NISS TV (Narodowy Instytut Studiów Strategicznych), "TI DAY. Cyfryzacja człowieka. Transhumanizm. Cywilizacja Człowieka nad przepaścią.", *YouTube*, 04.09.2022, accessed 06.09.2024, https://www.youtube.com/watch?v=16USMtJ995Q.

135 NISS TV (Narodowy Instytut Studiów Strategicznych), "Zbigniew Gelzok. 5G a Covid–19 na Śląsku. Czy telekomy zarabiają na pornografii i pedofilii?", *YouTube*, 12.05.2020, accessed 06.09.2024, https://www.youtube.com/watch?v=JoXs2yDb3–A.

136 NISS TV (Narodowy Instytut Studiów Strategicznych), "Dwie drogi do suwerenności Polski – synergiczne wizje. H. Kazahari, P. Jankowski, W. Dobrzyński", *YouTube*, 14.07.2020, accessed 06.09.2024, https://www.youtube.com/watch?v=PLws5FrplJM.

point of view. Many people appearing in NISS recordings are active propagators of Russian conspiracy and political narratives. The recordings from the organization's channel featured, among others, geopolitician and leader of the Polish anti-war movement Leszek Sykulski,[137] former chairman of the Jedność Narodu party and would-be candidate for the office of the President of the Republic of Poland Romuald Starosielec,[138] and publicists of "Myśl Polska" weekly Witold Modzelewski,[139] and Maria Szyszkowska.[140] However, recordings of this content are not only adjacent to the anti-scientific content presented on the channel. They also present numerous people who combine the propagation of anti-scientific narratives with those sympathizing with Russia, such as the leader of the so-called "Góralskie Veto!" anti-lockdown group Sebastian Pitoń,[141] Confederation party MEP Grzegorz Braun,[142] leaders of Polish monarchists Adam Wielomski[143] and farmers' protests Marcin Bustowski,[144] or the previously mentioned Mossakowski.[145]

The Polish anti-abortion movement is also active in spreading anti-vaccination conspiracy narratives (but not 5G). Mariusz Dzierżawski's Foundation,

---

137 NISS TV (Narodowy Instytut Studiów Strategicznych), "Nawigatorzy Państwa Polskiego obrali kurs na wojnę"– dr Leszek Sykulski specjalista ds. geopolityki", *YouTube*, 24.10.2022, accessed 06.09.2024, https://www.youtube.com/watch?v=MmuNNaMmyWQ.

138 NISS TV (Narodowy Instytut Studiów Strategicznych), "Kandydat na Prezydenta o NWO, o pieniądzu, o skradzionych podpisach, o bojkocie wyborów.", *YouTube*, 12.06.2020, accessed 06.09.2024, https://www.youtube.com/watch?v=52y_OncHWA.

139 NISS TV (Narodowy Instytut Studiów Strategicznych), "Kongres Pokoju. Pokój – różne poglądy jeden cel. Liberalizm – pokojowy kolonializm, pokojowy podbój.", *YouTube*, 11.03.2023, accessed 06.09.2024, https://www.youtube.com/watch?v=rhiezHyeyDw.

140 NISS TV (Narodowy Instytut Studiów Strategicznych), "Kongres Pokoju. Różne poglądy jeden cel ... POKÓJ! cz.1", *YouTube*, 13.03.2023, accessed 06.09.2024, https://www.youtube.com/watch?v=DIurXny7vhU.

141 NISS TV (Narodowy Instytut Studiów Strategicznych), "S. Pitoń – Globaliści a obóz azjatycki. Kryzysy nas straszą. Człowiek umarł ze strachu przed ZUS–em!", *YouTube*, 09.10.2022, accessed 06.09.2024, https://www.youtube.com/watch?v=XE1otsvRGDE.

142 NISS TV (Narodowy Instytut Studiów Strategicznych), "»Wirus« w służbie ponadnarodowych i ponadpaństwowych korporacji. Protesty w całej Polsce 10.10.2020!", *YouTube*, 10.10.2020, accessed 06.09.2024, https://www.youtube.com/watch?v=jdbKcMwcJgg.

143 NISS TV (Narodowy Instytut Studiów Strategicznych), "Piotr Szlachtowicz o zniewolonym umyśle".

144 NISS TV (Narodowy Instytut Studiów Strategicznych), "Koronawirus III Wojna Światowa – wojna biologiczna", *YouTube*, 03.03.2020, accessed 06.09.2024, https://www.youtube.com/watch?v=tCQz4hKGRPA.

145 NISS TV (Narodowy Instytut Studiów Strategicznych), "Wszystkim Polakom należy się Bezwarunkowa Dywidenda Narodowa! NISS w Centrum Edukacyjnym Polska.", *YouTube*, 30.04.2020, accessed 06.09.2024, https://www.youtube.com/watch?v=vSTAAIUPSsI.

Pro – Right to Life, created a special report to prove that the preparations used to immunize against viruses are produced using living cells obtained from aborted foetuses and called this procedure: experiments similar to those conducted in the concentration camps.[146] The leader of the Life and Family Foundation, Kaja Godek, described them similarly, stating that "all COVID-19 vaccines at some stage of production or testing benefitted from the crime of abortion" and that "biological material taken directly from a murdered person was used to release each of the COVID vaccines to the market."[147] Dzierżawski even criticized the consent for vaccinations issued by the Vatican.[148] As a result, both he and Godek demanded the rejection of laws aimed at introducing any coercion to COVID-19 vaccination.[149]

146 No author, "»To nie kwestia historyczna«. Raport o związkach szczepionek z przemysłem aborcyjnym", *Do Rzeczy*, 12.10.2021, accessed 06.09.2024, https://www.dorzeczy.pl/religia/210793/powstal-raport-o-zwiazkach-szczepionek-z-przemyslem-aborcyjnym.html.

147 Mikołaj Pietraszewski, "»Festiwal propagandy«, »oklaski od Konfederacji«. Znowu burza po słowach Kai Godek", *Radio Zet*, 05.01.2022, accessed 06.09.2024, https://www.wiadomosci.radiozet.pl/polska/Kaja-Godek-szczepionki-na-COVID-19-to-korzystanie-ze-zbrodni-aborcji.-Komentarze-ofensywa-antyszczepionkowcow.

148 Mariusz Dzierżawski, "Przymus szczepienia na COVID w Watykanie", *Do Rzeczy*, 30.09.2021, accessed 06.09.2024, www.dorzeczy.pl/tylko-na-dorzeczy/205807/watykan-przymus-szczepienia-na-covid-pracownikow-stolicy-apostolskiej.html.

149 Amk, "Godek na komisji zdrowia: Wszystkie szczepionki korzystają ze zbrodni aborcji", *Rzeczpospolita*, 05.01.2022, accessed 06.09.2024, www.rp.pl/polityka/art19258431-godek-na-komisji-zdrowia-wszystkie-szczepionki-korzystaja-ze-zbrodni-aborcji.

CHAPTER 5

Vaccination Macht Frei

The Political Representation of Anti-Vaccination and Anti-Fifth Generation Telephony Movements

Around 2015, the topic of vaccines and 5G, publicized by media associated with alternative medicine, new spirituality, and the far right, began to arouse broader interest among some politicians as well. Those who regularly appeared on these channels, together with activists of the vaccination and 5G telephony movements, began to publicize the topic beyond the previous circles of already interested supporters of technophobic conspiracy narratives and often also included it on their political agenda. Many also had in common with activists and "experts" of these movements, their joint activities in other protest groups, associations, and foundations, or even running from one party list for parliament or local or regional government. The most significant support for anti-scientific movements was provided by three formations: 1Polska.pl, Kukiz'15, and Confederation, and due to their publicizing these issues, the topic penetrated the mainstream parties. The presence of the message was also strengthened by the migration of politicians from these groups to other formations. Consequently, they also began to raise the issues of 5G and vaccines, using elements of conspiracy or technophobic narratives.

The first Polish strictly political group to add the topic of 5G and vaccines to its demands was the 1Polska.pl initiative, founded in 2015. According to the intention of its creators, it was to be an electoral platform for grassroots social initiatives. In practice, however, it turned out to be an association mainly of far-right activists and supporters of conspiracy narratives, including primary opponents of the introduction of 5G and mandatory vaccinations. Although the group existed for only four years, it contributed to the rapprochement of several groups, which, over time, began to represent supporters of anti-scientific conspiracy narratives while already active in other parties and organizations.

Future founders of the National Institute of Strategic Studies, such as Mateusz Jarosiewicz, Wojciech Edward Dobrzyński, Piotr Robert Jankowski,[1] and

1 Creator of the internet channel "NaviTy – Stacja Wolnych Ludzi" and the cryptocurrency NaviCoin.

 | DOI:10.1163/9789004747852_007

Marek Sas-Kulczycki,[2] were active in 1Polska.pl. Tobiasz Żuchiewicz became the link with the next group to become an essential political representative of anti-scientific movements in the mainstream of Polish politics, i.e., with the supporters of Grzegorz Braun. Żuchiewicz combined his activity in the Polish Association for the Prevention of Electrocontamination, "Right to Life," with his activity in 1Polska.pl, and later in NISS and Braun's foundation, Pobudka,[3] and also supported the movement opposing mandatory vaccinations.[4] Even though Żuchiewicz and the Braun parted ways, Confederation party is today an essential link in the network of opponents of 5G[5] telephony and mandatory vaccinations.[6]

Among Braun's supporters, Mossakowski,[7] Izabela Rostworowska, a later candidate from Grzegorz Braun's KWW list "God bless you!" in Kraków, and Małgorzata Łabędzka, an economist from Wrocław, were also active in the fight against 5G. Another group active in the battle against 5G, which started in 1Polska.pl, are the future MPs of the Kukiz'15 movement, Robert Majka and Paweł Szramka. It is also worth noting the presence of Hanna Kazahari, an opponent of mandatory vaccinations and 5G and a propagator of anti-Semitic conspiracy narratives, and Robert Jarosław Bogusławski, a nationalist from Wrocław, a former candidate for the office of the Mayor of the city, and a leader of protests against 5G telephony and mandatory vaccinations[8] in the region,[9] in the ranks of this party. Bogusławski shouted during a demonstra-

2 President of NISS, former columnist of the far-right blog *Nowy Ekran* and founder of the *Ekspedyt.org* portal.

3 LeszekBubelTV, "GRZEGORZ BRAUN – PSEUDOCHRZEŚCIJANIN I UZURPATOR, CZĘŚĆ 1 – TOBIASZ ŻUCHEWICZ", *BitChute*, 08.12.2019, accessed 06.09.2024, https://www.bitchute.com/video/Tpb15zEL1gRc.

4 Edyta Iwaniuk, "Zdejmij maseczkę, pogoń PiS. Antycovidowcy, antyszczepionkowcy i anty-5G wzywają do wyjścia na ulice", *Wyborcza.pl*, 07.10.2020., accessed 06.09.2024, https://wyborcza.pl/7,162657,26375707,zdejmij-maseczke-i-pogon-pis-antycovidowcy-antyszczepionkowcy.html.

5 Michał Świech, "Dumnie walczył z 5G. Teraz zostanie posłem", *Telepolis*, 19.07.2023, accessed 06.09.2024, https://www.telepolis.pl/wiadomosci/wydarzenia/konfederacja-5g-rafal-forys-teorie-spiskowe.

6 See further.

7 See previous chapters for more details.

8 Ml, "Kandydat na prezydenta Wrocławia, przeciwnik obowiązkowych szczepień gościem »Skandalistów« o 20:00", *Polsat News*, 06.10.2018, accessed 06.09.2024, https://www.polsatnews.pl/wiadomosc/2018-10-06/kandydat-na-prezydenta-wroclawia-przeciwnik-obowiazkowych-szczepien-gosciem-skandalistow-o-2000.

9 Piotr Zapotoczny, "Protest w centrum Opola. Straszono technologią 5G, lobby szczepionkowym i politykami na usługach koncernów", *Wyborcza.pl*, 25.01.2020, accessed 06.09.2024,

tion in Opole: "5G technology is very harmful, so why is no politician focusing on it? Some make business out of it, and we talk about big money. They lobby in parliaments and bribe politicians to keep quiet. Companies make much money from it, and our children suffer from it."[10] It is worth noting that, in addition to Mossakowski and the future founding members of NISS, Damian Bieńko from National Free Poland and the future influential activist of the Zmiana party, Nabil al-Malazi,[11] also participated in the creation of 1Polska.pl among activists sympathizing with Russian narratives.

However, the key factor in transferring anti-scientific narratives from the margins to the mainstream was the interest of, at that time, not a party but the right-wing populist Kukiz'15 movement. On the wave of high support for its leader in the presidential elections, a party emerged as a kind of "joint forces" of entities defining themselves as "anti-system."[12] As a result, the lists of this movement in the parliamentary elections included a diverse group consisting of libertarians, nationalists, supporters of direct democracy and religious orthodox. Among them were also opponents of mandatory vaccinations and 5G. As a result, these issues became the subject of interpellations, conferences, and discussions organized in the Parliament and began to be described more broadly by the most prominent Polish media. The activity of MP Paweł Skutecki proved crucial for propagating the narrative of opposition to mandatory vaccinations in the 8th term of the Sejm. In turn, for the anti-5G movement emphasizing the alleged threats resulting from its use, such a role was played by MP Jarosław Sachajko.

Skutecki's professional career was connected with media, politics, and business, not medicine. He was the press officer of the Voivodeship Labour Office in Warsaw, the Kuyavian-Pomeranian Voivode and Janusz Korwin-Mikke's far-right party, the Union of Real Politics (UPR), and a journalist and copywriter. He also ran his own business. He unsuccessfully ran for the European Parliament from the UPR list in 2009[13] and for the Sejm from the far-right

---

https://opole.wyborcza.pl/opole/7,35086,25632807,kontrowersyjny-protest-na-pl-wolnosci-narodowiec-traszyl-technologia.html.

10 Ibid.

11 See further: Przemysław Witkowski, *Partia rosyjska* (Warszawa, Arbitror, 2023),150–154.

12 Arkadiusz Lewandowski, "Antysystemowość ruchu Kukiz'15 w okresie kampanii wyborczej 2015 roku", in *Nowe partie i ruchy polityczne. Praktyka i idee współczesnej demokracji* ed. Katarzyna Walecka; Kinga Wojtas (Warszawa, Wydawnictwo Naukowe UKSW, 2017),179–199.

13 No author, "Wyniki głosowania: Komitet Wyborczy Unia Polityki Realnej", *Państwowa Komisja Wyborcza*, no data, accessed 06.09.2024, https://pe2009.pkw.gov.pl/PUE/PL/WYN/W110/2.htm.

Congress of the New Right list in 2011.[14] He only won the mandate in 2015 from his hometown of Bydgoszcz, starting from the first place on the Kukiz'15 list.[15] At that time, he was also an activist of the National Association of Vaccination Awareness "STOP NOP" (then led by Justyna Socha). Socha also ran for the Parliament from the Kukiz'15 list in 2015 but did not win a mandate. Despite not being elected, she began to regularly attend the Sejm, as Skutecki made her his assistant.

Already in the fourth month of his term (in February 2016), Skutecki, together with other MPs elected from the Kukiz'15 lists (Piotr Liroy Marzec, Robert Winnicki, Wojciech Bakun, and Jarosław Porwich), established the Parliamentary Team for the Safety of the Vaccination Program. The team members had no medical education, yet they considered the goals of their work to be establishing the actual balance of benefits and losses resulting from the vaccination programme, informing patients about all possible side effects and complications after using vaccines, and eliminating sanctions for not vaccinating a child.[16]

Skutecki also regularly participates in demonstrations of the movement opposing mandatory vaccinations every June and also gives speeches during this manifestation.[17] In 2018, he also collaborated on the draft bill on abolishing mandatory vaccinations, prepared by STOP NOP. According to the bill, vaccinating a child was a voluntary choice of parents, and vaccinations would be mandatory only in the event of an epidemiological threat or epidemic.[18] Activists opposing mandatory vaccinations collected 120,000 signatures for the bill, which allowed them to submit it to the Parliament for consideration.[19] In Sep-

---

14 No author, "Wybory 2011: Komitet Wyborczy Nowa Prawica – Janusza Korwin-Mikke", *Państwowa Komisja Wyborcza*, no data, accessed 06.09.2024, https://wybory2011.pkw.gov.pl/rfl/pl/393d3249f302f85c28fdca91340862b9.html.

15 Wb, "Wybory 2015. Znamy wyniki wyborów w okręgu bydgoskim", *Wyborcza.pl*, 26.10.2015, accessed 06.09.2024, https://bydgoszcz.wyborcza.pl/bydgoszcz/1,48722,19092194,wybory-2015-znamy-wyniki-wyborow-w-okregu-bydgoskim.html.

16 Łukasz Woźnicki, "Antyszczepionkowcy mają swój zespół w Sejmie. Poseł Kukiza: Ja nie szczepię syna", *Wyborcza.pl*, 08.02.2016, accessed 06.09.2024, https://wyborcza.pl/7,75398,19599598,antyszczepionkowcy-maja-swoj-zespol-w-sejmie-posel-kukiza.html.

17 Krystian Lurka, "Skutecki: Wielu Polaków wie więcej o szczepieniach niż lekarze", *TerMedia*, 11.06.2018, accessed 06.09.2024, https://www.termedia.pl/mz/Skutecki-Wielu-Polakow-wie-wiecej-o-szczepieniach-niz-lekarze,30337.html.

18 MS, "Antyszczepionkowcy chcą usunięcia obowiązku szczepień. Zebrali 120 tys. podpisów pod projektem ustawy", *Newsweek*, 12.07.2018, accessed 06.09.2024, https://www.newsweek.pl/polska/polityka/antyszczepionkowcy-zlozyli-w-sejmie-podpisy-pod-projektem-ustawy/7ze4pw2.

19 Medexpress, "Kukiz'15 nie ma nic wspólnego z projektem ustawy znoszącej obowiązek szczepień", *Medexpress.pl*, 16.07.2018, accessed 06.09.2024, https://www.medexpress.pl/

tember 2018, the Parliament rejected the bill, but ten MPs elected from the Kukiz'15 lists voted in favour of its adoption,[20] and sixteen abstained.[21]

The team appointed by Skutecki also invited foreign leaders of the anti-vaccination movement to the Parliament. They invited: Christopher Exley from Keele University, who tried to show that aluminum contained in vaccines causes autism (the results of many studies contradict this thesis);[22] Antonietta Gatti and Stefano Montanari, representing the Health, Law and Science Association, known primarily for their claims that vaccines contain toxic metals;[23] Robert F. Kennedy Jr., who believes that vaccines are responsible for causing ADD, ADHD, speech disorders, autism, food allergies and autoimmune diseases, and the HPV vaccine, in his opinion, causes uterine cancer[24] (in fact, it prevents 90% of cases of this disease[25]). At the same time, Kennedy was very critical of the American involvement in aid for Ukraine, attacked by Russia, calling the conflict "the United States' war against Russia" and claiming that the war aimed to "sacrifice the flower of Ukrainian youth in a slaughterhouse of death and destruction for the geopolitical ambitions of the neoconservatives".[26] Kennedy also opposes Ukraine's admission to NATO and has announced that as US president, he will consider admitting Russia

---

ochrona-zdrowia/kukiz-15-nie-ma-nic-wspolnego-z-projektem-ustawy-znoszacej-obowiazek-szczepien-71051.

20 These were the following Members of Parliament: Anna Siarkowska, Andrzej Kobylarz, Jarosław Sachajko, Paweł Skutecki, Magdalena Błeńska, Marek Jakubiak, Janusz Sanocki, Jacek Wilk, Robert Winnicki and Jarosław Porwich.

21 These were MPs Barbara Bubula and Jan Szyszko elected from the PiS lists, and MPs elected from the Kukiz'15 lists: Józef Brynkus, Barbara Chrobak, Paweł Grabowski, Jerzy Jachnik, Bartosz Józwiak, Norbert Kaczmarczyk, Paweł Kukiz, Maciej Masłowski, Robert Mordak, Stefan Romecki, Paweł Szramka, Agnieszka Ścigaj, Stanisław Tyszka and Jakub Kulesza.

22 Michael Simpson, "Autism and vaccines – 164 science articles say no link (UPDATED)", *Skeptical Raptor* (blog), 10.07.2017, accessed 06.09.2024, www.skepticalraptor.com/skepticalraptorblog.php/vaccines-and-autism-science-unrelated.

23 Co tam w Sejmie, "2019.02.19 Parlamentarny Zespół ds. Bezpieczeństwa Programu Szczepień Ochronnych Dzieci i Dorosłych", *Youtube*, 19.02.2019, accessed 06.09.2024, https://www.youtube.com/watch?v=Q4UrZntXAaI.

24 Robert F. Kennedy Jr, "What are we doing to our children?", *Twitter*, 21.03.2019, accessed 06.09.2024, http://www.twitter.com/RobertKennedyJr/status/1108734709041938432.

25 Heikki Peltola and Olli P. Heinonen, "Frequency of true adverse reactions to measles-mumps-rubella vaccine. A double-blind placebo-controlled trial in twins", *Lancet* 1, no. 8487 (1986):939–942.

26 Freddie Sayers, "Robert Kennedy Jr: America needs a revolution", *UnHerd*, 03.5.2023, accessed 06.09.2024, https://unherd.com/2023/05/robert-kennedy-jr-america-needs-a-revolution.

to the North Atlantic Alliance and de-escalating tensions with the People's Republic of China[27] (he declared, he intends to run for US president in 2024[28]).

Activity in the anti-vaccination movement is not Skutecki's only platform in which he expresses sympathy to Russian narratives or directly to Russia. In 2018, while serving in office, he also visited Lebanon and Syria,[29] where he met with the chairman of the Hezbollah parliamentary club, Mohammad Raad, and the Syrian dictator Bashar al-Assad. The US, UK, and Israel consider Hezbollah, a terrorist organization. In Poland, he spoke favourably about Hezbollah in a recording made public by Xportal, a media outlet run by the national Bolshevik and Falanga groups linked to the attacks in Ukraine.[30] Both the Syrian authorities and the leaders of Hezbollah, as well as Falanga, have close and amicable relations with Russia and actively participate in promoting Russian narratives.

In the 2018 local elections, Skutecki was a candidate for Kukiz'15 for the mayor of Bydgoszcz, taking the last 5th place. He also failed to win a mandate as a Member of the European Parliament in February 2019. As a result, in August 2019, he joined the newly formed Confederation party.[31] However, he failed to renew his parliamentary mandate from its lists.

Along with the disintegration of the Kukiz'15 parliamentary club, which had been ongoing since the first year of its existence, and the transfer of its members to other parties, critical opinions regarding mandatory vaccinations or 5G telephony began to gain new allies. Some of the politicians continued their former activities from the time of their activity in Kukiz'15, while others noticed this issue only after meeting new colleagues. The importance of vaccinations and 5G also varied for them. Some made it the central point of their activity. For others, it was present but marginal.

---

27 Ibid.

28 Michelle L. Price, "Anti–vaccine activist RFK Jr. challenging Biden in 2024", *Associated Press*, 06.04.2023, accessed 06.09.2024, https://apnews.com/article/robert-kennedy-jr-presidential-campaign-9fb5ed5c8e1fd31d2a4458d44b086593.

29 Barbara Burdzy, "Na kawie z dyktatorem Syrii. Niebezpieczne gry posła Kukiz'15", *Fakt*, 10.12.2018, accessed 06.09.2024, https://www.fakt.pl/wydarzenia/polityka/pawel–skutecki–kto–stoi–za–poslem–kukiz15/bxzrmq2.

30 Redakcja Xportal.pl, "Paweł Skutecki dla Xportalu: Hezbollah to najlepsza armia Bliskiego Wschodu", *Xportal.pl*, 23.08.2018, accessed 06.09.2024, http://xportal.press/?p=33464.

31 Rej, "Wybory parlamentarne 2019. Paweł Skutecki w Konfederacji. Do Sejmu wystartuje w Warszawie", *Wyborcza.pl*, 12.08.2019, accessed 06.09.2024, https://bydgoszcz.wyborcza.pl/bydgoszcz/7,48722,25081579,wybory–parlamentarne–2019–pawel–skutecki–w–konfederacji–do.html.

Regardless, from the Kukiz'15 movement, which brought the topic of opposition to mandatory vaccinations and 5G telephony to the Polish Parliament, they began to penetrate increasingly into other mainstream groups. At the same time, local protests organized by activists also influenced individual MPs, especially during their visits to their home districts. The combination of both influences – on the one hand, the influence of MPs – propagators in the Parliament, on the other – the grassroots-level pressure resulted in an increasing number of parliamentary speeches and interpellations regarding vaccines and 5G.

For the issue of 5G and the alleged threats from its use, an essential channel for the topic to penetrate the mainstream political discourse was the coalition parliamentary club Koalicja Polska – Polskie Stronnictwo Ludowe – Kukiz'15, and in it, above all, MP Jarosław Sachajko. He is not a telecommunications specialist. Sachajko studied physiotherapy, economics, and agronomy, and he has a doctorate based on the thesis entitled *Production, economic, and energy assessment of potato and spring triticale cultivation in the conditions of diversified agricultural technology*. He also completed postgraduate studies in the field of real estate valuation. Sachajko worked as an assistant professor at the Lublin Agricultural University and the Higher School of Management and Administration in Zamość. He was also a criminal analyst at the District Prosecutor's Office in Zamość and ran his own business in the training industry. Sachajko became politically involved in the Civic Movement for the Single-member districts. In 2015, he ran for the Parliament in the Chełm district from the first place on the Kukiz'15 list, obtaining the mandate of a member of parliament. In the Parliament of the 8th term, he became the chairman of the Agriculture and Rural Development Committee and a member of the Health Committee, and in the next term, vice-chairman of the Agriculture and Rural Development Committee.

He, like Skutecki, also opposes mandatory vaccinations. During the COVID-19 pandemic, he announced that he would accept vaccinations until a special compensation fund for alleged "vaccine victims" was established. On Radio Lublin, he demanded the establishment of such a fund and called for a significant public "debate about the real effects of vaccination."[32] "I believe that adults have the right to die or have complications. If someone wants to die and does so with full awareness, we cannot prevent them from doing so,"

32 Tomasz Piątek, "Sojusznik Białorusi i antyszczepionkowców na jednej liście z ministrem Kamińskim", *Wyborcza.pl*, 11.09.2023, accessed 06.09.2024, https://wyborcza.pl/7,75398,30175157,sojusznik-bialorusi-i-antyszczepionkowcow-na-jednej-liscie-z.html.

Sachajko said at the time. He did not get vaccinated against COVID-19.[33] As early as 2017, Sachajko submitted two interpellations to the Minister of Health, demanding explanations regarding alleged post-vaccination reactions[34] and information on vaccinations in Poland compared to other EU countries.[35] He also submitted other interpellations related to the subject, such as the one from 2020, supporting an open letter by Ryszard Rutkowski, a doctor who opposes mandatory vaccinations.[36]

However, the topic of 5G has become more important to Sachajko than vaccinations, as evidenced by the number of interpellations he has submitted on telephony and "electromagnetic fields" and the selection of his closest collaborators. Sachajko knows and cooperates with the anti-5G movement's Jadwiga Łopata and Julian Rose from his activity opposing the use of GMO in Polish agriculture.[37] Other activists of anti-5G movement – Tobiasz Żuchewicz, Agata Ratajczyk (both representatives of the Polish Association for the Prevention of Electrocontamination, "Right to Life,") and Barbara Gałdzińska-Calik (James Clerk Maxwell Institute for Electromagnetic Research Foundation) were social assistants to MP Sachajko during the 9th term of the Polish Parliament.[38] The MP is also working with the ufologist and opponent of 5G,

33 P.mal, "Poseł Kukiz'15 Jarosław Sachajko o obowiązku szczepień: Dorośli mają prawo umrzeć", *Rzeczpospolita*, 06.12.2021, accessed 06.09.2024, https://www.rp.pl/polityka/art19172541-posel-kukiz-15-jaroslaw-sachajko-o-obowiazku-szczepien-dorosli-maja-prawo-umrzec.

34 Jarosław Sachajko and Jerzy Kozłowski, "Interpelacja nr 15350 do ministra zdrowia w sprawie finansowania szczepionek i obowiązku szczepień w Polsce", *Sejm Rzeczpospolitej Polskiej*, 07.09.2017, accessed 06.09.2024, https://sejm.gov.pl/Sejm8.nsf/InterpelacjaTresc.xsp?key=399D7B34.

35 Jarosław Sachajko, "Interpelacja nr 16657 do ministra zdrowia w sprawie szczepień obowiązkowych oraz Programu Elektronicznego Nadzoru nad Dystrybucją Szczepionek", *Sejm Rzeczpospolitej Polskiej*, 23.10.2017, za: https://www.sejm.gov.pl/Sejm8.nsf/InterpelacjaTresc.xsp?key=3F6DFCFD.

36 Jarosław Sachajko, "Interpelacja nr 9600 do ministra zdrowia w sprawie badań nad COVID-19, sekcji zwłok osób zmarłych, opracowania krajowych standardów, lekarstw i badania szczepionek oraz zagadnień poruszanych w liście otwartym prof. dr hab. Ryszarda Rutkowskiego", *Sejm Rzeczpospolitej Polskiej*, 12.08.2020, accessed 13.04.2025, https://www.sejm.gov.pl/Sejm9.nsf/InterpelacjaTresc.xsp?key=BSN9ZH

37 Janek Cios, "Jarosław Sachajko (Kukiz'15) – przeciw GMO", *Życie Zamościa*, 30.11.2016, accessed 06.09.2024, https://zyciezamoscia.pl/informacje/jaroslaw-sachajko-kukiz15-przeciw-gmo.

38 Zbigniew Podjadek, "Anatomia polskiego ruchu anty5G. Kim są ludzie, którzy walczą z budową sieci nowej generacji?", *Wyborcza.pl*. 16.09.2020, za: https://wyborcza.biz/biznes/7,177150,26290907,anatomia-polskiego-ruchu-anty5g-kim-sa-ludzie-ktorzy-walcza.html.

Janusz Zagórski, whose YouTube channel Niezależna Telewizja is the primary channel for conspiracy narratives about 5G. Sachajko also filed seven interpellations on 5G.[39]

Other Kukiz'15 MPs were also active in the matter of opposition to 5G. MPs Maciej Masłowski, Józef Brynkus, and Agnieszka Ścigaj appeared with Sachajko at the Kraków conference,[40] where the aforementioned coalition Poland Free from 5G was formed.[41] In turn, another MP from the Kukiz'15 club at the time, Marek Jakubiak, asked from the parliamentary rostrum: "Are you really convinced that the 5G installation will not have any negative effect on the lives of our children, on the lives of our children's children? You are a young person. Are you all really convinced that the frequencies and power of these devices will not cause what everyone around you is talking about? The frequency affecting microwave ovens is harmful. If, at the moment, we sign and introduce with such joy a transmitter every few hundred meters, then I

39 Jarosław Sachajko et al., "Interpelacja nr 28560 do ministra cyfryzacji w sprawie wątpliwości związanych z budową sieci 5G w Polsce", *Sejm Rzeczpospolitej Polskiej*, 02.01.2019, accessed 06.09.2024, https://www.sejm.gov.pl/Sejm8.nsf/InterpelacjaTresc.xsp?key=B88JVU; Jarosław Sachajko, "Interpelacja nr 32367 w sprawie bezpieczeństwa wprowadzenia technologii 5G do Ministra Zdrowia", *Sejm Rzeczpospolitej Polskiej*, 01.07.2019, accessed 06.09.2024, https://www.sejm.gov.pl/Sejm8.nsf/InterpelacjaTresc.xsp?key=BE6C88; Jarosław Sachajko, "Interpelacja nr 32369 w sprawie bezpieczeństwa wprowadzenia technologii 5G do Ministra Obrony Narodowej ", *Sejm Rzeczpospolitej Polskiej*, 01.07.2019, accessed 06.09.2024, https://www.sejm.gov.pl/Sejm8.nsf/InterpelacjaTresc.xsp?key=BE6C8A; Jarosław Sachajko, "Interpelacja nr 32370 w sprawie bezpieczeństwa wprowadzenia technologii 5G do Ministra Infrastruktury", *Sejm Rzeczpospolitej Polskiej*, 01.07.2019, accessed 06.09.2024, https://www.sejm.gov.pl/Sejm8.nsf/InterpelacjaTresc.xsp?key=BE6C8B; Jarosław Sachajko, "Interpelacja nr 32371 w sprawie bezpieczeństwa wprowadzenia technologii 5G do Ministra Sprawiedliwości i Ministra Cyfryzacji", *Sejm Rzeczpospolitej Polskiej*, 01.07.2019, accessed 06.09.2024, https://www.sejm.gov.pl/Sejm8.nsf/InterpelacjaTresc.xsp?key=BE6C8C; Jarosław Sachajko, "Interpelacja nr 32373 w sprawie bezpieczeństwa wprowadzenia technologii 5G do Ministra Środowiska", *Sejm Rzeczpospolitej Polskiej*, 01.07.2019, accessed 06.09.2024, https://www.sejm.gov.pl/Sejm8.nsf/InterpelacjaTresc.xsp?key=BE6C8E; Jarosław Sachajko, Paweł Szramka "Interpelacja nr 970 w sprawie rozwoju sieci 5G i tworzenia spółki celowej "Polskie 5G" do Ministra Cyfryzacji", *Sejm Rzeczpospolitej Polskiej*, 18.12.2019, accessed 06.09.2024, https://www.sejm.gov.pl/sejm9.nsf/InterpelacjaTresc.xsp?key=BKCE7Q.

40 Administrator, "Czy 5G to wojskowa broń światowych elit do selektywnego zabijania ludzi?! Na to wygląda!", *Warszawska Gazeta*, 02.03.2019, accessed 06.09.2024, https://gazetawarszawska.com/index.php/pugnae/3151-czy-5g-to-wojskowa-bron-swiatowych-elit-do-selektywnego-zabijania-ludzi-na-to-wyglada.

41 Ada Chojnowska, "Kontrola umysłu i eksperymenty na ludziach ... Koalicja przeciw 5G obradowała w Krakowie", *Wyborcza.pl*, 28.02.2019, accessed 06.09.2024, https://krakow.wyborcza.pl/krakow/7,44425,24504806,kontrola-umyslu-i-eksperymenty-na-ludziach-koalicja-przeciw.html.

ask where we will live. People, think about it."[42] One of Sachajko's interpellations regarding 5G was also signed by members of the Sejm of the 8th term, Stefan Józef Romecki, Józef Brynkus and Bartosz Józwiak and Barbara Chrobak. The latter additionally, in her interpellation to the Minister of Health and the Minister of the Environment (Interpellation No. 23562), asked about the spraying of so-called chemtrails over Poland, repeating another conspiracy narrative unsupported by scientific facts.[43]

Also, in the case of MP Sachajko, the element of connections with the regimes of Russia and Belarus is repeated once again. In 2016, Sachajko opposed sanctions against Russia in an interview with the Russian portal Life.ru,[44] and the portal Regnum, currently blocked in Poland by the Internal Security Agency, publicized Sachajko's struggle against the so-called Istanbul Convention on Combating Domestic Violence.[45] However, the MP has closer ties with Belarus. Since 2016, he has been the vice-chairman of the Poland-Belarus Association, one of the organizations established after the liquidation of the Polish-Soviet Friendship Society. The association considers Lukashenka's rule to be legal and the Belarusian elections to be unrigged. The chairman of the organization to which Sachajko belongs is Bolesław Borysiuk, a former post-communist party Democratic Left Alliance activist and national-communist Samoobrona party MP. In the communist times, he was secretary of the Supreme Council of the Polish Students' Association and the Polish-Soviet Friendship Society and deputy editor-in-chief of the weekly "Rzeczywistość," a magazine of the most pro-Soviet part of the communist party, whose political tone was set, from behind the scenes, by the former head of the Ministry of Internal Affairs, strongly associated with the KGB, General Mirosław Milewski. In 2002, Borysiuk became an advisor to Samoobrona, headed its city authorities in Lublin, and 2005, was elected a member of parliament from its lists. After his parliamentary mandate expired, Borysiuk founded the Party of Regions and 2009 was a candidate on the Eurosceptic

---

42 Marek Jakubiak, "Technologia 5G", *Facebook*, 07.07.2019, accessed 06.09.2024, https://www.facebook.com/watch/?ref=external&v=539788803222400.

43 Barbara Chrobak, "Interpelacja nr 23562 do ministra zdrowia, ministra środowiska w sprawie rozpylania w atmosferze szkodliwych substancji", *Sejm Rzeczpospolitej Polskiej*, 29.06.2018, accessed 06.09.2024, https://www.sejm.gov.pl/Sejm8.nsf/InterpelacjaTresc.xsp?key=3323CEFA.

44 Наталья Корчмарек, "С нас хватит! Даже в Польше задумались об отмене санкций против России", *Life*, 18.08.2016, accessed 06.09.2024, https://life.ru/p/892033#S.embed_link–K.C–B.1–L.2.zw.

45 Tomasz Piątek, "Sojusznik Białorusi i antyszczepionkowców".

pan-European Libertas party lists.[46] In 2015 Party of Regions merged with the pro-Russian Zmiana party, led by mentioned earlier, alleged Russian agent, Mateusz Piskorski. Sachajko social assistants Żuchewicz and Ratajczyk were also actively supporting Minsk dictatorship – they participated in a rally supporting Alexander Lukashenka on August 21, 2020, in front of the Belarusian embassy.[47] Also, in the 9th term, primarily due to the activities of MP Sachajko, the topic of 5G found a wider field of influence, drawing PSL politicians into the activities combating this generation of telephones. Sachajko submitted further interpellations on this matter.[48] Jarosław Rzepa, Urszula Nowogórska, and Paweł Szramka also submitted their interpellations, repeating conspiracy narratives about 5G.

Supporters of the struggle against 5G are also present in the Polish People's Party (PSL) outside the Parliament. In the 2018 local elections, Well-known anti-5G activist Rafik Ennaoui ran for the regional council on the PSL list in the West Pomeranian Voivodeship.[49] In Kraków, Marta Patena ran for the Sejm from the PSL lists in 2019, and in 2023 from the Trzecia Droga[50] lists. She is one of the better-known anti-5G activists. From 1998 to 2018, she was a Krakow City Councilor and, in 2014, a candidate of the Civic Platform (PO) for the position of Mayor of Kraków.[51] She also ran for the Parliament from the lists of this party in 2001 and 2007 and to the European Parliament in 2009.[52]

---

46 Sławomir Skomra, "SLD z sierotami po Lepperze", *Wyborcza.pl*, 20.08.2010, accessed 06.09.2024, https://wyborcza.pl/7,75398,8273052,sld-z-sierotami-po-lepperze.html.

47 Podjadek, "Anatomia polskiego ruchu anty5G".

48 Jarosław Sachajko and Paweł Szamka, "Interpelacja nr 970" oraz Jarosław Sachajko and Paweł Szramka, "Interpelacja nr 969 w sprawie ewentualnych praktyk korupcyjnych prowadzonych przez koncern Ericsson do Ministra Sprawiedliwości", *Sejm Rzeczypospolitej Polskiej*, 18.12.2019, accessed 06.09.2024, http://www.sejm.gov.pl/sejm9.nsf/InterpelacjaTresc.xsp?key=BKCE7P.

49 He is also an active participant in the protest movement against mandatory vaccinations. His sister is Sofia Ennaoui, a Polish athlete specializing in middle- and long-distance running. Finalist of the 2016 Olympic Games in Rio de Janeiro in the 1500 meters and vice-champion of the European Championships in 2018 in this distance.

50 The Third Way – a Polish political coalition of a centrist and Christian Democratic-liberal nature, established before the parliamentary elections in 2023, founded by the Polish People's Party and Poland 2050.

51 Maciej Skowronek, "Marta Patena kandydatką PO na prezydenta Krakowa", *Radio Kraków*, 31.08.2014, accessed 06.09.2024, https://www.radiokrakow.pl/aktualnosci/krakow/marta-patena-kandydatka-po-na-prezydenta-krakowa.

52 Maria Patena, "O mnie", *Patena.pl*, no data, accessed 06.09.2024, http://www.patena.pl/pl/o-mnie.

In 2016, she left the PO club[53] and supported her recent rival Jacek Majchrowski, becoming his social advisor for bioelectromagnetic issues, replacing Gałdzińska-Calik in this function.[54] She also spoke as a 5G Free Zone at the anti-5G Kraków International Conference. Ultimately, Patena joined the PSL and became a social assistant to the party chairman Władysław Kosiniak-Kamysz.[55] It was most likely due to his cooperation with Patena that the PSL leader publicly addressed the technophobic narrative regarding 5G during his presidential campaign in 2020. Commenting on the statement of the presidents of Poland and the USA on cooperation in the development of technological solutions, including 5G, at his election rally in Słupca in June 2020, he asked: "Is it true that President Duda reached an agreement with the Americans on 5G, which raises far-reaching doubts in many circles in Poland? Do we have safe research that could already confirm it?".[56]

Sachajko paved the way for activists fighting against 5G to speak in the Parliament. One of the forums for presenting views was the meetings of the Parliamentary Team for Children's Cybersecurity in 2019, which Sachajko headed during the 8th term. In the first of them, in addition to activists fighting against the introduction of 5G to Poland,[57] scientific experts related to telecommunications also took part.[58] They presented statistics indicating that EMF is not linked to cancer cases in Poland and countries with higher permitted radiation levels.[59] In response, activists claimed that 5G is a mili-

53 Olo, "Marta Patena odchodzi z Platformy Obywatelskiej, kruszeje koalicja w magistracie", *Wyborcza.pl*, 08.06.2016, accessed 06.09.2024, https://krakow.wyborcza.pl/krakow/7,44425,20207354,marta-patena-odchodzi-z-platformy-obywatelskiej-kruszeje-koalicja.html.

54 Łukasz Jeżak, "Doradczyni Jacka Majchrowskiego: Sieć 5G posłuży przede wszystkim rozrywce kosztem zdrowia innych osób [Rozmowa]", Love Kraków, 09.06.2020, accessed 06.09.2024, https://lovekrakow.pl/aktualnosci/siec-5g-posluzy-przede-wszystkim-rozrywce-i-zabawie-kosztem-zdrowia-innych-osob-rozmowa_36157.html.

55 Podjadek, "Anatomia polskiego ruchu anty5G".

56 Władysław Kosiniak-Kamysz, "Czy to prawda, że @AndrzejDuda dogadał się z Amerykanami o 5G", *Twitter*, 25.06.2020, accessed 06.09.2024, https://twitter.com/KosiniakKamysz/status/1276054582515453953.

57 Such activists as: Barbara Gałdzińska-Calik, Ewa Pawela, Zbigniew Gelzok.

58 Such experts as: prof. dr hab. inż. Krawczyk, head of the Institute of Communications dr inż. Jerzy Żurek from the Institute of Communications, prof. dr hab. n. med. Aleksander Sieroń, representatives of the Ministry of Health, Ministry of Digital Affairs and operators.

59 Arkadiusz Dziermański, "Pod przykrywką cyberbezpieczeństwa dzieci, sejm odwiedzili przeciwnicy łączności komórkowej", *Telepolis*, 20.12.2019, accessed 06.09.2024, https://www.telepolis.pl/artykuly/felietony/zmiany-norm-pem-5g-sejm-sachajko-braun.

tary technology that is used for mind control, causes cancer, works using "torsion fields," and will be used to transmit pornography to children.[60] The next meeting was attended only by opponents of the introduction of 5G technology[61] and their allies in the Parliament, such as Sachajko and Braun. From that moment on, these meetings became primarily a presentation of conspiracy narratives related to 5G. As a result of these meetings, the Polish Coalition Free from 5G demanded stricter EMF standards, the withdrawal of Wi-Fi from schools, nurseries, and hospitals and their replacement with cable Internet, as well as the construction of a fibre-optic Internet network in Poland[62] and the rejection of the act on supporting the development of telecommunications services and networks (popularly called the "5G Mega-Act"). On July 4, 2019, twenty-one MPs voted against the bill's adoption, and twenty-six abstained. Notably, most of them, i.e., twenty-eight, were from the Kukiz'15[63] and PSL[64] clubs. Both groups were already getting closer to each other at that time, and a month later, PSL and Kukiz'15 concluded an electoral agreement.[65]

At the end of the 8th term of the Polish Parliament, the Kukiz'15 parliamentary club significantly reduced its representation. Composed of several groups (libertarians, nationalists, supporters of direct democracy, religious integralists, and regionalist politicians) connected by a leader and declared "antisystemic," it could not develop a stable structure. In 2015–2019, it went through a series of splits, and its MPs number dropped from forty-two on day one to sixteen at the end of the term, and six after the 2019 elections.[66] At

60 Ibid.

61 Among others: Ewa Pawela, Witold Hake, Zofia Filipiuk, Diany Wojtkowiak, Sławomir Dul, Zbigniew Gelzok, Barbara Gałdzińska-Calik, Paweł Wypychowski, Rafał Górski,

62 Arkadiusz Dziermański, "Protest przeciwko sieci 5G w Polsce był smutny. Teorie o szkodliwym PEM wybrzmiały w sejmie", *Telepolis*, 01.05.2019., https://www.telepolis.pl/artykuly/wydarzenia/siec-5g-w-polsce-protest-szkodliwe-pem-sejm.

63 The following MPs voted against: Józef Brynkus, Andrzej Kobylarz, Robert Mordak, Jarosław Sachajko, Paweł Skutecki, while the following MPs abstained: Grzegorz Długi, Norbert Kaczmarczyk, Paweł Kukiz, Błażej Parda, Stefan Romecki, Paweł Szramka, Agnieszka Ścigaj, Stanisław Tyszka and Elżbieta Zielińska, as well as an MP elected from the Kukiz'15 list, and non-attached at the time of the vote, Janusz Sanocki.

64 The following MPs voted against: Paweł Bejda, Krystian Jarubas, Michał Kamiński, Mieczysław Kasprzak, Kazimierz Kotowski, Stefan Niesiołowski, Urszula Pasławska, Krzysztof Paszyk, Marek Sawicki and Piotr Walkowski, while the following MPs abstained: Eugeniusz Kłopotek, Jan Łopata, Genowefa Tokarska and Piotr Zgorzelski.

65 Paweł Wroński, "Kukiz jednak z PSL? »Rozmowy są na dobrej drodze«", *Wyborcza.pl*, 04.08.2019, accessed 06.09.2024, https://wyborcza.pl/7,75398,25058036,kukiz-jednak-z-psl-rozmowy-sa-na-dobrej-drodze.html.

66 These were the following MPs: Paweł Kukiz, Jarosław Sachajko, Paweł Szramka, Agnieszka Ścigaj, Stanisław Tyszka and Stanisław Żuk.

the same time, several former members of parliament of the group (primarily associated with the nationalist party National Movement and the milieu of Janusz Korwin-Mikke) established an umbrella political party called the Confederation of Freedom and Independence at the turn of 2018 and 2019. And this group became essential for promoting technophobic narratives, including conspiratorial ones.

Initially, the leading promoter of this type of content among the members of this party's circle was Rafał Majka. Since the end of the eighties, he has been active in various right-wing and extreme right-wing political parties.[67] He found himself in the Parliament at the end of the 8th term, obtaining a mandate as the next on the list after the Kukiz'15 MP, Wojciech Bakun, elected president of Przemyśl. Earlier, in 2015, he did not receive sufficient votes. Despite this, in the same year, he was appointed, on the recommendation of Kukiz'15, as a member of the State Tribunal. Despite his activities related to the mainstream of Polish politics, Majka was primarily active in anti-Semitic circles and those promoting conspiracy narratives. He believes the long-time suspended (1944) Constitution of 1935 is a legal law in Poland, and the legal president is Jan Zbigniew Potocki, a self-proclaimed head of the state and a businessman from small Polish city Kalisz. Majka claims that he "has evidence" that the Vatican[68] is implanting microchips in people and that Polish children are being sold to the West as a result of secret trade agreements concluded by a satanic cult.[69] He also believes that Jews rule Poland and publicly recommended the book "Jewish Supremacy" by David Duke, a former leader of the Ku-Klux-Klan.[70] However, he became best known for raising the issue of "mind control through vaccines and 5G technology" in the Parlia-

67 Such parties as: Fighting Solidarity and the Freedom Party of Kornel Morawiecki and the League of Polish Families, and he also unsuccessfully ran for the Podkarpackie regional assembly in 2006 from the Law and Justice list and in 2014 from the National Movement list.

68 Kamil Sikora, "Sędzia Trybunału Stanu "ma dowody" na czipowanie ludzi przez Watykan i zdradę państwa przez prezydenta Komorowskiego", *NaTemat*, 24.05.2016, accessed 06.09.2024, https://natemat.pl/180625,sedzia-trybunalu-stanu-pisze-o-chipowaniu-ludzi-przez-watykan-i-zdradzie-panstwa-przez-komorowskiego.

69 Ssd, "Polski poseł odpłynął. Mówi o depopulacji Polaków!", *Fakt*, 25.01.2019, accessed 06.09.2024, https://www.fakt.pl/polityka/posel-robert-majka-wierzy-w-teorie-spiskowe-siec-5g-to-depopulacja/9rn0sxt.

70 Wiktor Ferfecki, "Ostrzeżenie przed poglądami posła Roberta Majki", *Rzeczpospolita*, 25.09.2019, accessed 06.09.2024, https://www.rp.pl/polityka/art1114851-ostrzezenie-przed-pogladami-posla-roberta-majki.

ment.[71] In his public speeches, he referred to information provided by activists opposing mandatory vaccinations and the introduction of 5G telephony – Jerzy Zięba and Ewa Pawela.[72] Former MP Majka is also a regular guest of Janusz Zagórski's Niezależna Telewizja. Majka sent also an interpellation to Prime Minister Mateusz Morawiecki, in which he appealed for scientific expertise to be conducted on the harmful effects of radiation related to the future operation of the 5G network in Poland. He also raised questions about responsibility for the adverse effects of implementing the new technology, claiming that the matter was of national importance and would affect the health of Poles. Majka also claimed that the technology was introduced to Poland to reduce the population of Poles.[73] During his term in office, Majka was also a frequent guest on programmes on the YouTube channels Centrum Edukacyjne Polska, Niezależna Telewizja, and wRealu24.pl, described earlier in this work in the context of their promotion of conspiracy narratives, including those of Russian origin.

Majka did not win re-election in 2019 and currently holds no public office. His role as the main propagator of conspiracy and technophobic narratives was taken over by another Confederation MP, Grzegorz Braun (now also MEP). He does not believe in vaccinations, which he considers not to be based on "confirmable, scientifically or statistically verifiable facts."[74] In his opinion, it is a "grand medical experiment" and a "lottery" carried out "by progressive-totalitarian rabid hordes," which undermines "the sovereignty of parental authority" and "strikes our civilizational, free traditions."[75] He is also one of the prominent advocates of the struggle against 5G telephony. In December 2019, opponents of mobile communications visited Parliament, participating in a session of the team for children's cybersecurity. Braun and Sachajko invited them. During the meeting, Braun stated, "5G looks like dan-

71 Tomasz Domański, "Poseł pyta o biologiczny wpływ 5G na populację Polaków. Spider's Web odpowiada", *Spider's Web*, 25.01.2019, accessed 06.09.2024, https://spidersweb.pl/2019/01/posel-robert-majka-5g-teorie-spiskowe.html.

72 Ibid.

73 Robert Majka, "Interpelacja nr 28857 do prezesa Rady Ministrów w sprawie realizacji planu wcielenia w życie w Polsce w 2020 r. technologii nowej generacji 5G w kontekście realizacji ustawy z druku nr 1066 i jej wpływu na populację Polek i Polaków", *Sejm Rzeczypospolitej Polskiej*, 13.01.2019, accessed 06.09.2024, http://www.sejm.gov.pl/sejm8.nsf/InterpelacjaTresc.xsp?key=B8PJWS.

74 Grzegorz Braun, "Przymus szczepień? Owszem – przeciw totalniackiej wściekliźnie", *Grzegorz Braun*, 16.06.2017, accessed 06.09.2024, https://www.grzegorzbraun.pl/2017/06/16/przymus-szczepien-owszem-przeciw-totalniackiej-wsciekliznie.

75 Ibid.

gerous military technology and an invention of the Eurokolkhoz.[76]" The MP also recorded several videos he posted on his YouTube channel, which he considered the "dangers" of the new technology. In the context of 5G, however, he did not negate the idea of introducing the technology to develop the ICT network. Still, he questioned how the act concerning it was being processed, declaring that it was unclear who ordered the research on network security. He also drew attention to the issue of changing the standards for exposure to electromagnetic fields announced in connection with 5G.[77]

Braun was even more active in the anti-vaccination movement. He not only led many anti-vaccination and anti-restriction demonstrations in Poland[78] but also travelled to similar demonstrations in Germany[79] and created an international network of those fighting the "COVID conspiracy",[80] and also publicly threatened the Polish Minister of Health with the death penalty.[81] He appeared, among others, on January 15, 2022, at an anti-vaccination demonstration in Rzeszów under the slogan "Stop sanitary segregation"[82] and at similar ones in Warsaw and other cities. When a doctor who opposed mandatory vaccinations, Zbigniew Martyka, was brought before the Medical Court at the District Medical Chamber in Kraków, a demonstration was organized before

76 An offensive term referring to the word "kolkhoz" used by the far right in Poland on the EU.

77 Anna Mierzyńska, "O co walczy Grzegorz Braun? Zbudował radykalny polityczny ekosystem i ma coraz większe wpływy", *OKO.Press*, 29.04.2023, accessed 06.09.2024, https://www.oko.press/co-robi-grzegorz-braun.

78 BJS, AGIE, SC, "Marsz przeciwko restrykcjom. Wśród uczestników poseł Braun [WIDEO]", *TVP Info*, 20.03.2021, accessed 06.09.2024, https://www.tvp.info/52885484/pandemia-koronawirusa-polska-warszawa-marsz-przeciwko-restrykcjom-epidemicznym-wsrod-uczestnikow-posel-konfederacji-grzegorz-braun.

79 No author, "Koronasceptycy próbowali wtargnąć do Reichstagu. W proteście w Berlinie uczestniczył Grzegorz Braun", *Wprost*, 31.08.2020, accessed 06.09.2024, https://www.wprost.pl/swiat/10358116/berlin-protest-koronasceptykow-i-proba-wtargniecia-do-reichstagu-posel-grzegorz-braun-uczestniczyl-w-demonstracji.html.

80 Mierzyńska, "O co walczy Grzegorz Braun?".

81 Dorota Roman, Weronika Fabjańska, "Braun do ministra Niedzielskiego: Będziesz pan wisiał! Elżbieta Witek składa zawiadomienie do prokuratury", *Wyborcza.pl*, 16.09.21, accessed 06.09.2024, https://www.wyborcza.pl/7,82983,27578646,braun-do-ministra-niedzielskiego-bedziesz-pan-wisial.html.

82 Bp, "Agata Osiniak z Kolbuszowej: Ponad 50 procent naszego społeczeństwa została oszukana? Lekarz ostro o szczepionkach [WIDEO]", *Korso Kolbuszowskie*, 20.01.22, accessed 06.09.2024, https://korsokolbuszowskie.pl/strefa-covid-19/dr-agata-osiniak-ponad-50-procent-naszego-spoleczenstwa-zostala-oszukana-lekarz-ostro-o-szczepionkach-wideo/tGlgcEuYwybpu2P9mqrH.

the building where the trial was taking place. There were scuffles, and Braun tried to forcefully enter the building to conduct a parliamentary intervention.[83] He participated in debates organized by the anti-vaccination movement, among others in Jasło[84] and Rzeszów, and often appeared in public with "experts" or movement.[85] His party, the Confederation, organized the "Congress of Freedom" under the slogan "Threats to freedom and opportunities for the future." The participants demanded the elimination of anti-pandemic restrictions and guaranteeing the absence of compulsory vaccinations, especially among children and youth.[86] He also repeatedly introduced "experts" and activists of the movement opposing mandatory vaccinations to the Parliament, including to the deliberations of the Parliament Health Committee.[87] Together with other Confederation politicians (Jacek Wilk, Andrzej Zapałowski, and Włodzimierz Skalik), he also established a "committee of inquiry" as part of the Parliamentary Team for Abuse and Violations of Law in Connection with COVID-19,[88] which he founded, and which was boycotted by representatives of other parties. The committee invited numerous movement representatives opposing mandatory vaccinations to its "hearings" and enabled them to propagate conspiratorial narratives in the Parliament.

Braun also promotes conspiracy narratives, especially anti-Semitic ones, on issues other than vaccines and 5G. There are also many similarities between the narratives he promotes and those spread by the Russian Federation. According to Braun, the PiS government has led to the Germanization

83 Agnieszka Pitoń, "Dr Martyka stracił na rok prawo wykonywania zawodu. Awantura z udziałem posłów Konfederacji przed Izbą Lekarską w Krakowie", *Wyborcza.pl*, 19.12.2022, accessed 06.09.2024, https://www.krakow.wyborcza.pl/krakow/7,44425,29280014,protest-antyszczepionkowcow-pod-okregowa-izba-lekarska-w-krakowie.html.

84 ADMINIWJKK, "Kolejna konferencja zorganizowana przez Wolne Wybory! Lekarze w Jaśle", *Wolne Wybory*, 06.11.2021, accessed 06.09.2024, https://www.wolnewybory.org/kolejna-konferencja-zorganizowana-przez-wolne-wybory-niezalezni-lekarze-w-jasle.

85 Posoborowe absurdy, "Wykład »Fałszywa pandemia« – Grzegorz Braun, dr Anna Martynowska, dr Tomasz Wołoszyn i Roman Szostek.", *Gloria.tv*, 17.11.2020, accessed 06.09.2024, https://www.gloria.tv/post/ypQg84EEbg2b1xMesFgDfzTZA#2080.

86 Ryszard Zajączkowski et al., "MANIFEST LUBELSKI ŚRODOWISK WOLNOŚCIOWYCH", *Ordo Medicus*, 04.10.2021, accessed 06.09.2024, https://www.ordomedicus.org/manifest-lubelski-srodowisk-wolnosciowych.

87 No author, "Polscy lekarze bronią swoich praw – Konferencja prasowa w Sejmie RP", *Polskie Stowarzyszenie Niezależnych Lekarzy i Naukowców*, 12.04.2022, accessed 06.09.2024, https://www.psnlin.pl/video,polscy-lekarze-bronia-swoich-praw-konferencja-prasowa-w-sejmie-rp,80.html.

88 Wiktor Ferfecki, "Fałszywa komisja posła Brauna", *Rzeczpospolita*, 28.03.2022, accessed 06.09.2024, https://www.rp.pl/polityka/art35961971-falszywa-komisja-posla-brauna.

and Judaization of the country.[89] For Braun, the future that threatens Poland is Ukraine – or "New Khazaria"[90] ruled by Jewish President Volodymyr Zelynsky, Jewish Prime Minister Volodymyr Hroysman, with the money of Jewish oligarch Ihor Kolomyskyi, and all under the control of the Jewish religious group Chabad Lubavitch from its headquarters in Dnipro.[91] Their goal, according to Braun, is to build "Jewish island sovereignty" in the region from Polish Szczecin to Ukrainian Odessa. In his opinion Israel, Germany, and Polish traitors are striving to develop Mitteleuropa[92] and place a "Judeopolonia"[93] in it.[94] The goal is, according to Braun, to place Poles in reservation camps,[95] where, according to Braun, Jews are to engage in the slave trade, just like in the Middle Ages.[96] And the US military is to watch over everything. The vice-chairman of the Confederation called the bringing of American troops to the territory of Poland "treason".[97] After the ratification in August 2020 of the agreement between the governments of Poland and the US

89 BS, "TV Republika "amerykańską i żydowską agenturą", czyli Grzegorz Braun w formie", *Strajk.eu*, 09.05.2016, accessed 06.09.2024, https://strajk.eu/tv-republika-amerykanska-i-zydowska-agentura-czyli-grzegorz-braun-w-formie.

90 Lech Keller, "Nowa Chazaria, czyli tajny plan Putina-Szarona?", *Wolna Palestyna*, 07.06.2014, accessed 06.09.2024, https://palestyna.wordpress.com/2014/06/07/nowa-chazaria-czyli-tajny-plan-putina-szarona.

91 Martyna Pięta, "Braun o Żydach – rabini Chabad-Lubawicz to antychrześcijańska sekta", *Media Narodowe*, 15.01.2019, accessed 06.09.2024, https://medianarodowe.com/braun-o-zydach-rabini-chabad-lubawicz-to-antychrzescijanska-sekta.

92 Grzegorz Braun, "TYLKO U NAS! Grzegorz Braun: Sojusznicza fatamorgana", *Polska Niepodległa*, 07.10.2018, accessed 06.09.2024, https://polskaniepodlegla.pl/opinie/item/17598-tylko-u-nas-grzegorz-braun-sojusznicza-fatamorgana.

93 Judeopolonia (or Judeo-Polonia, Judeopolska, Judeo-Polska, Judeo-Polin) – a conspiracy narrative about an alleged threat to Polishness and the Polish state, attributing to Jews' plans to dominate and create their state on the territory of Poland. A variation of this is the concept of "Ukropolin," that is a conspiracy narrative of the joint Polish-Ukrainian state ruled by Jews.

94 Tajne Komplety, "Pan Grzegorz Braun – Czarny scenariusz judeopolonia cz.4d", *YouTube*, 01.08.2012, accessed 06.09.2024, https://www.dlapolski.pl/pan-grzegorz-braun-czarny-scenariusz-judeopolonia.

95 PCh24TV · Polonia Christiana, "Grzegorz Braun: jesteśmy świadkami próby ostatecznego rozwiązania kwestii polskiej!", *YouTube*, 08.02.2018, accessed 06.09.2024, https://www.pch24.pl/grzegorz-braun--jestesmy-swiadkami-ostatecznego-rozwiazania-kwestii-polskiej--video,58119,tv.html.

96 Wojciech Karpieszuk, "Konwencja Konfederacji przed wyborami do europarlamentu. »Zatrzymamy zalew imigrantów«, *Wyborcza.pl*, 06.04.2019, accessed 06.09.2024, http://warszawa.wyborcza.pl/warszawa/7,54420,24622911,konwencja-konfederacji-przed-wyborami-do-europarlamentu-zatrzymamy.html.

97 Redakcja, "Wczoraj Braun nazwał »zdradą« wojska USA w Polsce. Dziś jego wystąpienie zamieścił Sputnik", *Niezalezna.pl*, 20.11.2019, accessed 06.09.2024, https://niezalezna.pl/

on enhanced defence cooperation, he called this document "Colonization plus" and "Occupation plus"[98] and called for the recognition of the US ambassador as *persona non grata.*[99] In his opinion, it is also necessary to change Polish alliances. He believed that the Republic of Poland should leave NATO and ally with Belarus, Iran, Turkey, and China, and build a military fort named after Xi Jinping on Polish territory.[100] Braun also calls for the overthrow of the Polish system by force. "To put it bluntly – first a coup d'état and then a complete change of the political system. I am a monarchist, so I pray for the return of the King,"[101] he informed publicly. He wants to punish women with imprisonment for abortion and publicly whip gays. Robert Biedroń, as a "public, professional sodomite," should, according to Braun, go to prison.[102] Following the conspiracy narrative of so-called "cultural Marxism," he declares that "the world revolution, mafias, services and lodges, Jews and freemasons, (…) treat sodomites as the lost lumpenproletariat of their revolution."[103]

Since the invasion of the Russian Federation into the territory of Ukraine, Braun has been actively opposing the aid provided to both Ukrainian refugees and their state.[104] He publicly declared "the incredible privilege of new-

---

polska/wczoraj-braun-nazwal-zdrada-wojska-usa-w-polsce-dzis-jego-wystapienie-zamiescil-sputnik/298238.

98 PAP, "Braun uderza w ustawę o współpracy z USA: Kolonizacja plus, a może nawet Okupacja plus", *Dziennik.pl*, 15.09.2020, accessed 06.09.2024, https://wiadomosci.dziennik.pl/polityka/artykuly/7823966,grzegorz-braun-konfederacja-usa-umowa-wspolpraca-polska-wojsko-armia.html.

99 Laf, Ml, "Konfederacja chce uznania Mosbacher za osobę niepożądaną w Polsce", *Polsat News*, 15.07.2020, accessed 06.09.2024, https://www.polsatnews.pl/wiadomosc/2020-07-15/mosbacher-persona-non-grata-postulat-konfederacji.

100 Tkwl, "A tyle mówił o zdradzie! Braun: »Fort Xi jest moją odpowiedzią na koncepcję Fort Trump wysuniętą przez polityków warszawskich«", *wPolityce.pl*, 22.11.2019, accessed 06.09.2024, https://wpolityce.pl/polityka/474174-braun-fort-xi-jest-moja-odpowiedzia-na-fort-trump.

101 Przemysław Witkowski, "Żydzi, masoni i niskie podatki. Braun i Korwin-Mikke: ukryte twarze Konfederacji", *Polityka.pl*, 13.07.2023, accessed 06.09.2024, https://www.polityka.pl/tygodnikpolityka/kraj/2219070,1,zydzi-masoni-i-niskie-podatki-braun-i-korwin-mikke-ukryte-twarze-konfederacji.read.

102 Adm, "Grzegorz Braun o Robercie Biedroniu: Powinien trafić do więzienia", *Rzeczpospolita*, 21.03.2019, accessed 06.09.2024, https://www.rp.pl/Polityka/190329834-Grzegorz-Braun-o-Robercie-Biedroniu-Powinien-trafic-do-wiezienia.html.

103 Ibid.

104 Wprost.pl, "Skandaliczny wpis Grzegorza Brauna po agresji Rosji na Ukrainę. »NIE otwierać terytorium RP na przestrzał«, »nadchodźcy«", *Wprost*, 24.02.2022, accessed 06.09.2024, https://www.wprost.pl/polityka/10634365/skandaliczny-wpis-grzegorza-brauna-po-agresji-rosji-na-ukraine-nie-otwierac-terytorium-rp-na-przestrzal-nadchodzcy.html.

comers"[105] and that "Poland should not spend a single dime on aid to Ukraine and refugees."[106] In his opinion, the war is being used as a pretext to change the population structure of the Republic of Poland[107] and the operation of mass resettlement of the population.[108] Braun was the only Polish MP who did not vote in favour of the adoption by the Parliament of the resolution condemning the crime of genocide in Ukraine because, according to Braun, Parliament should not prejudge in advance who was the perpetrator there.[109] He also protested in the Parliament against the location of the Ukrainian flag in the Sejm[110] and called the pouring of red paint on the Russian ambassador in Warsaw "barbaric."[111] When a man was murdered on Nowy Świat Street in Warsaw and rumours appeared that the perpetrator was a Ukrainian, Braun organized a press conference and announced that opening Poland to uncontrolled migration "in the nature of mass resettlement" would result in severe consequences in the social and political dimension.[112] A week later, law enforcement authorities established that the perpetrators were Sebastian W. and Łukasz G. – Poles.[113] In the Parliament, during the session of the Parliamentary

105 Wojciech Borakiewicz, "Konfederacja: Państwo ma Ukraińcom pomagać w drugiej kolejności. Ks. Kneblewski: Słusznie", *Wyborcza.pl*, 09.03.2022, accessed 06.09.2024, https://bydgoszcz.wyborcza.pl/bydgoszcz/7,48722,28200951,konfederacja-panstwo-ma-ukraincom-pomagac-w-drugiej-kolejnosci.html.

106 Piotr Bakselerowicz, "Antyukraińskie spotkanie Grzegorza Brauna z wyborcami. Pomoc uchodźcom porównuje do grzechu", *Wyborcza.pl*, 19.03.2022, accessed 06.09.2024, https://zielonagora.wyborcza.pl/zielonagora/7,35182,28238667,spotkanie-grzegorza-braun-a-z-wyborcami-pomaganie-uchodzcom.html.

107 Agnieszka Piwar, "»Musimy zachować zdrowy rozsądek« – wywiad Agnieszki Piwar z Grzegorzem Braunem", *Myśl Polska*, 21.03.2022, accessed 06.09.2024, https://myslpolska.info/2022/03/20/grzegorz-braun-musimy-zachowac-zdrowy-rozsadek.

108 Ibid.

109 Anna Mierzyńska, "Poseł Braun wykorzystuje Sejm, by nakręcać antyukraińskie nastroje. To spodoba się w Rosji", *OKO.Press*, 19.07.2022, accessed 06.09.2024, https://www.oko.press/posel-braun-wykorzystuje-sejm-by-nakrecac-antyukrainskie-nastroje-to-spodoba-sie-w-rosji.

110 Ibid.

111 No author, "Braun: To jest akt barbarzyństwa", *DoRzeczy*, 09.05.2022, accessed 06.09.2024, https://dorzeczy.pl/opinie/298624/braun-skomentowal-incydent-z-andriejewem.html.

112 Anna Mierzyńska, "Tak skrajna prawica wykorzystała fake newsa o morderstwie, by szczuć Polaków przeciwko Ukraińcom", *OKO.Press*, 14.05.2022, accessed 06.09.2024, https://oko.press/tak-skrajna-prawica-wykorzystala-fake-newsa-o-morderstwie-by-szczuc-polakow-przeciwko-ukraincom.

113 Katarzyna Lipka, "Nie, podejrzanymi o zabójstwo na Nowym Świecie nie są obcokrajowcy", *FakeNews.pl*, 20.05.2022, accessed 06.09.2024, https://fakenews.pl/spoleczenstwo/nie-podejrzanymi-o-zabojstwo-na-nowym-swiecie-nie-sa-obcokrajowcy.

Team for International Relations, and later during the Warsaw march "Dmowski to the Castle," Braun presented the "Appeal for Peace in Central Europe," in which he stated that the war in Ukraine broke out because Russia feared NATO expansion and Ukraine was introducing changes to the language law that discriminated against the Russian minority.[114] During the Parliamentary Team for International Relations and the Interests of Poland and the Polish Diaspora (which includes only three Confederation MPs, including Braun), he presented a brochure entitled "Stop the Ukrainization of Poland".[115] The Confederation activist Halszka Bielecka,[116] who accompanied him, announced in the Parliament that "Ukraine has been the largest exporter of prostitutes to Europe for years," "prostitution is popular there among girls and every fourth high school girl dreams of becoming a prostitute," so "Polish girls must be protected so that this model does not take root",[117] and the document stated that "the presence of a significant Ukrainian minority in Poland will negatively result in the import of various pathologies of social life from beyond the eastern border; from corruption and crime to demands for easier access to abortion."[118] Braun's Confederation party also wants to remove the Ukrainian minority from the list of national minorities in Poland, close the Polish–Ukrainian border, and reduce the level of social assistance provided to refugees.[119] Braun is the only MP to abstain from voting on Poland's consent to Sweden and Finland joining the North Atlantic Treaty Alliance[120] and is one of two MPs to abstain in the vote condemning the annexation of four Ukrainian

114 Anna Mierzyńska, "Poseł polskiego Sejmu promuje w Warszawie prorosyjskie oświadczenie obwiniające Ukrainę i NATO", *OKO.Press*, 26.06.2022, accessed 06.09.2024, https://oko.press/posel-polskiego-sejmu-promuje-prorosyjskie-oswiadczenie.

115 Wprost.pl, "Grzegorz Braun zaprezentował skandaliczną inicjatywę, w sieci wrze. »Świadomie działa przeciwko naszemu bezpieczeństwu«", *Wprost*, 15.07.2022, accessed 06.09.2024, https://www.wprost.pl/polityka/10784413/grzegorz-braun-zaprezentowal-skandaliczna-inicjatywe-w-sieci-wrze-swiadomie-dziala-przeciwko-naszemu-bezpieczenstwu.html.

116 Previously a member of Janusz Korwin-Mikke's party, the Congress of the New Right, a former collaborator of the religious orthodox Institute for Legal Culture Ordo Iuris Foundation, leader of a small group associated with the far right, the Drivers' Party.

117 Wprost.pl, "Grzegorz Braun zaprezentował skandaliczną inicjatywę".

118 Zespół ekspertów Korony, *Stop ukrainizacji Polski*, *Konfederacja Korony Polskiej*, July 2022, accessed 06.09.2024, https://stopukrainizacjipolski.pl/fileadmin/user_upload/SUP_4_podglad.pdf.

119 Ibid.

120 Another Confederation MP, the chairman of the National Movement, Robert Winnicki, voted against Finland's entry into NATO, but he later explained that he had cast such a vote by mistake, see: Mikołaj Pietraszewski, "Jako jedyny zagłosował przeciw Finlandii w NATO. Teraz podał powód", *Radio Zet*, 07.07.2022, accessed 06.09.2024, https://wiadomo

oblasts to Russia – Kherson, Luhansk, Donetsk, and Zaporizhia.[121] He was also against the resolution commemorating the Holodomor – the Great Famine in Ukraine.[122] He was also the only MP to oppose the recognition of Russia as a state sponsor of terrorism,[123] the appeal to stop further construction of Nord Stream 2, and the call for EU and NATO countries to support Ukraine in the face of Russian aggression.[124] Together with politicians from the Alternative for Germany and Flemish Interest, Braun also called for lifting sanctions imposed on Russia, abolishing subsidies for wind and solar energy, and eliminating taxation of fossil fuels.[125]

While Braun's initiatives do not enjoy much support in Poland, and the attendance at events he organizes is low,[126] they are widely noticed in the

---

sci.radiozet.pl/polska/polityka/Robert-Winnicki-zaglosowal-przeciw-Finlandii-w-NATO.-Posel-Konfederacji-podal-powod.

121 Mikołaj Małecki, "Sejm niemal jednogłośnie wezwał UE do niewydawania wiz Rosjanom. Dwóch posłów wstrzymało się od głosu", *Dziennik. Gazeta Prawna*, 29.09.2022, accessed 06.09.2024, https://www.gazetaprawna.pl/artykuly/8558510,wizy-dla-rosjan-ue-sejm-uchwala-braun-berkowicz.html.

122 Elżbieta Witek, "UCHWAŁA SEJMU RZECZYPOSPOLITEJ POLSKIEJ z dnia 1 grudnia 2022 r. w sprawie upamiętnienia 90. rocznicy Wielkiego Głodu na Ukrainie", *Sejm Rzeczypospolitej Polskiej*, 01.12.2022, accessed 06.09.2024, https://www.sejm.gov.pl/media9.nsf/files/MPRA-CLPUK9/%24File/Uchwa%C5%82a%20Sejmu%20w%20sprawie%20upami%C4%99tnienia%2090.%20rocznicy%20Wielkiego%20G%C5%82odu%20na%20Ukrainie.pdf.

123 Kamila Cieślik, "Sejm miał uznać Rosję za państwo sponsorujące terroryzm. "Braun zgłosił sprzeciw", obrady przerwano", *Gazeta.pl*, 01.12.2022, accessed 06.09.2024, https://wiadomosci.gazeta.pl/wiadomosci/7,114884,29207224,sejm-mial-zajac-sie-ustawa-uznajaca-rosje-za-panstwo-sponsorujace.html .

124 Wg, "Grzegorz Braun głosował przeciw uchwałom ws. Nord Stream 2 i Rosji. Solidarna Polska: »Trzeba deputinizacji polskiej polityki«", *Niezalezna.pl*, 25.02.2023, accessed 06.09.2024, https://niezalezna.pl/476421-grzegorz-braun-glosowa-przeciw-uchwaom-ws-nord-stream-2-i-rosji-solidarna-polska-trzeba-deputinizacji-polskiej-polityki.

125 Wojciech Szymański, "Sankcje na Rosję. Populiści chcą powrotu do normalności", *DW*, 26.09.2022, accessed 06.09.2024, https://www.dw.com/pl/sankcje-na-rosj%C4%99-populi%C5%9Bci-apeluj%C4%85-o-powr%C3%B3t-do-normalno%C5%9Bci/a-63246522.

126 Nobody showed up at the demonstration in Sanok (Karolina Kraushar, "Sanok. Konfederacja zorganizowała protest »Stop ukrainizacji Polski«. I nikt na niego nie przyszedł", *Gazeta.pl*, 26.09.2022, accessed 06.09.2024, https://wiadomosci.gazeta.pl/wiadomosci/7,114883,28955056,sanok-stop-ukrainizacji-polski-na-protescie-organizowanym.html) and in Szczecin (Sd, "Na placu Adamowicza: »Stop ukrainizacji Polski« [GALERIA, FILM]", *Kurier Szczeciński*, 07.10.2022, accessed 06.09.2024, https://24kurier.pl/aktualnosci/wiadomosci/na-placu-adamowicza-stop-ukrainizacji-polski), and about 300 people came to the demonstration in Warsaw (Mateusz Kaługa, "Marsz »Stop ukrainizacji Polski« w Warszawie z udziałem radnego z Szydłowca. Arkadiusz Soko-

Russian media and positioned as a significant opposition to Poland's policy of supporting Ukraine. "Poland intends to stop the creation of a common state with Ukraine. In Poland, forces are emerging that are categorically opposed to creating a Polish-Ukrainian federal state and cooperation with Russia," wrote the Pravda portal about Braun's initiatives. After his "Dmowski to the Castle" march, Politnavigator published an article, "Stop feeding Ukrainians! – Polish nationalists marched in Warsaw," and after the presentation of the "Stop Ukrainization of Poland" initiative, another one – "The Polish Sejm is outraged: Ukrainian refugees are turning Poles into second-class citizens."[127] Braun's pamphlet was discussed by the News Front channel, which Western countries officially recognize as part of the assets of Russia's secret services.[128] "Poland is being transformed into Ukropolonia with the help of Bandera's descendants," "[Ukrainian] immigrants are transforming Poland into a binational state," "Poles are being forced to tolerate Bandera and Bandera's supporters," "The tragedy of Volhynia is leading to clashes with Ukrainians in Poland" – these are the headlines that have been appearing since 2019 in the Polish version of the Kaliningrad-based RuBaltic portal.[129] The media from Belarus, an ally of Russia, provide similar coverage of Braun's activities. There, alongside theses about Poland's interest in the partition of Ukraine, one could also hear praise for the "Stop Ukrainization of Poland" campaign.[130] Braun's statements and those of the staff from the demonstration he organized were also featured in the *60 Minutes* programme, broadcast on the first channel of Russian public television. "Citizens who disagree with the government's actions believe that the support for Ukraine is hitting their pockets. They demand that the authorities stop masking their theft with aid for Ukraine,"[131] said the programme

łowski pokazany … w rosyjskiej telewizji", *Echodnia.eu*, 03.10.2022, accessed 06.09. 2024, https://echodnia.eu/radomskie/marsz-stop-ukrainizacji-polski-w-warszawie-z-udzialem-radnego-z-szydlowca-arkadiusz-sokolowski-pokazany-w-rosyjskiej-telewizji/ar/c1-16928313).

127 Mierzyńska, "Poseł Braun wykorzystuje Sejm".

128 Julia Dauksza and Anna Gielewska, "Ukrainizacja w kolorze braun", *Frontstory*, 05.08.2022, accessed 06.09.2024, https://frontstory.pl/ukrainizacja-grzegorz-braun-propaganda-rosja-ukraina-dezinformacja.

129 Ibid.

130 SL, KOAL, "Prokremlowski propagandysta o posłach Konfederacji: »To moi koledzy«", *TVP.info*, 28.07.2022., za https://web.archive.org/web/20220729142309/https://www.tvp.info/61539675/bialorus-prokremlowski-propagandysta-chwali-politykow-konfederacji.

131 PAP, jd, "Niedzielski o występie Brauna w rosyjskiej telewizji: staje się stronnikiem Putina niszczącego niepodległą Ukrainę", *Wirtualne Media*, 03.03.2022, accessed 06.09. 2024, https://www.wirtualnemedia.pl/artykul/grzegorz-braun-w-rosyjskiej-telewizji-krytyka.

host, Olga Skabieyeva, wife of a deputy from Putin's United Russia party, Yevgeny Popov. When Braun and his supporters marched through the streets of Warsaw on November 11 as participants in the Independence March, the banner he held, "Stop Ukrainization of Poland," was presented in the Russian media as the official slogan of this largest cyclical demonstration in Poland.[132]

Braun's social assistant Paweł Skutecki[133] failed to renew his parliamentary mandate from Confederation lists. He also did not become its presidential candidate, despite his efforts. However, he continued his activities in the anti-5G and anti-vaccine movements. With the help of Braun, and Janusz Korwin-Mikke, he reactivated the Parliamentary Team for the Safety of Vaccinations of Children and Adults. The goals of the Team included: "establishing the actual balance of benefits and losses of the vaccination program," "informing patients about all possible side effects and complications after using vaccines," and "eliminating sanctions for not vaccinating a child".[134] Its activities were limited to several meetings and inviting representatives of the movement opposing mandatory vaccinations to the Parliament.

In 2020, Skutecki founded the Zjednoczeni party to unite groups harmed by the "system," among others, in courts and hospitals[135] and work together with the Confederation.[136] In the party's leaflet distributed during the demonstration against mandatory vaccinations and anti-pandemic restrictions on June 6, 2020, one could read that the coronavirus crisis is "used for the mass and constant expansion of global control," "global media terror" with "totalitarian measures".[137] The "21 United Demands," i.e., the programmatic demands published by Skutecki in the Facebook group Zjednoczeni on May 3, also includes the demand: "Stop 5G.[138]"

In addition to Skutecki and Socha, another social assistant to Braun was Piotr Jawornik, vice president of the STOP NOP association, who actively pro-

---

132 The real slogan of this demonstration in 2022 was "Strong Nation – Great Poland".

133 Anna Mierzyńska, "»Czwarta noga Konfederacji«. Antyszczepionkowcy budują ugrupowanie Zjednoczeni", *OKO.Press*, 05.06.2020, accessed 06.09.2024, https://www.oko.press/antyszczepionkowcy-maja-byc-czwarta-noga-konfederacji.

134 No author, "REGULAMIN Parlamentarnego Zespołu ds. Bezpieczeństwa Programu Szczepień Ochronnych Dzieci i Dorosłych", *Sejm Rzeczypospolitej Polskiej*, no data, accessed 06.09.2024, https://orka.sejm.gov.pl/opinie8.nsf/nazwa/zesp_szczepienochronnych/$file/zesp_szczepienochronnych.pdf .

135 PAP, "Byli współpracownicy Kukiza sceptyczni ws. budowy przez niego nowej partii", *Dzieinnik.pl*, 07.03.2020, accessed 06.09.2024, https://wiadomosci.dziennik.pl/polityka/artykuly/6453215,kukiz-nowa-partia-byli-wspolpracownicy-sceptycyzm.html.

136 Mierzyńska, "»Czwarta noga Konfederacji«".

137 Ibid.

138 Ibid.

moted conspiracy narratives about 5G.[139] Both Socha and another leader of the anti-vaccination movement, Jerzy Zięba, expressed their support for the Confederation in the elections to the Sejm and the European Parliament in 2019. Socha herself was encouraged by the Confederation to run on its lists but ultimately decided not to do so. Her candidacy was mentioned by the media again in 2023,[140] but this year, she was not ultimately included on the lists either. However, a soon as in 2019, opponents of mandatory vaccinations in large, organized numbers found themselves on the Confederation lists. In practically every district, 1–2 candidates ran – supporters of abolishing mandatory vaccinations.[141]

A significant role in spreading the narrative of the anti-vaccine movement was played by scientists who were not associated with medicine but they are in the same time politicians of Confederation party.[142] Mirosław Piotrowski of the Catholic University of Lublin is a former LPR and PiS MEP and a candidate in the 2020 presidential election. 2018, he founded his party, the True Europe Movement – Europa Christi. In 2019, PiS did not support him in the European Parliament elections, and the list of his new party was crossed out before the voting day. He also failed to register in Lublin as a candidate for senator. When he announced his candidacy for president, Telewizja Trwam and Radio Maryja[143] ended their cooperation with him. As a result, he won only 0.11% of the votes and took the last, eleventh place. Since then, he has been gradually approaching the Confederation, and Grzegorz Braun, whom he supported in 2021 when he ran for the position of the president of Rzes-

139 Piotr Grabiec, "Nadajnik Play podpalony. Przeciwnicy 5G świętują, a razem z nimi asystent jednego z posłów", *Spider's Web*, 26.05.2020, accessed 06.09.2024, https://spidersweb.pl/2020/05/play-stacja-bazowa-pozar.html.

140 Tomasz Nyczka, "Zamieszanie w Konfederacji w Poznaniu. Socha ostatecznie skreślona, uderza w Mentzena i Bosaka", *Wyborcza.pl*, 29.07.2023, accessed 06.09.2024, https://poznan.wyborcza.pl/poznan/7,36001,30018814,zamieszanie-w-konfederacji-w-poznaniu-socha-ostatecznie-skreslona.html.

141 In Greater Poland, the president of the Kalisz district, Andrzej Kuświk; in Mazovia, the president of the Association for Natural Childbirth and Feeding, Ewa Nitecka; in Warsaw, the councillor of the Stare Babice commune, Marta Pamięta; in Podkarpacie, Marzena Kardasińska; in Silesia, the then Confederation MP, Jacek Wilk; and in other regions, the activists of STOP NOP: Marta Gontarska, Emilia Jawornik and Joanna Mazur, as well as the anti-vaccine doctor, Hubert Czerniak.

142 Among academics not in any way connected with medicine and not involved in the activities of political parties, the following were or are still active in the anti-vaccination movement: Dr Hab. Marek Szczepaniec (economist), Dr Hab. Wojciech Dobiński (geomorphologist), Dr Eng. Józef Okulewicz (logistician), Dr Marta Ułaszewska-Żuk (psychologist), Dr Mariusz Żuk (psychologist), Dr Krystyna Ziemian (mathematician).

143 Because they supported PiS candidate Andrzej Duda.

zów.[144] Also during the pandemic, Piotrowski supported Braun's[145] anti-vaccination views.[146] At that time, together with other Confederation politicians (Jacek Wilk, Andrzej Zapałowski, and Włodzimierz Skalik), he became a member of the so-called "investigative committee," which Braun established as part of the Parliamentary Team for Abuse and Violations of Law in Connection with COVID-19.[147]

Another scientist and politician of the Confederation supporting the movement against mandatory vaccinations is Ryszard Zajączkowski.[148] The Osuchowa Foundation, founded by Braun, published a book for Zajączkowski entitled *Global Change: What We Must Defend and How to Act?*.[149] The author calls for "creating resistance forces that will oppose the small handful of people obsessed with the vision of enslaving all of humanity".[150] He claims that changes comparable to those that occurred after the birth of Jesus will soon take place on Earth and that "then a new era began, and now another one is beginning, which essence will be the clash between anti-establishment movements and globalists, who for now operate in conspiracy."[151] According to Zajączkowski, in the context of vaccinations, we are dealing with the "Great Reset," the goal of which is "mass control," "depopulation",[152] the introduction of "globalist communism," which wants to bring nations to their knees and

144 Red., "Prof. Piotrowski popiera Brauna i mówi o Ferencu: »Na starość zdziwaczał«", *CzytajRzeszów.pl*, 29.04.2021, accessed 06.09.2024, https://czytajrzeszow.pl/?idd=26&id=17059&poz=mn1&show=-1&pagebda1_mobile=0.

145 Ferfecki, "Fałszywa komisja posła Brauna".

146 Rafał Pankowski et al., *Wirus nienawiści:"Brunatna Księga" czasu epidemii*, Stowarzyszenie "Nigdy Więcej", May 2020,, accessed 06.09.2024, https://www.nigdywiecej.org/docstation/com_docstation/172/wirus_nienawisci_brunatna_ksiega_czasu_epidemii.pdf.

147 Ferfecki, "Fałszywa komisja posła Brauna".

148 Tomasz Maciuszczak, "Profesor KUL gwiazdą antycovidowego marszu w Lublinie. Co na to uczelnia?", *Dziennik Wschodni*, 12.10.20, accessed 06.09.2024, https://www.dziennikwschodni.pl/lublin/profesor-kul-gwiazda-antycovidowego-marszu-w-lublinie,n,1000276120.html.

149 See further: Ryszard Zajączkowski, *Globalna zmiana. Co musimy obronić i jak działać?* (Warsaw: Fundacja Osuchowa, 2022).

150 Mira Piłaszewicz, "Debata dziesięciu uczestników »Nie ma tego złego ... Czyli świat płonie i oczyszcza się w ogniu« u Miry Piłaszewicz [WIDEO]", *Wolność.tv*, 25.04.2022, accessed 06.09.2024, https://www.wolnosc.tv/debata-dziesieciu-uczestnikow-nie-ma-tego-zlego-czyli-swiat-plonie-i-oczyszcza-sie-w-ogniu-u-miry-pilaszewicz-wideo.

151 No author, "Deklaracja warszawska przeciw nowemu totalitaryzmowi", *Norymberga 2.0*, November 2021,, accessed 06.09.2024, https://www.norymberga2.pl/page/view/7/deklaracja-warszawska.html.

152 No author, "Dokąd zmierzamy jako ludzkość – prof. Ryszard Zajączkowski Lubartów 29.08.2022 r.", *Polskie Stowarzyszenie Niezależnych Lekarzy i Naukowców*, 07.09.2022, accessed 06.09.2024, https://www.psnlin.pl/video,dokad-zmierzamy-jako-ludzkosc-prof-ryszard-zajaczkowski-lubartow-29-08-2022-r,109.html.

enslave people,[153] and we are standing at the beginning of "the greatest genocide in the history of the world."[154] His involvement in propagating narratives consistent with the Russian ones is also noticeable. He is involved in the Polish Anti-War Movement of Leszek Sykulski and Sebastian Pitoń, organizes collections for anti-war posters,[155] and believes that Poland is not accepting refugees but is carrying out a planned resettlement operation aimed at changing the ethnic composition of Poland.[156]

It would be possible to discuss a larger number of Confederation politicians (scientists and non-scientists) involved in the anti-vaccination movement, but generally during the 9th term of the Polish Parliament, all Confederation MPs were involved in the activities of the anti-vaccination movement. In their public statements, they presented a negative attitude towards sanitary restrictions related to the COVID-19 pandemic and a negative attitude towards the coronavirus vaccine. They also questioned the very qualification of the events as a state of pandemic,[157] often directly claiming that the threat posed by the virus is exaggerated,[158] COVID-19 is a "typical disease,"[159] which can also be treated with already known drugs, e.g., amantadine.[160]

---

153 Tomasz Kowalewicz, "Profesor KUL przeciw powszechnej akcji szczepień. »Bóg przystosował człowieka do życia w wirusach«", *Wyborcza.pl*, 15.12.20, accessed 06.09.2024, https://lublin.wyborcza.pl/lublin/7,48724,26610185,profesor-kul-w-szczepionce-na-koronawirusa-widzi-wieksze-zagrozenie.html.

154 PSNLiN, "Dokąd zmierzamy jako ludzkość – prof. Ryszard Zajączkowski Lubartów 29.08.2022", *Rumble*, 29.08.2022, accessed 06.09.2024, https://rumble.com/v1iyxp1-dokd–zmierzamy–jako–ludzko–prof.–ryszard–zajczkowski–lubartw–29.08.2022.html.

155 Klaudia Kowalczyk, "Pacyfizm czy rosyjska racja stanu? Profesor KUL zbiera na antywojenne bilbordy", *Jawny Lublin*, 17.04.23, accessed 06.09.2024, https://jawnylublin.pl/pacyfizm-czy-rosyjska-racja-stanu-profesor-kul-zbiera-na-antywojenne-bilbordy/.

156 Łukasz Jakubowski and Anna Tatar, *Jak policzyć nienawiść? Hejterzy o Ukraińcach*, Warszawa, Stowarzyszenie "Nigdy Więcej", February 2023, accessed 06.09.2024, https://www.nigdywiecej.org/docstation/com_docstation/172/jak_policzyc_nienawisc._hejterzy_o_ukraincach.pdf. Kindle.

157 Magda Kuczyńska, "SPRAWOZDANIE Z BERLIŃSKIEGO PROTESTU PRZECIW PLANDEMII", *Konfederacja Korony Polskiej*, no data, accessed 06.09.2024,https://konfederacjakoronypolskiej.pl/sprawozdanie-z-berlinskiego-protestu-przeciw-plandemii.

158 Karol Woliński, "Poseł Grzegorz Braun prowokuje: »Jesteśmy znów w stanie wojny«", *CzytajRzeszów.pl*, 01.09.2020, accessed 06.09.2024, https://czytajrzeszow.pl/?idd=10&id=12614.

159 Konfederacja, "Tego potrzebuje teraz PolskaPL! Jeśli się zgadzasz, podaj dalej!!", *Twitter*, 13.09.2021, accessed 06.09.2024, https://twitter.com/KONFEDERACJA_/status/1437334131550167040.

160 Konfederacja, "UWAGA, koniec z leczeniem #COVID–19 amantadyną!", *Twitter*, 03.12.2020, accessed 06.09.2024, https://twitter.com/KONFEDERACJA_/status/1334424821720109056.

They also claimed that attempts to limit the pandemic were tantamount to blocking citizens' access to healthcare, and compared the restrictions on civil liberties related to attempts to stop the spread of the virus to the times of Nazism and the Third Reich[161] and presented them as lawlessness and a brutal violation of the rule of law by the government.[162] They maintained that they were illegal and called for not accepting tickets for failure to comply with the government's anti-pandemic regulations.[163] Their statements included comparisons of the Polish government's treatment of unvaccinated people to the situation of Jews in the Third Reich or people of colour during apartheid in South Africa[164] or even to the solutions used in extermination camps. They presented[165] people who did not comply with the government's restrictions as heroes and simultaneously victims of the government's actions, the main goal of which was to break the spirit of resistance defending civil liberties.[166]

Confederation politicians questioned the effectiveness of vaccines against COVID-19.[167] They called vaccination a "medical experiment," suggesting that it could cause serious side effects, affect genes, and cause dangerous "complications, as well as fatalities."[168] They also cited medical opinions that differed

161 Dominika Sitnicka, "Transparent »Szczepienie czyni wolnym« to nie wypadek przy pracy. Tak Konfederacja pogrywa epidemią", *OKO.Press*, 18.12.2021, accessed 06.09.2024, https://www.oko.press/jak-konfederacja-politycznie-pogrywa-epidemia.

162 Zew, 'Konfederacja: Segregacja sanitarna jest bezprawna i nieskuteczna', *Rzeczpospolita*, 18.11.2021, accessed 06.09.2024, https://www.rp.pl/polityka/art19114141-konfederacja-segregacja-sanitarna-jest-bezprawna-i-nieskuteczna.

163 Mk, "Konfederacja: PiS po cichu wprowadza państwo policyjne", *Tygodnik Bydgoski*, 12.01.2021, accessed 06.09.2024, https://www.tygodnikbydgoski.pl/aktualnosci/konfederacja-pis-po-cichu-wprowadza-panstwo-policyjne.

164 Mieczysław Rudy, "Konfederacja w sprawie projektu dot. weryfikacji covidowej: Potencjalnie przestępczy", *Dziennik. Gazeta Prawna*, 05.01.2022, za: https://www.gazetaprawna.pl/wiadomosci/kraj/artykuly/8327759,konfederacja-w-sprawie-projektu-dot-weryfikacji-covidowej-potencjalnie-przestepczy.html.

165 Ww, "»Szczepienie czyni wolnym« na banerze posłów Konfederacji przed Sejmem. Fala komentarzy", *Onet*, 1512.2021, za: https://www.wiadomosci.onet.pl/kraj/szczepienie-czyni-wolnym-na-banerze-poslow-konfederacji-fala-komentarzy/57p4339.

166 Anna Mierzyńska, "O co walczy Grzegorz Braun?".

167 Konfederacja, "Badania nad szczepionką Pfizer-BioNTech trwały tylko kilka miesięcy", *Twitter*, 07.12.2020, accessed 06.09.2024, https://www.twitter.com/KONFEDERACJA_/status/1335968878355226626.

168 Prz, "Konfederacja żąda pełnej dobrowolności szczepień na Covid-19. »Sceptycyzm rośnie«", *Polsatnews.pl*, 07.12.2020, accessed 06.09.2024, https://www.polsatnews.pl/wiadomosc/2020-12-07/konfederacja-zada-pelnej-dobrowolnosci-szczepien-na-covid-19-sceptycyzm-rosnie.

from the medical consensus, suggesting far-reaching consequences of vaccination for individuals and entire populations. They often emphasized the right of parents to decide on medical procedures concerning their offspring. They also used the term "sanitary segregation" to describe attempts to introduce restrictions for unvaccinated people.[169] They also suggested that their introduction was the result of succumbing to pressure from abroad, from pharmaceutical companies, the European Commission, and the World Health Organization, which, under the guise of caring for public health, want to introduce mechanisms of control over countries such as Poland, and also strive to create an informal, non-democratic instance of power, whose scope would include managing populations by pressuring states to introduce increasingly far-reaching control over the health, movement and digitization of citizens' everyday lives, along with the collection of medical data.[170]

A constant element of their public statements was presenting anti-pandemic restrictions as harmful to the national and global economy.[171] In their social media and public statements, they publicized cases of entrepreneurs who went bankrupt due to government policy and presented them as victims of the authorities. Those who protested against mandates or bans were described as heroes.

Although Confederation politicians focused primarily on vaccines during the 9th term of the Sejm, they also raised the issue of fifth-generation telephony. Active propagators of conspiracy narratives about 5G were candidates from the Confederation lists, such as Hubert Czerniak,[172] former party MP Jacek Wilk,[173] and farmers' leader supporting the party Marcin Bustowski.[174] All of them either took part in demonstrations against 5G or publicly propagated conspiracy narratives regarding the use of this technology. Despite the

169 No author, "STANOWISKO KONFEDERACJI KORONY POLSKIEJ W SPRAWIE OBOWIĄZKOWYCH SZCZEPIEŃ DZIECI", *Konfederacja Korony Polskiej*, no data, accessed 06.09.2024, https://www.konfederacjakoronypolskiej.pl/stanowisko-konfederacji-korony-polskiej-w-sprawie-obowiazkowych-szczepien-dzieci.

170 Sitnicka, "Transparent »Szczepienie czyni wolnym«".

171 Konfederacja, "Musimy natychmiast zatrzymać ten chory lockdown, całkowicie otworzyć gospodarkę", *Twitter*, 13.02.2021, accessed 06.09.2024, https://www.twitter.com/KONFEDERACJA_/status/1360609171050270725.

172 No author, "Hubert Czerniak – fake newsy o 5G i szczepieniach przeciw HPV", *Demagog*, 09.12.2021, accessed 06.09.2024, https://demagog.org.pl/fake_news/hubert-czerniak-fake-newsy-o-5g-i-szczepieniach-przeciw-hpv.

173 Polskie Forum, "Jacek Wilk o szczepieniach, WHO, cenzurze, 5G, Billu Gatesie i Szumowskim! Kto korzysta na pandemii?", *YouTube*, 28.04.2020, accessed 06.09.2024, https://www.youtube.com/watch?v=DWfidWGKPJo.

174 Zapotoczny, "Protest w centrum Opola".

assurances of the party leader, Sławomir Mentzen, about the accidental and temporary presence of supporters of conspiracy narratives on the Confederation lists,[175] the electoral lists of this group in subsequent elections seem to contradict this thesis. Many active propagators of such views can be found in them and are also in very exposed positions.

Such narratives were propagated by many of the party's candidates for mayors of large cities in the local and municipal elections of 2024. Rafał Foryś, the Confederation's candidate for mayor of Radom, demanded a ban on the introduction of 5G telephony, considered the pandemic to be planned by global elites,[176] made public anti-Semitic statements,[177] recommended treating the coronavirus with thyme and refused to get vaccinated against COVID-19, considering the vaccine to be a "medical experiment."[178] Similarly, Karolina Pikuła, the Confederation's candidate for mayor of Rzeszów, who also opposed the expansion of 5G telephony on her social media, warned against the lethal COVID-19 vaccine, additionally promoting anti-Ukrainian content and suggested hanging political opponents.[179] Jacek Budniok, the Confederation's candidate for mayor of Katowice, shares even more conspiratorial and anti-scientific narratives. His social media includes claims that mountains are ancient trees,[180] that the world is to be ruled by Masons and Illuminati, who are creating a New World Order,[181] and who have HAARP weapons that cause earthquakes.[182] Budniok also publishes posts in which he claims that "there is only one disease," and that diabetes and cancer have the exact causes and

---

175 No author, "Mentzen tłumaczy się z szurów na listach Konfederacji i nie przeszkadzałby mu poseł wierzący w planetę Nibiru", *Donald*, 11.10.2023, accessed 06.09.2024, https://www.donald.pl/artykuly/wwMn66Y2/mentzen-tlumaczy-sie-z-szurow-na-listach-konfederacji-i-nie-przeszkadzalby-mu-posel-wierzacy-w-planete-nibiru.

176 Nocny Pingwin, "Rafał Foryś – kandydat Konfederacji na prezydenta Radomia", *Twitter*, 16.02.2024, accessed 06.09.2024, https://twitter.com/Nocny_Pingwin/status/1758496975870197875.

177 Nocny Pingwin, "Po akcji Grzegorza Brauna z gaśnicą", *Twitter*, 16.02.2024, accessed 06.09.2024, https://twitter.com/Nocny_Pingwin/status/1758496979833720916.

178 Nocny Pingwin, "Kandydat Konfederacji przyznał, że wolałby", *Twitter*, 16.02.2024, accessed 06.09.2024, https://twitter.com/Nocny_Pingwin/status/1758496973936595272.

179 Nocny Pingwin, "Karolina Pikuła – numer 30 na liście Konfederacji", *Twitter*, 10.10.2023, accessed 06.09.2024, https://twitter.com/Nocny_Pingwin/status/1711631396803141796/photo/2.

180 Nocny Pingwin, "Oto przykładowe fragmenty filmów, jakie udostępnia kandydat", *Twitter*, 01.03.24, accessed 06.09.2024, https://twitter.com/Nocny_Pingwin/status/1763575697560174674.

181 Nocny Pingwin, "Kandydat Konfederacji udostępnia dużo postów", *Twitter*, 01.03.2024, accessed 06.09.2024, https://twitter.com/Nocny_Pingwin/status/1763575689884639491.

182 Ibid.

the same treatment,[183] that COVID-19 vaccinations are harmful, and that the pandemic is planned.[184]

There were even more such candidates in the 2023 parliamentary elections. Diana Ruchniewicz – fourth on the Konfederacja list in Warsaw, announces on her social media the dangers of 5G, the supposedly planned nature of the pandemic, believes in the concept of chemtrails, forced implanting microchips in people and using by superpowers the electromagnetic weapons, as well as in the narrative in which Bill Gates, George Soros and Klaus Schwab want to cause cancer in people by adding worms to food. Ruchniewicz places this content in an anti-Semitic conspiracy narrative, in which all subsequent Polish presidents since 1989 have been Jews.[185] Żaneta Nochowicz and Danuta Grzyb, in addition to the narratives mentioned above, also believe that the Earth is not a geoid but a flat disk. Nochowicz also promotes the conspiracy concept of the QAnon movement, in which global elites rejuvenate themselves with the blood of tortured children and numerous narratives about the conflict in Ukraine, consistent with Russian ones.[186] Grzyb, in turn, adds to the above narratives the belief in lizard-like aliens ruling the Earth – reptilians.[187] Tomasz Walczuk, number 6 on the Confederation list in district 5, adds to the belief in the harmfulness of 5G and the existence of chemtrails the assumption that legal abortion there is responsible for the great fires in Australia.[188] Krystyna Jarocka, the Confederation's candidate for the Senate from district 9, in addition to the quite typical set of ideas mentioned before (harmful 5G, vaccines, "plandemic," "implanting microchips in people," chemtrails), believes in the existence of the planet Niburu and the resurrection of a

183 Nocny Pingwin, "Na profilu Budnioka znajdziemy też sporo antynaukowych bzdur", *Twitter*, 01.03.2024, accessed 06.09.2024, https://twitter.com/Nocny_Pingwin/status/1763575693810565291.

184 Nocny Pingwin, "Standardem na profilach Konfederatów są fałszywe informacje", *Twitter*, 01.03.2024, accessed 06.09.2024, https://twitter.com/Nocny_Pingwin/status/1763575685828739414.

185 Nocny Pingwin, "Diana Ruchniewicz – czwarta na liście Konfederacji", *Twitter*, 11.09.2023, accessed 06.09.2024, https://twitter.com/Nocny_Pingwin/status/1701229831160045646.

186 Nocny Pingwin, "Żaneta Nochowicz – płaskoziemka na liście Konfederacji", *Twitter*, 22.09.2023, accessed 06.09.2024, https://twitter.com/Nocny_Pingwin/status/1705069820704641406.

187 Nocny Pingwin, "Danuta Grzyb – numer 12 na liście Konfederacji", *Twitter*, 15.09.2023, accessed 06.09.2024, https://twitter.com/Nocny_Pingwin/status/1702618470762877030.

188 Nocny Pingwin, "Tomasz Walczuk, numer 6 na liście Konfederacji", *Twitter*, 16.09.2023, accessed 06.09.2024, https://twitter.com/Nocny_Pingwin/status/1703003120152854903.

child as a result of the sacrament of baptism.[189] Other candidates add their belief in curing cancer with frozen lemon (Danuta Leheńka–Das, number 12 on the Confederation list in district 2)[190] nuts (Marek Banyś, number 11 on the Confederation list in district 8),[191] cinnamon (Stanisław Bąk, number 8 on the Confederation list in district 7),[192] cloves (Sylwia Siłuszyk, number 15 on the Confederation list in district 12),[193] and drinking urine, amulets protecting against energy vampirism and clairvoyance (Dorota Rycak, number 20 on the Confederation list in district 7),[194] the World Trade Center blew up with an atomic bomb, (Krzysztof Olędzki – number 21 on the Confederation list in district 19),[195] portals to the "outer Earth" in Antarctica (Rafał Czubak – number 5 on the Confederation list in district 17),[196] and the Martian civilization located there (Mariusz Bożyk – number 16 on the Confederation list in district 14).[197]

The fact that the first representative of anti-vaccine movement, who started his activities in its ranks, entered the Sejm from the lists of this party in the 2023 parliamentary elections can be considered a summary of the Confeder-

---

189 Nocny Pingwin, "Krystyna Jarocka – kandydatka Konfederacji do senatu z okręgu 9", *Twitter*, 12.09.2023, accessed 06.09.2024, https://twitter.com/Nocny_Pingwin/status/1701607205152370748.

190 Nocny Pingwin, "Danuta Leheńka–Das – numer 12 na liście Konfederacji w okręgu 2", *Twitter*, 14.09.2023, accessed 06.09.2024, https://twitter.com/Nocny_Pingwin/status/1702323945763164293.

191 Nocny Pingwin, "Marek Banyś – numer 11 na liście Konfederacji w okręgu 8", *Twitter*, 20.09.2023, accessed 06.09.2024, https://twitter.com/Nocny_Pingwin/status/170441 7778814382386.

192 Nocny Pingwin, "Stanisław Bąk – numer 8 na liście Konfederacji w okręgu 7", *Twitter*, 21.09.2023, accessed 06.09.2024, https://twitter.com/Nocny_Pingwin/status/170478127 9680450629.

193 Nocny Pingwin, "Sylwia Siłuszyk – numer 15 na liście Konfederacji w okręgu 12", *Twitter*, 02.10.2023, accessed 06.09.2024, https://twitter.com/Nocny_Pingwin/status/17087234 90902200362.

194 Nocny Pingwin, "Dorota Rycak – numer 20 na liście Konfederacji w okręgu 7", *Twitter*, 19.09.2023, accessed 06.09.2024, https://twitter.com/Nocny_Pingwin/status/170405930 8583841795.

195 Nocny Pingwin, "Krzysztof Olędzki – numer 21 na liście Konfederacji w okręgu 19", *Twitter*, 09.10.2023, accessed 06.09.2024, https://twitter.com/Nocny_Pingwin/status/1711252650979664234.

196 Nocny Pingwin, "Rafał Czubak – numer 5 na liście Konfederacji w okręgu 17", *Twitter*, 08.10.2023, accessed 06.09.2024, https://twitter.com/Nocny_Pingwin/status/17108516 81775161403.

197 Nocny Pingwin, "Mariusz Bożyk – numer 16 na liście Konfederacji w okręgu 14", *Twitter*, 29.09.2023, accessed 06.09.2024, https://twitter.com/Nocny_Pingwin/status/1707584784 396935423.

ation's cooperation with activists of the movement opposing the vaccination obligation. Previously, politicians took up the issues of opposition to the vaccination obligation or other anti-scientific conspiracy narratives and added these issues to their agenda. This time, however, the opposite happened for the first time. In district no. 31, winning 18648 votes, wedding photographer Grzegorz Płaczek entered the Parliament. As MP, he sits on the parliamentary health committee and the education committee. During the pandemic, he was very active in the movement opposing the vaccination obligation and government anti-pandemic restrictions.[198] At the same time, he provided false information on his social media regarding vaccinations[199] and child mortality during the pandemic.[200] In his case, like in the case of many other activists opposing the mandatory vaccination and politicians of the Confederation, the presence of narratives consistent with the Russian ones can be noticed, as he mixed his messages about COVID-19 and the pandemic with false and manipulated information about the so-called retirement tourism[201] and free apartments for Ukrainians[202] and against Poland's aid for Ukraine and accepting refugees from that country.[203]

198 Anna Mierzyńska, "Ujawniamy prawdziwą »piątkę Konfederacji«. Czego nienawidzą kandydaci skrajnej prawicy?", *OKO.Press*, 26.07.2023, accessed 06.09.2024, https://oko.press/prawdziwa-piatka-konfederacji.

199 No author, "Szczepienie na COVID-19 zagraża dzieciom? Fałszywe informacje!", *Demagog*, 10.11.2022, accessed 06.09.2024, https://demagog.org.pl/fake_news/szczepienie-na-covid-19-zagraza-dzieciom-falszywe-informacje.

200 No author, "Dementujemy fałszywe informacje w »odkryciach« Grzegorza Płaczka", *Demagog*, 13.10.2021, accessed 06.09.2024, https://demagog.org.pl/fake_news/dementujemy-falszywe-informacje-w-odkryciach-grzegorza-placzka.

201 Stanisław Mroszczak, "Manipulacje o »turystyce emerytalnej« ukraińskich seniorów", *Wojownicy Klawiatury*, 04.05.2023, accessed 06.09.2024, https://wojownicyklawiatury.pl/manipulacje-o-turystyce-emerytalnej-ukrainskich-seniorow.

202 Michał Istel, "Krakowskie rodziny »mieszkań nie dostaną«, bo zasiedlą w nich Ukraińców? To fałszywy przekaz", *Konkret24*, 11.03.2022, accessed 06.09.2024, https://konkret24.tvn24.pl/polska/krakowskie-rodziny-mieszkan-nie-dostana-bo-zasiedla-w-nich-ukraincow-to-falszywy-przekaz-ra1098931-ls5791257.

203 Jan Kunert, "Konta antyszczepionkowe pracują dla Kremla. »Potwierdziło się, że paliwo było ze Wschodu«", *Konkret24*, 05.04.2022, accessed 06.09.2024, https://konkret24.tvn24.pl/polityka/konta-antyszczepionkowe-pracuja-dla-kremla-potwierdzilo-sie-ze-paliwo-bylo-ze-wschodu-ra1100795-ls5791231.

CHAPTER 6

Covidians and Corona-Sceptics

Mainstream Parties and the Anti-vaccination and Anti-5G Telephony Movements

The attitude of Law and Justice politicians towards the movements opposing mandatory vaccinations and fifth-generation telephony has changed over time. In addition, politicians from different parties that were part of Mateusz Morawiecki's, de facto, coalition governments[1] also showed different attitudes towards these issues. The intensity of interest in these narratives also varied, and their subsequent turns, weakening and peaks of interest were noticeable. Additionally, the attempt to clearly define this attitude is made more difficult by the fact that PiS formed the government during the COVID-19 pandemic and, as a result, influenced the limitation of its politicians' activity in the movement opposing mandatory vaccinations, due to the contradiction of its postulates with the government's position on the fight against the pandemic. Hence, there are specific problems in obtaining a clear statement of PiS's attitude towards anti-scientific narratives.

However, it is possible to distinguish critical groups in this political milieu that are particularly active in promoting narratives consistent with conspiracy narratives. In the matter of strengthening doubts about the safety of the 5G network for people, politicians from the right wing of the party were particularly active, essentially religious integralists, associated primarily with the religious Radio Maryja, such as Antonii Macierewicz, Bartosz Kownacki, Gabriela Masłowska, Anna Sobecka and Jan Szyszko, Czesław Hoc, Tomasz Rzymkowski, Bartłomiej Dorywalski, Patryk Wicher, Jerzy Wilk, Terasa Hałas, Władysław Kurowski, Barbara Bubula, and Andrzej Zybertowicz. In the case of the latter two, an additional influence had their connection with the non-governmental organization Institute of Civil Affairs, which promotes numerous technophobic conspiracy narratives. A similar correlation could be seen in the matter of mandatory vaccinations. First of all, the position against

1 In addition to Law and Justice, these governments were formed by Polish Solidarity (Solidarna Polska, from 2023 Sovereign Poland/Suwerenna Polska), and for various periods of time also by groups such as Agreement (Porozumienie) party, The Right Wing of the Republic (Prawica Rzeczypospolitej) party, Republican Party (Partia Republikańska), OdNowa association, Polish Matters (Polskie Sprawy) association,

 | DOI:10.1163/9789004747852_008

introducing mandatory vaccinations, and over time also against broader pandemic restrictions, was demonstrated by politicians who are religious integralists, supported by Radio Maryja, with particular emphasis on politicians of the Suwerenna Polska party. Their group gained specific importance with the decreasing pro-government majority in the 9th term of the Polish parliament and the increasing tension within the coalition. As a result, the support of MPs opposing mandatory vaccinations was necessary to maintain the parliamentary majority, which influenced the slowdown in the introduction or even abandonment of some anti-pandemic restrictions by Mateusz Morawiecki's government. Therefore, it can be assumed that despite their small representation, the influence of MPs opposing mandatory vaccinations was disproportionately high. An additional factor that could influence the views of these MPs could be the fact that they were frequent guests on media channels promoting conspiratorial narratives about vaccinations and 5G, such as wRealu24.pl, Media Narodowe and others cooperated with organizations, spreading narratives contrary to the scientific consensus.

The first well-known PiS politician to publicly discuss conspiracy narratives about the anti-5G movement was then-Minister of National Defense Antoni Macierewicz. In 2016, during the conference "Problems of contemporary politics. Armed conflicts and terrorism" organized at the Higher School of Social and Media Culture in Toruń (a university associated with Tadeusz Rydzyk), he was asked by a participant of the meeting, "Does the Polish state have strategies for illegal research, i.e., testing electromagnetic weapons on Polish citizens?" He replied that over the past fifteen years, people from Poland (at least ten people) had contacted him, claiming that they had been subjected to such research and could not find help and protection anywhere. He declared that analyses were being conducted but did not specify what they were.[2] The minister's response was accepted by supporters of the conspiracy narrative about electromagnetic weapons (from which, among other things, the belief in the dangers of the 5G network evolved) as confirmation of their story.[3] In this context, members of the STOP Organized Electronic Torture Associa-

2 Maksymilian Dura, "Ataki elektromagnetyczne na polskich obywateli? »Nieporozumienie«", *Defence24*, 16.03.2016, accessed 5.09.2024, https://defence24.pl/geopolityka/ataki-elektromagnetyczne-na-polskich-obywateli-nieporozumienie.

3 Jacek Harłukowicz, "Macierewicz u Rydzyka o eksperymentach z bronią elektromagnetyczną na Polakach", *Wyborcza.pl*, 15.03.2016, accessed 5.09.2024, https://wroclaw.wyborcza.pl/wroclaw/7,35771,19770604,macierewicz-u-rydzyka-o-eksperymentach-z-bronia-elektromagnetyczna.html.

tion began to demand an investigation into the matter from the Prosecutor General.[4]

As the activity of the movement against the 5G network grew, interpellations and statements began to appear from PiS MPs referring to the conspiracy narratives or fuelling anxiety related to its introduction. Again, these issues were raised primarily by religious integralists, including those associated with the Radio Maryja milieu, such as MPs Czesław Hoc,[5] Jerzy Wilk,[6] Bartłomiej Dorywalski, Filip Kaczyński, Władysław Kurowski, and Patryk Wicher.[7]

For some PiS MPs, who raised doubts about the 5G network, the religious factor was also added to the connections with activist fighting against this generation of telephony. Examples of such people include Władysław Kurowski and Teresa Hałas, who not only expressed doubts about the safety of introducing 5G[8] but also supported the postulates of the movement opposing mandatory vaccinations.[9] Yet another group of PiS politicians involved in the opposition to 5G and mandatory vaccinations were directly involved in the activities of non-governmental organizations, undermining the scientific consensus on these issues. Such links between these two milieus were primarily

---

4 RAV, "Ofiary broni elektromagnetycznej piszą do Ministerstwa Sprawiedliwości", *Newsweek*, 26.10.2016, accessed 5.09.2024, https://www.newsweek.pl/polska/ofiary-broni-elektromagnetycznej-pisza-do-ministerstwa-sprawiedliwosci/eq65vj9.

5 Czesław Hoc, "Interpelacja nr 1736 do ministra cyfryzacji w sprawie lokowania stacji bazowych telefonii komórkowej w strefie uzdrowiskowej A w oparciu o przepisy tzw. megaustawy", *Sejm Rzeczpospolitej Polskiej*, 27.01.2020, accessed 5.09.2024, http://www.sejm.gov.pl/sejm9.nsf/InterpelacjaTresc.xsp?key=BLFJWP.

6 Jerzy Wilk, "Interpelacja nr 8310 do ministra środowiska w sprawie informacji o skutkach dla zdrowia, jakie powoduje funkcjonowanie anten telefonii komórkowej", *Sejm Rzeczpospolitej Polskiej*, 23.06.2020, accessed 5.09.2024, http://www.sejm.gov.pl/sejm9.nsf/InterpelacjaTresc.xsp?key=BR3D7M oraz Jerzy Wilk, "Interpelacja nr 8311 do ministra cyfryzacji w sprawie informacji o skutkach dla zdrowia, jakie powoduje funkcjonowanie anten telefonii komórkowej", *Sejm Rzeczpospolitej Polskiej*, 23.06.2020, accessed 5.09.2024, http://www.sejm.gov.pl/sejm9.nsf/InterpelacjaTresc.xsp?key=BR3D7N.

7 Patryk Wicher et al., "Zapytanie nr 254 do ministra zdrowia w sprawie wpływu technologii 5G na zdrowie i życie ludzkie", *Sejm Rzeczpospolitej Polskiej*, 04.02.2020, accessed 5.09.2024, https://www.sejm.gov.pl/sejm9.nsf/InterpelacjaTresc.xsp?key=BLRC4T oraz Patryk Wicher et al., "Zapytanie nr 255 do ministra klimatu w sprawie wpływu technologii 5G na środowisko naturalne", *Sejm Rzeczpospolitej Polskiej*, 04.02.2020, accessed 5.09.2024, https://www.sejm.gov.pl/Sejm9.nsf/InterpelacjaTresc.xsp?key=BLRC4U.

8 Teresa Hałas, "Zapytanie nr 1659 do prezesa Rady Ministrów w sprawie instalacji sieci 5G", *Sejm Rzeczpospolitej Polskiej*, 12.10.2020, accessed 5.09.2024, http://www.sejm.gov.pl/sejm9.nsf/InterpelacjaTresc.xsp?key=BUJB4K.

9 Łukasz Szpyrka, Jakub Szczepański, "Kulisy antyszczepionkowego buntu w PiS", *Interia*, 23.11.2021, accessed 5.09.2024, https://wydarzenia.interia.pl/autor/lukasz-szpyrka/news-kulisy-antyszczepionkowego-buntu-w-pis,nId,5663360.

Barbara Bobula and Andrzej Zybertowicz. The former is a member of the board of the Association of Catholic Families and a retired teacher at a Catholic high school. She also became a journalist for Radio Maryja. At the same time, from 2000 to 2004, she was a member of the editorial board of the *Obywatel* magazine and chairwoman of the programme council of the previously discussed Institute of Civil Affairs foundation. Andrzej Zybertowicz, social advisor to the President of the Republic of Poland Andrzej Duda and advisor to the head of the National Security Bureau, can also be considered another "transmission belt" between activists opposing the introduction of 5G technology in Poland and the PiS. At the same time, Zybertowicz was a member of the Honorary Council of the predecessor of *Nowy Obywatel*, the *Obywatel* magazine, and appeared several times at festivals organized by this magazine. These magazines were published, among others, by the Institute of Civil Affairs leaders.

Additionally, in 2015, Zybertowicz and his team published a book entitled *Suicide of Enlightenment? How Neuroscience and New Technologies Are Devastating the Human World*. In the book, he and his colleagues put forward the thesis that "when technology became networked – and the next phase of this dissemination will be 5G and the Internet of Things – it stopped being our servant, it began to develop according to autonomous rules. It became so complex that fewer and fewer people understand it, fewer and fewer institutional entities control it, it governs our lives, adapts us to itself."[10]

As a result of these contacts, activists opposing the introduction of the 5G network in Poland persuaded the government to create and publish the so-called White Paper explaining the operation of the planned infrastructure and launching the accompanying website gov.pl/5G. Opponents of EMF (i.e., so-called electromagnetic radiation) also reached the Chancellery of the President in 2016 and held a meeting with Anna Streżyńska, the then Minister of Digital Affairs. They proposed a law to link the possibility of building new mobile networks with examining the electromagnetic fields.[11] Thanks to their contacts in government circles, they also began to participate more and more often in the deliberations of various parliamentary committees. Since November 2019, over thirty interpellations regarding 5G have been submitted

10 Jan Zygmuntowski, "Prof. Andrzej Zybertowicz: smartfon to nie narzędzie, to władca", *Polskie Radio*, 28.10.2021, accessed 5.09.2024, https://trojka.polskieradio.pl/artykul/2836312,prof-andrzej-zybertowicz-smartfon-to-nie-narzedzie-to-wladca.

11 Sylwia Czubkowska, "Komu zależy, by płonęły maszty 5G? Ponad 140 ataków w całej Europie!", *Spider's Web*, 04.06.2020, accessed 5.09.2024, https://spidersweb.pl/plus/2020/06/5g-potesty-covid-ataki-teorie-spiskowe.

to the ministries of health, climate, justice and digital affairs, and the Prime Minister. A particular government portal showing whether the standards of such radiation have been exceeded also cost PLN 11 million.[12]

The support of PiS politicians also helped activists opposed to mandatory vaccinations to enter the mainstream of Polish politics. Before the COVID-19 pandemic, such actions included public statements by then-Prime Minister Beata Szydło, PiS MP, and later MEP Patryk Jaki. In September 2017, Szydło stated on Radio Maryja: "There are parents who are convinced that vaccination can harm their child, but there are also those who are convinced that vaccination will protect against the disease. I cannot assess this in terms of its merits. I am not a doctor. We will consider how to reconcile these two positions, which will not be easy."[13]

Jaki made more far-reaching statements in September 2015 when he stated at a press conference he organized with Socha that "forcing a parent to take on such a sensitive issue as vaccination, where it concerns the child's safety when the parent does not want to have the vaccination, is an abuse of the Polish state and we want to combat this abuse."[14] He also publicly urged the President of the Republic of Poland Andrzej Duda to veto the act extending the vaccination requirement, and wrote to the Minister of Health: "the procedure forcing vaccinations creates discrimination against Polish citizens concerning other residents of the European Union." He withdrew from his position, opposing the vaccination requirement, in October 2018.[15]

Andrzej Duda provided additional support for opponents of the vaccination requirement during the 2020 election campaign. "If anyone wants to, if there is a vaccine, let them get vaccinated, but if anyone does not want to, it is

12 Sylwia Czubkowska, "Strach przed elektrosmogiem już blokuje inwestycje telekomom. Czy może opóźnić wprowadzenie 5G?", *Wyborcza.pl*, 12.04.2019, accessed 5.09.2024, https://wyborcza.biz/biznes/7,177150,24642674,strach-przed-elektrosmogiem-juz-blokuje-inwestycje-telekomom.html.

13 Łukasz Sakowski, "Premier Szydło nie wie czy szczepionki są dobre dla dzieci, czy złe. To wsparcie antyszczepionkowców", *OKO.Press*, 24.09.2017, accessed 5.09.2024, https://oko.press/premier-szydlo-wie-szczepionki-sa-dobre-dla-dzieci-zle-wsparcie-antyszczepionkowcow.

14 Michał Danielewski, "Oto pożytek z koronawirusa w Polsce: PiS docenił standardy WHO. Już nie są antypolskim spiskiem", *OKO.Press*, 11.03.2020, accessed 5.09.2024, https://www.oko.press/oto-pozytek-z-koronawirusa-w-polsce.

15 akw, tmw, rzw, "Jaki dwa razy o szczepieniach. Rok temu przeciw »przymuszaniu«, teraz obiecuje »promocję«, *TVN24*, 03.10.2018, accessed 5.09.2024, https://www.tvn24.pl/wiadomosci-z-kraju,3/patryk-jaki-o-obowiazkowych-szczepieniach-w-sejmie-projekt,873150.html.

their personal decision (...) I am not a supporter of mandatory vaccinations," he said during his public speech in Końskie in July 2020.[16]

During the pandemic, however, the leadership of the PiS parliamentary club considered supporting the vaccination requirement.[17] However, despite the rejection by PiS leaders of the anti-vaccination narrative, it was in this party that one of the movement's most notorious representatives continued to operate for several years – MP Anna Siarkowska. She began her political career in the zero years of the 21st century in the ranks of the All-Polish Youth, managing its structures in Siedlce. She ran for the Sejm from the ultra-religious-nationalist party League of Polish Families lists in 2005 but did not manage to become an MP. The party sent her to work on the TVP Polonia[18] programme council, where she became the vice-chair of that body. Siarkowska entered the Parliament for the first time elected from the Kukiz'15 lists in 2015, leaving the movement's parliamentary club a year later. In 2017, she became the leader of the Republican Party, closely cooperating with PiS, and Siarkowska became a member of the Law and Justice parliamentary club. In 2019, she was re-elected, running from PiS list to the Polish Parliament.

Siarkowska combines religious integralism and opposition to mandatory vaccinations in her political views. She opposes the legality of abortion and the in vitro procedure, and is very hostile towards the LGBT community. At the same time, she opposes compulsory vaccinations and the introduction of anti-pandemic restrictions on mandatory vaccinations for COVID-19. She has also expressed scepticism about vaccinations in general, including in the case of other infectious diseases. She believes that politicians are using the COVID pandemic to implement the fourth industrial revolution, which is to minimize travel and interpersonal contacts switch to remote work and education, and all to finally eliminate the barriers between humans and computers and lead to implanting chips in people and abolishing private property. In her opinion, this is to lead to the division of humanity into those who will surrender in exchange for privileges and the excluded and is reminiscent of the dystopias of Nazism and Soviet communism.[19]

---

16 Dominika Sitnicka, "W imię wolności? Jak politycy dla zysku podsycają lęk przed szczepionkami", *OKO.Press*, 30.05.2021, accessed 5.09.2024, https://oko.press/w-imie-wolnosci-jak-politycy-dla-zysku-podsycaja-lek-przed-szczepionkami.

17 Danielewski, "Oto pożytek z koronawirusa".

18 Polish satellite television targeted to the people of Polish descent leaving abroad.

19 Piotr Głuchowski, "W PiS antyszczepionkowcy górą: Kiedyś, nie daj Boże, czip będzie konieczny, żeby wyjść z domu", *Wyborcza.pl*, 20.01.2022, accessed 5.09.2024, https://wyborcza.pl/duzyformat/7,127290,28022089,w-pis-antyszczepionkowcy-gora-kiedys-nie-daj-boze-czip-bedzie.html.

In the 9th term of the Polish Parliament, Siarkowska became the chairwoman of "the parliamentary team for the sanitarization issue," which brought together MPs opposing mandatory vaccinations. It consisted primarily of MPs from the Confederation, Suwerenna Polska and PiS. In addition to Siarkowska from the PiS parliamentary club, the team included eight more MP s. As in the case of the opposition to 5G telephony, these were primarily former members of the Kukiz'15 movement and Catholic integralists. Thus, the secretary of the team was Małgorzata Janowska, an accountant elected from the Kukiz'15 lists, and the presidium also included a former catechist, member of the Radio Maryja Family,[20] and from 2019 to 2022 the Secretary General of Suwerenna Polska,[21] Mariusz Kałużny. He claimed that: "Vaccinated people infect in the same way as unvaccinated people. All the stories about them infecting less (...) can be dismissed as fairy tales because no one has proven it. In parliament [there are many] MPs who think similarly, but are afraid to talk about it." In another interview, he added that: "Almost 40 percent of Poles have not been vaccinated, so we represent this Poles."[22]

Out of the six MP s from the PiS club in this group, five had direct links to Radio Maryja. Zbigniew Girzyński was a journalist for this medium for three years. Teresa Hałas regularly appears on the radio, and owned by the same owner, TV Trwam. Sławomir Zawiślak, through his parliamentary office in Zamość, collected signatures in favour of placing "TV Trwam" on the so-called digital multiplex. In 2021, Maria Kurowska and Janusz Kowalski publicly opposed the activities of the Watchdog Polska association, which demanded the disclosure of the finances of the Lux Veritatis foundation, whose president is Tadeusz Rydzyk, Radio Maryja owner and CEO.

Of this group, Kowalski from Sovereign Poland party was particularly intensely involved in opposing the vaccination requirement. He criticized the Minister of Health of the government co-founded by his own party, stating that "I do not know in whose name he is encouraging vaccinations since even the President of the Republic of Poland is against such a requirement" and claimed that the consequence of this "compulsion" would be "thousands of doctors leaving the profession and the actual paralysis of the health service."[23] Additionally, with Siarkowska, he conducted a parliamentary inter-

20 The Radio Maryja Family is a community of listeners as well as people identifying with the Radio Maryja station. The association operates both in Poland and abroad. The community has its own statute, approved by the Polish Episcopal Conference in 2006.

21 At that time party has been using the name Solidarna Polska.

22 Głuchowski, "W PiS antyszczepionkowcy górą".

23 Ibid.

vention in a children's care home in Nowy Dwór Gdański, where, as the two MP s claimed, children were to be vaccinated forcibly.[24]

On June 29, 2021, the parliamentary team for the sanitarization issue members submitted a draft bill entitled "Stop sanitary segregation," which stated that "no one may be discriminated against in political, social or economic life due to (...) not being vaccinated against COVID-19." They prohibited the requirement to provide information about vaccination to participate in any cultural, sports or educational events and purchase goods and services."[25] Due to meetings with PiS leaders (Jarosław Kaczyński and Ryszard Terlecki) and Prime Minister Mateusz Morawicki, the team members achieved partial success. After meeting with the team, Terlecki informed the PiS members of parliament that the act imposing compulsory COVID-19 vaccinations would not be processed and that mandatory vaccination are unnecessary.[26] Due to this decision by the PiS club, thirteen out of seventeen members of the Prime Minister's medical council resigned.[27] However, the MP s – opponents of mandatory vaccinations – did not manage to obtain the parliamentary majority needed to adopt the bill submitted by the team. Despite this, as many as thirteen MPs from the ruling coalition voted in favour of the bill on October 29, 2021, including Minister of Justice Zbigniew Ziobro, his deputy Michał Woś, and Deputy Minister of Agriculture Norbert Kaczmarczyk, while another eight abstained.[28] Most of them were members of the Sovereign Poland party. Additionally, Ziobro declared himself an opponent of COVID passports and mandatory vaccinations, and Kaczmarczyk stated that an additional problem in the pandemic is "medical staff," and for years, "unfortunately, only elites had access to this profession, and people with talent could not break into this club."[29] Another MP of Sovereign Poland and Ziobro's legal representative, Piotr Sak, publicly stated that the act on mandatory vaccinations may open

24 Julia Theus, "Tak wyglądało najście Kowalskiego i Siarkowskiej na dom dziecka. »Grozili, byli nieprzyjemni«", *OKO.Press*, 15.07.2021, accessed 5.09.2024, https://oko.press/kowalski-i-siarkowska-nachodza-dom-dziecka.

25 Głuchowski, "W PiS antyszczepionkowcy górą".

26 Ibid.

27 Barbara Bodalska, "Polska: Koniec bezradnej Rady Medycznej przy premierze", *Euractiv.pl*, 17.01.2022, accessed 5.09.2024, https://www.euractiv.pl/section/zdrowie/news/polska-pandemia-koronawirus-zdrowie-rada-medyczna-rezygnacja-premier-morawiecki-niedzielski.

28 The following members of the PiS club abstained: Teresa Glenc, Mariusz Gosek, Sebastian Kaleta, Jan Kanthak, Bartosz Kownacki, Władysław Kurowski, Beata Mateusiak, Jacek Ozdoba, Piotr Uruski, while Dariusz Olszewski, Agnieszka Górska and Paweł Lisiecki were also in favor of adopting the act submitted by PZdS.

29 Głuchowski, "W PiS antyszczepionkowcy górą".

up other expectations for employers. "Maybe the employer should know if the employee has the HPV or HIV. There has to be a stopgap somewhere,"[30] he stated.

The significant influence of MP s opposing mandatory vaccinations on the decisions of the PiS club and the government resulted, among other things, from the small parliamentary majority on which Morawicki's government was based in the 9th term of the Sejm. Since the exit of the Porozumienie party from the coalition in August 2021, the government has had majority based only on a few votes and, therefore, could not remove from the parliamentary club MPs opposing mandatory vaccinations without the government falling simultaneously.

As a result of repeatedly opposing the government's line, Siarkowska concluded that, as she stated, "Kaczyński gets a white fever at the sound of my name" and, assuming that she would not receive a place on his party's electoral list in the 2023 Sejm elections, she began to look for a new place for herself on the Polish political scene. Initially, in September 2021, she became involved in the creation of the Poland Is One (Polska Jest Jedna) movement,[31] which was created by journalist Jan Pospieszalski, the president of Siemianowice Śląskie Rafał Piech, former MP and presidential candidate Jan Łopuszański, activists of the movement opposing the mandatory vaccination Zbigniew Hałat,[32] Piotr Witczak[33] and future MP of the Confederation Grzegorz Płaczek.[34] As the creator of the movement, Piech, says, the only postulate of the new organization was to fight mandatory vaccination and lockdowns.[35] In May 2022, Siarkowska joined the Sovereign Poland party. In July 2023, she joined the National Movement party and the parliamentary group of the Con-

---

30 Ibid.

31 Since 2023 it is a political party.

32 Paweł Pawlik, "Antyszczepionkowcy tworzą nowy ruch. Wśród założycieli znane osoby", *Onet*, 27.09.2021, accessed 5.09.2024, https://wiadomosci.onet.pl/slask/polska-jest-jedna-piech-pospieszalski-i-poslanka-pis-tworza-ruch-antyszczpionkowy/ct26yry.

33 No author, "POWSTAŁ RUCH POLSKA JEST JEDNA SPRZECIWIAJĄCY SIĘ SEGREGACJI SANITARNEJ. APELUJE DO PREZYDENTA O »WETOWANIE AKTÓW PRAWNYCH, KTÓRE GODZĄ W PRAWA I WOLNOŚCI OBYWATELSKIE«", *Radio Maryja,* 26.09.2021, accessed 5.09.2024, https://www.radiomaryja.pl/informacje/powstal–ruch–polska–jest–jedna–sprzeciwiajacy–sie–segregacji–sanitarnej–apeluje–do–prezydenta–o–wetowanie–aktow–prawnych–ktore–godza–w–prawa–i–wolnosci–obywatelskie.

34 Ibid.

35 Aneta Kaczmarek, "Prezydent Siemianowic na czele ruchu. Chce walczyć z segregacją sanitarną", *PortalSamorządowy.pl,* 6.10.2021, accessed 5.09.2024, https://www.portalsamorzadowy.pl/ochrona-zdrowia/prezydent-siemianowic-na-czele-ruchu-chce-walczyc-z-segregacja-sanitarna,317519.html

federation party. On the lists of this party, she ran in the elections to the Sejm in 2023 but did not win re-election.

In her case, there are also noticeable connections with groups sympathetic to Russian narratives. She was a journalist for the quarterly "Myśl.pl," by Sylwester Chruszcz and Krzysztof Tenerowicz. The first is a long-time collaborator of Mateusz Piskorski, accused by the prosecutor's office of espionage for Russia and China and a participant in the election observation missions he organized in the territories of the former USSR. For years, Polish and foreign media have indicated him as a politician with solid pro-Russian sympathies.[36] Tenerowicz, in turn, is a close associate of Adam Andruszkiewicz. The press, in turn, accused the deputy minister of pro-Belarusian sympathies.[37]

The attitude of parliamentary groups other than the parties discussed above[38] towards 5G telephony and mandatory vaccinations was not hostile, and the statements of their politicians were practically devoid of elements of technophobic conspiracy narratives. They did not have close connections with activists of these movements, and their positions differed from those postulated by technophobic organizations. However, there is a clear difference in opinion on these issues between the opposition parties (in the 2015–2023 period), and the positions of the Left and the Civic Platform can be considered opposite poles.

On the issue of mandatory vaccinations, the Left can be considered to be vehemently opposed to the position of the movement opposing mandatory vaccinations. It was the members of this club who, during the 9th term of the Sejm, which fell during the COVID-19 pandemic, called for the introduction of mandatory vaccinations against this disease, tightening restrictions for unvaccinated people and strengthening the control of "COVID passports,"[39] i.e., They wanted to introduce precisely what the parliamentary sanitary team led by Siarkowska wished to block.

---

36 Anna Mierzyńska, "Zmienny jak poseł Chruszcz. Kolejny narodowiec dołącza do PiS. Czy na pewno nie lubi już Putina?", *OKO.Press*, 03.04.2019, accessed 5.09.2024, https://oko.press/zmienny-jak-posel-chruszcz-kolejny-narodowiec-dolacza-do-pis-czy-na-pewno-nie-lubi-juz-putina.

37 Reset Obywatelski, "Adam Andruszkiewicz działał w grupie parlamentarnej sympatyków Łukaszenki", *YouTube*, 22.03.2023, accessed 5.09.2024, https://www.youtube.com/watch?v=TAhxMl5K6jo.

38 That is: 1polska.pl, Kukiz'15 movement, Confederation, One Poland, Sovereign Poland, Polish People's Party, law and Justice.

39 Martyna Śmigiel, "TAK dla obowiązkowych szczepień! – Lewica z radykalnym pomysłem na walkę z pandemią", *Wyborcza.pl*, 05.12.2021, accessed 5.09.2024, https://warszawa.wyborcza.pl/warszawa/7,54420,27878096,tak-dla-obowiazkowych-szczepien-lewica-z-radykalnym-pomyslem.html.

The Civic Platform did not support the Left's idea. Its leaders, including Donald Tusk[40] and Tomasz Siemoniak,[41] spoke critically of the idea, pointing to persuasion, not coercion, as a means of increasing vaccination rates and questioning the possibility of the PiS government implementing such a project, pointing to its problems in implementing other measures to combat the pandemic. Regardless, the party supported vaccinations, and its politicians did not question their effectiveness.

The Left, as a parliamentary club, also did not support views against 5G telephony based on conspiratorial narratives. However, among the members of this club, one could also find supporters of conspiratorial narratives related to this generation of telephony. Such a person was Robert Obaz, a member of the Sejm of the 9th term, who, however, limited himself to posts on the subject on social media without moving on to any more formal actions against the 5G network or closer cooperation with the movement opposing it. MP Andrzej Szejna took up broader activity related to 5G in the Sejm, submitting three interpellations on the subject.[42] His interpellations concerning 5G, however, did not put forward arguments based on conspiracy narratives but emphasized primarily the USA's financial issues and political influence on introducing this generation of networks and the issues of participation in the legislative process of the Eriksson company. The only other interpellation by members of the parliamentary club of the Left on using the 5G network to reduce digital exclusion in Poland maintained a favourable tone to fifth-generation telephony.[43]

---

40 No author, "D.Tusk o ewentualnym obowiązku szczepień: jestem zwolennikiem perswazji",. *PolitykaZdrowotna.com*, 09.12.2021, accessed 5.09.2024, https://politykazdrowotna.com/artykul/dtusk-o-ewentualnym-obowiazku-szczepien-jestem-zwolennikiem-perswazji/825020.

41 No author, "Lewica składa projekt ustawy o obowiązkowych szczepieniach", *Dziennik. Gazeta Prawna*, 06.12.2021, accessed 5.09.2024, https://www.gazetaprawna.pl/wiadomosci/kraj/artykuly/8307534,lewica-sklada-projekt-ustawy-obowiazkowe-szczepienia.html.

42 Andrzej Szejna, "Interpelacja nr 4795 do prezesa Rady Ministrów w sprawie rozporządzenia dotyczącego budowy sieci 5G", *Sejm Rzeczpospolitej Polskiej*, 20.04.2020, accessed 5.09.2024, http://www.sejm.gov.pl/sejm9.nsf/InterpelacjaTresc.xsp?key=BP3F2D; Andrzej Szejna, "Interpelacja nr 4796 do ministra cyfryzacji w sprawie rozporządzenia dotyczącego budowy sieci 5G", *Sejm Rzeczpospolitej Polskiej*, 20.04.2020, accessed 5.09.2024, http://www.sejm.gov.pl/sejm9.nsf/InterpelacjaTresc.xsp?key=BP3F2E oraz Andrzej Szejna, "Interpelacja nr 4797 do ministra spraw zagranicznych w sprawie rozporządzenia dotyczącego budowy sieci 5G", *Sejm Rzeczpospolitej Polskiej*, 20.04.2020, accessed 5.09.2024, http://www.sejm.gov.pl/sejm9.nsf/InterpelacjaTresc.xsp?key=BP3F2F.

43 Marek Rutka et al., "Interpelacja nr 7110 do ministra cyfryzacji w sprawie wykorzystania sieci 5G do zmniejszenia wykluczenia cyfrowego w Polsce", *Sejm Rzeczpospolitej Polskiej*,

The Civic Platform as a party did not promote conspiratorial narratives related to 5G telephony. However, even in this party, individual politicians spoke out in their interpellations and used motifs present in technophobic opinions about this generation of networks. Such statements were made by three MPs of this party: Waldemar Sługocki, Marek Sowa, and Bogusław Sonik. The latter submitted two interpellations related to this topic,[44] in which he raised this issue, repeating conspiratorial narratives about the harmfulness of using this generation of networks for Polish citizens' health. The interpellations of Sługocki[45] and Sowa[46] were constructed similarly. However, the party itself did not take up the subject more broadly, and in 2023, Sonik left the Civic Platform and did not run in the elections for the 10th term of the Sejm. Sługocki and Sowa, on the other hand, did not continue to speak out on the 5G network. The interpellation on 5G, submitted by the then-PO MP Joanna Mucha[47] to the Minister of Digital Affairs, the Minister of Internal Affairs and Administration, and the Minister-Coordinator of Special Services, was different, focusing on economic policy issues.[48] Her interpellation was primarily concerned with the problem of Chinese capital and Huawei devices, and she did not use the conspiratorial narratives of activists opposed to introducing this generation of telephony. In the context of 5G, it is also worth noting the vote against the adoption of the draft act amending the act on supporting

---

30.05.2020, accessed 5.09.2024, http://www.sejm.gov.pl/sejm9.nsf/InterpelacjaTresc.xsp?key=BQ9DPW.

44 Bogusław Sonik, "Interpelacja nr 2356 do ministra cyfryzacji w sprawie budowy sieci 5G", *Sejm Rzeczpospolitej Polskiej*, 14.02.2020, accessed 5.09.2024, https://www.sejm.gov.pl/sejm9.nsf/InterpelacjaTresc.xsp?key=BLYCLY oraz Bogusław Sonik, "Interpelacja nr 23 58 do ministra zdrowia w sprawie wpływu na zdrowie sieci 5G", *Sejm Rzeczpospolitej Polskiej*, 14.02.2020, accessed 5.09.2024, https://www.sejm.gov.pl/sejm9.nsf/InterpelacjaTresc.xsp?key=BLYCM2.

45 Waldemar Sługocki, "Interpelacja nr 8744 do ministra cyfryzacji w sprawie wprowadzenia w Polsce technologii 5G", *Sejm Rzeczpospolitej Polskiej*, 09.07.2020, accessed 5.09.2024, https://www.sejm.gov.pl/sejm9.nsf/InterpelacjaTresc.xsp?key=BRJA3J&view=null.

46 Marek Sowa, "Interpelacja nr 2737 do ministra zdrowia w sprawie wpływu stacji bazowy ch telefonii komórkowej na zdrowie człowieka", *Sejm Rzeczpospolitej Polskiej*, 26.02.2020, accessed 5.09.2024, https://www.sejm.gov.pl/sejm9.nsf/InterpelacjaTresc.xsp?key=BMDHP6.

47 In 2021, she joined the Polska 2050 movement founded by tv celebrity Szymon Hołownia.

48 Joanna Mucha, "Interpelacja nr 9556 do ministra cyfryzacji, ministra spraw wewnętrznych i administracji, ministra – członka Rady Ministrów, koordynatora służb specjalnych w sprawie sieci telekomunikacyjnej i instalacji urządzeń firmy Huawei", *Sejm Rzeczpospolitej Polskiej*, 10.08.2020, accessed 5.09.2024, http://www.sejm.gov.pl/sejm9.nsf/InterpelacjaTresc.xsp?key=BSN9Y5.

the development of telecommunications services and networks in the wording from paper no. 3545 (popularly called the "5G Mega-Act"), submitted by Marek Biernacki. However, he, like Sonik, has ceased to be a member of the Civic Platform and currently holds a mandate obtained from the Third Way coalition[49] lists and does not continue activities that could be considered as implementing the postulates of the movement as opposed to fifth-generation telephony.

49 The Third Way (Polish: Trzecia Droga, TD) is a political alliance in Poland that was formed before the 2023 parliamentary election. The coalition was created by centrist Poland 2050 of Szymon Hołownia and the agrarian Polish People's Party.

CHAPTER 7

You'll Be Hanged When Poland Returns

Protests, Radicalization, and Extremism in Movements against 5G and Mandatory Vaccinations

The issue of introducing the 5G network to Poland or discussing mandatory vaccinations and anti-pandemic restrictions has been among the most motivating factors for public activity in Polish political life in the last decade. Almost one hundred demonstrations and marches related to this topic have taken place in the Republic of Poland. It is difficult to estimate the total number of participants precisely. Still, based on data provided by local media, it can be assumed that over 100,000 people have participated in all demonstrations, marches, pickets, and rallies related to these topics since 2014. Additionally, these protests were not concentrated in one region of Poland but spread throughout its territory, not limited to larger cities but often also in smaller centres. This wave shows that movements opposing the 5G network and mandatory vaccinations can be considered nationwide and exceed the metropolitan-smaller centre division visible in other political issues (e.g., during parliamentary elections).[1]

Their intensity also often exceeded that seen in other protest movements. Not only were marches, pickets, and demonstrations related to the topics of the issue of introducing the 5G network to Poland, mandatory vaccinations, or anti-pandemic restrictions, but there were also physical clashes, attempts to enter institutions and private properties illegally, acts of vandalism (from graffiti to permanent destruction of property), as well as threats, including death threats, against experts and politicians opposing the postulates of the movement. This scale indicates the significant radicalization of groups opposing mandatory vaccinations and the introduction of the 5G network in Poland, in some cases turning into extremism. Considering this and the links indicated in this work, with groups spreading narratives similar to Russian ones, it is not surprising that Polish counterintelligence is interested in the movements

1 Łukasz Kobeszko, "Metropolie i prowincja. Polityczna wojna światów", *Rzeczpospolita*, 07.08.2020, accessed 30.08.2024, https://www.rp.pl/plus-minus/art585231-metropolie-i-prowincja-polityczna-wojna-swiatow.

 | DOI:10.1163/9789004747852_009

fighting against 5G telephony and mandatory vaccinations. At the beginning of 2024, the case of Polish counterintelligence officers entering the premises of activists of the movement became widely reported in the media. Aggressive behaviour is also displayed by the movement of Compatriots and Comrades (Rodacy Kamraci) associated with opponents of mandatory vaccinations. Violent language and behaviour, ties with pro-Russian circles, and hostility toward political and scientific elites mean that this dimension of the activities of these circles also deserves separate analyses.

Since the announcement of the pilot introduction of the 5G network in Poland in 2018 until the first half of 2020, protests against this generation of telephone networks have occurred in many Polish cities. Demonstrations, pickets, marches, and rallies took place not only in larger cities such as Kraków,[2] Gdańsk,[3] Radom,[4] Radomsko,[5] Koszalin,[6] Kołobrzeg,[7] Gliwice,[8] Bieruń,[9]

2 Pog, "Protest przeciwników 5G na Rynku Głównym. »Zabójcza technologia«", *Kraków. Nasze Miasto*, 19.06.2020, accessed 30.08.2024, https://krakow.naszemiasto.pl/protest-przeciwnikow-5g-na-rynku-glownym-zabojcza/ar/c3-7765204.

3 RIRM, "W kilku miastach Polski odbędą się protesty przeciwników technologii 5G", *Radio Maryja*, 25.01.2020, accessed 25.11.2025, https://www.radiomaryja.pl/informacje/w-kilku-miastach-polski-odbeda-sie-protesty-przeciwnikow-technologii-5g.

4 Adam Bednarek, "Maszt jak sklep monopolowy. To już poszło za daleko", *Spider's Web*, 28.11.2023, accessed 30.08.2024, https://spidersweb.pl/2023/11/radom-maszt-protest-odleglosc-od-dzialek.html.

5 No author, "Radomszczanie nie chcą masztu z nadajnikiem sieci telefonii komórkowej", *TVP3 Łódź*, 13.07.2020, accessed 30.08.2024, https://lodz.tvp.pl/48949780/radomszczanie-nie-chca-masztu-z-nadajnikiem-sieci-telefonii-komorkowej.

6 Wiesław Miller, Marek Łagocki, Dawid Macal, "W Koszalinie powstanie 5G? Mieszkańcy protestują! Mamy stanowisko ratusza", *KoszalinInfo.pl*, 20.04.2020, accessed 30.08.2024, https://koszalininfo.pl/w-koszalinie-powstanie-5g-mieszkancy-protestuja-mamy-stanowisko-ratusza.

7 Miastokolobrzeg, "Protest przeciwko sieci 5G", *YouTube*, 26.04.2020., accessed 30.08.2024, https://www.youtube.com/watch?v=73a8483itqE.

8 Telepolis, "Relacja z protestu przeciwko sieci 5G w Gliwicach TELEPOLIS.PL", *YouTube*, 15.10.2018, accessed 30.08.2024, https://www.youtube.com/watch?v=y4QLF20HMMM.

9 Bartosz Bednarczuk, "Śląskie: Bunt przeciwko sieci 5G! Nadajnika nie chcą u siebie mieszkańcy Bierunia. Głos zabrał burmistrz!", *TVS*, 12.06.2020, accessed 30.08.2024, https://tvs.pl/informacje/slaskie-bunt-przeciwko-sieci-5g-nadajnika-nie-chca-u-siebie-mieszkancy-bierunia-glos-zabral-burmistrz.

Bolesławiec,[10] Kielce,[11] Opole,[12] Bielsko-Biała,[13] Warsaw,[14] Kraśnik,[15] Słupsk,[16] Kazimierz Dolny,[17] Olkusz,[18] Olsztyn[19] but also in smaller towns and villages such as Lutomia Górna (Świdnica commune),[20] Gniechowice (Kąty Wrocławskie commune),[21] Stryszów,[22] Rzyki (Andrychów commune),[23] Izdebnik

10 Redakcja, "Kilkanaście osób przed jednostką wojskową na proteście przeciw 5G", *Istotne.pl*, 25.04.2020, accessed 30.08.2024, https://istotne.pl/boleslawiec/wiadomosc/Xdf/kilkanascie-osob-przed-jednostka-wojskowa-na-protescie-przeciw-5g.

11 Kamil Markiewicz, "Protestują przeciw budowie masztu 5G. Konstrukcja ma stanąć w okolicach ulicy Prostej", *TVP3 Kielce*, 13.07.2020, accessed 30.08.2024, https://kielce.tvp.pl/48955032/protestuja-przeciw-budowie-masztu-5g-konstrukcja-ma-stanac-w-okolicach-ulicy-prostej.

12 Piotr Zapotoczny, "Protest w centrum Opola. Straszono technologią 5G, lobby szczepionkowym i politykami na usługach koncernów", *Wyborcza.pl*, 25.01.2020, accessed 30.08.2024, https://opole.wyborcza.pl/opole/7,35086,25632807,kontrowersyjny-protest-na-pl-wolnosci-narodowiec-traszyl-technologia.html.

13 Raz, "Mieszkańcy Hałcnowa obawiają się budowy masztu 5G", *Bielsko.info*, 30.06.2020, accessed 30.08.2024, https://www.bielsko.info/wiadomosci/23785-mieszkancy-halcnowa-obawiaja-sie-budowy-masztu-5g-bielsko-biala.

14 Telepolis, "Relacja z protestu przeciwko sieci 5G w Warszawie TELEPOLIS.PL", *YouTube*, 01.05.2019, accessed 30.08.2024, https://www.youtube.com/watch?v=SxT5ztBnjWs.

15 Adam Bednarek, "Mapa wstydu. W tych polskich miastach protestowano przeciwko 5G", *Dobre Programy*, 24.09.2020, accessed 30.08.2024, https://www.dobreprogramy.pl/mapa-wstydu-w-tych-polskich-miastach-protestowano-przeciwko-5g,6628692371286145a.

16 Ibid.

17 Ibid.

18 Ibid.

19 Ibid.

20 No author, "[FOTO, WIDEO] »Lutomia mówi NIE, dla wieży 5G!«", *Doba.pl*, 27.04.2019, accessed 30.08.2024, https://doba.pl/dsw/artykul/-foto-wideo-lutomia-mowi-nie-dla-wiezy-5g-/25194/0.

21 Zespół wGospodarce.pl, "W Gniechowicach lęk przed 5G. Mieszkańcy nie chcą masztu", *wGospodarce.pl*, 27.06.2020, accessed 30.08.2024, https://wgospodarce.pl/informacje/81577-w-gniechowicach-lek-przed-5g-mieszkancy-nie-chca-masztu.

22 No author, "Petycja »Stop zagrożeniu zdrowia i życia«", *Urząd Gminy Stryszów*, 23.03.2020, accessed 30.08.2024, https://bip.malopolska.pl/ugstryszow,a,1752178,petycja-stop-zagrozeniu-zdrowia-i-zycia.html.

23 MG, "Mimo protestów, starostwo wydaje pozytywne opinię dla nowych stacji bazowych", *WadowiceOnline.pl*, 23.07.2020, accessed 30.08.2024, https://wadowiceonline.pl/biznes/16612-mimo-protestow-starostwo-wydaje-pozytywne-opinie-dla-nowych-stacji-bazowych.

(Wadowice commune),[24] Stara Wieś (Limanowa commune),[25] Myślec (Stary Sącz commune),[26] Dębogóra (Czerwonak commune),[27] Złocieniec,[28] Łuków,[29] Puchły (Narew commune),[30] Korczyna[31] or Lutoryż (Boguchwała commune).[32]

In many cases, demonstrators raised the issue of alleged threats to human health allegedly stemming from the installation of a fifth-generation telephone mast in a locality, often using justifications derived from conspiracy or technophobic narratives. According to the demonstrators, the 5G network was "unexplored," "harmful," and was supposed to "fry organs" and "kill." Arguments describing the introduction of new technology as undemocratic and unconsulted ("local governments have no say"), emphasizing the freedom of residents to decide on investments in their area or their health ("they have no

---

24 Marcin Płaszczyca, "Czy ludzie w Izdebniku dostaną raka od 5G? Ministerstwo Zdrowia odpowiada: »Nie mamy dowodów«', *Wadowice24*.pl, 03.03.2020, accessed 30.08.2024, https://www.wadowice24.pl/nowe/pieniadze/13055-czy-ludzie-w-izdebniku-dostana-raka-od-5g.html.

25 PAN, "Wraca temat budowy stacji bazowej. Okoliczni mieszkańcy boją się oddziaływania 5G", *Limanowa.in*, 11.07.2020, accessed 30.08.2024, https://limanowa.in/aktualnosci/wraca-temat-budowy-stacji-bazowej-okoliczni-mieszkancy-boja-sie-oddzialywania-5g/49749.

26 Ewa Stachura, "60-metrowy kolos? Mieszkańcy Myślca nie zgadzają się na budowę przekaźnika GSM", *Sądeczanin*, 16.06.2020, accessed 30.08.2024, https://sadeczanin.info/wiadomosci/60-metrowy-kolos-mieszkancy-myslca-nie-zgadzaja-sie-na-budowe-przekaznika-gsm.

27 Marta Danielewicz, "Maszt 5G będzie górował nad Puszczą Zielonką? Mieszkańcy Dębogóry nie chcą takiej inwestycji pod swoimi oknami przy wejściu do lasu", *Gazeta Pomorska*", 19.06.2020, accessed 30.08.2024, https://plus.pomorska.pl/maszt-5g-bedzie-gorowal-nad-puszcza-zielonka-mieszkancy-debogory-nie-chca-takiej-inwestycji-pod-swoimi-oknami-przy-wejsciu-do/ar/c1-15031340.

28 Kira Czarczyńska, "Nadajnik Play stanie mimo protestów mieszkańców", *Komórkomania*, 05.04.2010, accessed 30.08.2024, https://komorkomania.pl/18372,nadajnik-play-stanie-mimo-protestow-mieszkancow.

29 Lukow24, "Prywatny inwestor chce budować maszt telefonii 5G", *Lukow24*.pl, 28.02.2020, accessed 30.08.2024, https://www.lukow24.pl/wiadomosci/prywatny-inwestor-chce-budowac-maszt-telefonii-5g/XjaoyKLZibKAZL3fq3JB.

30 Źmicier Kościn, "Mieszkańcy wsi Puchły nie chcą wieży 5G w swojej miejscowości", *Polskie Radio Białystok*, 15.07.2020, accessed 30.08.2024, https://www.radio.bialystok.pl/wiadomosci/index/id/186742.

31 Andrzej Józefczyk, "Starosta mówi NIE dla nadajnika 5G w Korczynie", *TerazKrosno.pl.* 02.07.2020, accessed 30.08.2024, https://www.terazkrosno.pl/wiadomosci/10260-starosta-mowi-nie-dla-nadajnika-5g-w-korczynie.

32 Marcin Piecyk, "W Lutoryżu nie chcą masztu 5G. Mieszkańcy protestują, gmina Boguchwała ma związane ręce", *Nowiny24*, 18.07.2020, accessed 30.08.2024, https://nowiny24.pl/w-lutoryzu-nie-chca-masztu-5g-mieszkancy-protestuja-gmina-boguchwala-ma-zwiazane-rece/ar/c1-15083339.

right to decide on our health"), as well as accusations against politicians and doctors of serving the interests of telecommunications networks, pharmaceutical companies, and international organizations, were also frequently used during demonstrations, in slogans, speeches and on banners. An example of such a statement, combining all the threads mentioned above, is a fragment of a speech by Robert Jarosław Bogusławski[33] from a demonstration against 5G in Opole: "5G technology is very harmful, so why is no politician looking into it? Some do business on it, and we are talking about big money. They – the corporations – lobby in parliaments and bribe politicians to keep quiet about it. Corporations make vast amounts of money on this, and our children suffer. We have to talk about it, talk to our loved ones."[34]

As a result of such numerous protests, activists who created petitions, collected signatures, put up posters in the area, and publicized the issue in local media often managed, especially in smaller towns, to block the construction of a telecommunications mast of the network, but also of earlier generations, as well as to interest politicians in the subject and move the topic into the mainstream of Polish politics, as evidenced by numerous articles about the network in the national press and statements by MP s in the Polish Parliament.

At the same time, protests against the introduction of the 5G network to Poland began to die down shortly after their peak at the turn of 2019 and 2020. The COVID-19 epidemic started in November 2019 in the Chinese city of Wuhan. The first case of the disease was recorded in Poland on March 4, 2020, and from March 20, 2020, to May 15, 2022, by the regulation of the Minister of Health, a state of the epidemic was announced in Poland. Demonstrations against mandatory vaccinations and anti-pandemic restrictions replaced protests against 5G. This was not a beginning of protests connected to vaccinations, as they started much earlier. They had been taking place since 2012,[35] and a nationwide march of opponents of mandatory vaccinations had been held in Warsaw since 2016. The first march, organized on the so-called Inter-

---

33 Bogusławski was a candidate for the mayor of Wrocław in 2014 and 2018 (supported by the nationalist National Movement and the anti-vaccination National Association of Vaccination Awareness "STOP Vaccine adverse events"), a former activist of the far-right party 1Polska.pl, a candidate for the Sejm from the list of Grzegorz Braun's Szczęść Boże committee, spreading numerous conspiracy narratives, an associate of Piotr Rybak, sentenced to ten months of unconditional imprisonment for burning an effigy of a Jew on the Wrocław Market Square.

34 Zapotoczny, "Protest w centrum Opola".

35 Adriana Rozwadowska, "Półwiejska: Protest przeciw obowiązkowym szczepieniom [AKTUALIZACJA]", *Głos Wielkopolski*, 21.04.2012, accessed 30.08.2024, https://gloswielkopolski.pl/polwiejska-protest-przeciw-obowiazkowym-szczepieniom-aktualizacja/ar/558845.

national Day of Victims of Post-Vaccination Reactions, was attended by approximately 10,000 people.[36] However, before 2020, demonstrations of the anti-vaccination movement occurred primarily in large Polish cities. Protests took place, among others, in Łódź,[37] Inowrocław,[38] Warsaw,[39] and Poznań,[40] and their location was mainly related to the activity of local leaders of the movement. At the turn of 2019 and 2020, both movements operated in parallel, often mixing their conspiratorial narratives, hybridizing them into stories combining 5G, pandemics, and vaccinations.[41] However, since mid-2020, one has noticed a clear dominance of vaccinations and opposition to lockdowns in technophobic demonstrations over issues related to electromagnetic fields or the fifth-generation network. Since 2021, the protests have spread to other regions capitals, such as Katowice,[42] Wrocław,[43] Szczecin,[44] Opole,[45]

36 Joanna Opas, "Protest Przeciw Przymusowi Szczepień", *Polskie Radio Katowice*, 02.06.2018, accessed 30.08.2024, https://www.radio.katowice.pl/zobacz,36901,Protest–Przeciw–Przymusowi–Szczepien.html#.WxedZzSFOUk.

37 Redakcja, "Protest antyszczepionkowców w Łodzi. Bronili »swojego« lekarza [FILM, ZDJĘCIA]", *Dziennik Łódzki*, 21.12.2017, accessed 30.08.2024, https://dzienniklodzki.pl/protest–antyszczepionkowcow–w–lodzi–bronili–swojego–lekarza–film–zdjecia/ar/12787207.

38 Dominik Fijałkowski, "Szczepionki? Chcą mieć wybór. Protest rodziców w Inowrocławiu [zdjęcia]", *Express Bydgoski*, 24.08.2017, accessed 30.08.2024, https://expressbydgoski.pl/szczepionki–chca–miec–wybor–protest–rodzicow–w–inowroclawiu–zdjecia/ar/12415800.

39 Bartłomiej Leśniewski, "Marsz antyszczepionkowców w Warszawie", *Termedia*, 04.06.2017, accessed 30.08.2024, https://www.termedia.pl/mz/Marsz–antyszczepionkowcow–w–Warszawie,26348.html.

40 Rozwadowska, "Półwiejska: Protest przeciw obowiązkowym szczepieniom".

41 CEPowiśle, "Hanna Kazahari (Tokio): Plandemia – zaplanowana przez globalistów i syjonistów fałszywa flaga wirusa", *BitChute*, 08.04.2020, accessed 30.08.2024, https://www.bitchute.com/video/ccEa9MRbElOQ.

42 Krzysztof Konopka, "Protest antyszczepionkowców w Katowicach. Część osób znów w mundurach. »My jesteśmy czwartą falą«", *Gazeta.pl*, 07.08.2021, accessed 30.08.2024, https://wiadomosci.gazeta.pl/wiadomosci/7,114883,27426268,protest–antyszczepionkowcow–w–katowicach–stop–segregacji.html.

43 Beata Makowska, "Protest antyszczepionkowców we Wrocławiu", *Radio Wrocław*, 04.12.2021, accessed 30.08.2024, https://www.radiowroclaw.pl/articles/view/114727/Protest–antyszczepionkowcow–we–Wroclawiu.

44 Michał Kaczmarek, "Protest antyszczepionkowców przed gabinetem prof. Parczewskiego. »Stop segregacji sanitarnej«", *wSzczecinie.pl*, 18.05.2021, accessed 30.08.2024, https://wszczecinie.pl/index.php/protest-antyszczepionkowcow-przed-gabinetem-prof-parczewskiego-stop-segregacji-sanitarnej/37812.

45 Zapotoczny, "Protest w centrum Opola".

Gdańsk,[46] Olsztyn[47] and Lublin,[48] as well as to smaller cities such as Wałbrzych,[49] Zielona Góra,[50] Konin,[51] Stalowa Wola,[52] Siedlce[53] and Bielsko-Biała.[54]

Since the beginning of the anti-vaccination protests, Warsaw has been the focal point of these activities. Since 2016, an annual demonstration has been held there in June on the so-called International Day of Victims of PostVaccination Reaction (June 2–3).[55] On other dates (such as June 2017,[56] May 2020,[57]

---

46 Agnieszka Kamińska, "Marsz antyszczepionkowców ulicami Gdańska. Na transparentach: »My się nie szczepimy«, »Stop segregacji sanitarnej«", *Nowiny24*, 23.10.2021, accessed 30.08.2024, https://nowiny24.pl/marsz-antyszczepionkowcow-ulicami-gdanska-na-transparentach-my-sie-nie-szczepimy-stop-segregacji-sanitarnej/ar/c1-15867695.

47 SK, "Antyszczepionkowcy zaatakowali Olsztyn [ZDJĘCIA]", *Gazeta Olsztyńska*, 31.01.2021, accessed 30.08.2024, https://gazetaolsztynska.pl/690896,Antyszczepionkowcy-zaatakowali-Olsztyn-ZDJECIA.html.

48 Pk, "Marsz antyszczepionkowców na ulicach Lublina. Krzyczeli: »Cała Polska śpiewa z nami, nie jesteśmy królikami«", *Wyborcza.pl*, 17.09.2021, accessed 30.08.2024, https://lublin.wyborcza.pl/lublin/7,48724,27583860,marsz-antyszczepionkowcow-na-ulicach-lublina-krzyczeli-cala.html.

49 Kaki, "Demonstracja antyszczepionkowców pod domem prezydenta Wałbrzycha. Nazwali go doktorem Mengele", *Wyborcza.pl*, 03.05.2021, accessed 30.08.2024, https://wroclaw.wyborcza.pl/wroclaw/7,35771,27045236,antyszczepionkowcy-demonstruja-w-walbrzychu-poszli-pod-dom.html.

50 Maja Sałwacka, "Antyszczepionkowcy pod ratuszem straszyli za zgodą prezydenta: »Szczepionki są robione z płodów ludzkich«", *Wyborcza.pl*, 17.11.2021, accessed 30.08.2024, https://zielonagora.wyborcza.pl/zielonagora/7,35182,27817685,antyszczepionkowcy-pod-ratuszem-straszyli-szczepionki-sa-robione.html.

51 Bartłomiej Bolczyk, "W Koninie odbędzie się marsz »Z miłości do wolności«. Będą protestować przeciwko obostrzeniom antyepidemicznym", *Konin. Nasze Miasto*. 21.03.2021, accessed 30.08.2024, https://konin.naszemiasto.pl/w-koninie-odbedzie-sie-marsz-z-milosci-do-wolnosci-beda/ar/c1-8160855.

52 Magdalena Rodecka, "W Stalowej Woli protest przeciwko restrykcjom związanym z koronawirusem", *Stalowka.net*, 10.10.2020, accessed 30.08.2024, https://www.stalowka.net/wiadomosci.php?dx=23777.

53 Tygodnik, "W Siedlcach będzie pikieta »Zakończyć Plandemię«", *Tygodnik Siedlecki*, 06.10.2020, accessed 30.08.2024, https://www.tygodniksiedlecki.com/siedlce/t58287-w-siedlcach-bedzie-pikieta-zakonczyc-plandemie-htm.

54 Bielsko TV, "Protest antyszczepionkowców w obronie lekarki", *YouTube*, 17.11.2021, accessed 30.08.2024, https://www.youtube.com/watch?v=ztU3WxpYgwQ.

55 Joanna Opas, "Protest Przeciw Przymusowi Szczepień".

56 Leśniewski, "Marsz antyszczepionkowców w Warszawie".

57 Anna Mierzyńska, "Przedsiębiorcy, bezmaskowcy i antyszczepionkowcy nawołują do antyrządowej rewolucji. Zaskakujący miks", *OKO.Press*, 11.05.2020, accessed 30.08.2024, https://oko.press/przedsiebiorcy-bezmaskowcy-i-antyszczepionkowcy-nawoluja-do-antyrzadowej-rewolucji-zaskakujacy-miks.

August 2020,[58] September 2020,[59] May 2021,[60] September 2021,[61] December 2021,[62] and January 2022,[63]) there were also protests related to the government's decisions regarding the fight against the pandemic, the introduction of lockdowns or other anti-pandemic restrictions.

The demonstrations reached their peak in January 2022, when, in a coordinated manner, protesters marched in cities such as Warsaw,[64] Kalisz,[65] Szczecin,[66] Wrocław,[67] Lublin,[68] Olsztyn,[69] Częstochowa,[70] Sieradz and Zduńska

---

58 Luiza Pastuszka, "Warszawa. Protest antyszczepionkowców. »Zakończyć pandemię«", *WP*, 16.08.2020, accessed 30.08.2024, https://wiadomosci.wp.pl/warszawa/warszawa-protest-antyszczepionkowcow-zakonczyc-pandemie-6543709232253568a.

59 Radosław Opas, "Protest antyszczepionkowców w Warszawie. »Nie pozwolimy na kolejny lockdown«", *WP*, 11.09.2020, accessed 30.08.2024, https://wiadomosci.wp.pl/protest-antyszczepionkowcow-w-warszawie-nie-pozwolimy-na-kolejny-lockdown-6553025609075616a.

60 Magdalena Patryła, "Protest antyszczepionkowców w Warszawie. Przepychanki z policją", *RMF 24*, 22.05.2021, accessed 30.08.2024, https://www.rmf24.pl/raporty/raport-koronawirus-z-chin/polska/news-protest-antyszczepionkowcow-w-warszawie-przepychanki-z-polic,nId,5248913.

61 Katarzyna Redmerska, "Antyszczepionkowcy protestowali przed Sejmem" *Prawo.pl*, 15.09.2021, accessed 30.08.2024, https://www.prawo.pl/zdrowie/antyszczepionkowcy-protestowali-przed-sejmem,510721.html.

62 Aml, "Warszawa. Protest antyszczepionkowców przed Sejmem. Przynieśli transparent nawiązujący do Auschwitz", *Polsat News*, 15.12.2021, accessed 30.08.2024, https://www.polsatnews.pl/wiadomosc/2021-12-15/warszawa-protest-antyszczepionkowcow-przed-sejmem-przyniesiono-transparent-nawiazujacy-do-obozow.

63 No author, "Protest antyszczepionkowców przed Sejmem. »Koronawirus mutuje w faszystowską dyktaturę«", *Dziennik.pl*, 05.01.2022, accessed 30.08.2024, https://gospodarka.dziennik.pl/news/galeria/8327958,protest-antyszczepionkowcy-sejmem.html.

64 Ibid.

65 Agnieszka Walczak, "Kaliski protest antyszczepionkowców z antyrządowymi hasłami. Narodowcy mówili o »bandziorach w rządzie«", *Wyborcza.pl*, 15.01.2022, accessed 30.08.2024, https://kalisz.wyborcza.pl/kalisz/7,181359,28005858,kaliski-protest-antyszczepionkowcow-z-antyrzadowymi-haslami.html.

66 Aleksander Stanuch, "Antyszczepionkowcy pod eskortą policji przeszli przez Szczecin, by głosić, że »pandemia to ściema«", *Wyborcza.pl*, 15.01.2022, accessed 30.08.2024, https://szczecin.wyborcza.pl/szczecin/7,34939,28004524,protest-antyszczepionkowcow-w-szczecinie-szczepionki-na-covid-19.html.

67 Koz, UJ, "Protest antyszczepionkowców we Wrocławiu. Z pl. Solnego musiał odjechać szczepciobus [ZDJĘCIA]", *Wyborcza.pl*, 15.01.2022, accessed 30.08.2024, https://wroclaw.wyborcza.pl/wroclaw/7,35771,28004636,atyszczepionkowcy-we-wroclawiu.html.

68 Anna Opasiak, "Lublin/Covid-19 – manifestacje przeciwników szczepień", *Radio Plus*, 22.01.2022, accessed 30.08.2024, https://www.radioplus.pl/lublin/lublin-covid-19-manifestacje-przeciwnikow-szczepien-aa-uVd4-aa7t-1cVG.html.

69 Redakcja, "»Stop przymusowi szczepień, biedzie i bezprawiu«. Protest w Olsztynie [ZDJĘCIA, WIDEO]", *Olsztyn.com.pl*, 23.01.2022, accessed 30.08.2024, https://www.olsz

Wola,[71] Łódź,[72] Chełm,[73] Gorzów Wielkopolski,[74] Tarnów,[75] Kalisz,[76] Koszalin[77] Elbląg,[78] Toruń,[79] Bydgoszcz,[80] Kielce[81] Rzeszów,[82] Legnica[83] and Kłodzko.[84] However, on May 16, 2022, the state of the epidemic was changed by the government to a state of epidemic threat, so the protests began to

tyn.com.pl/artykul,stop-przymusowi-szczepien-biedzie-i-bezprawiu-protest-w-olsztynie-zdjecia-wideo,34158.html.

70 Bs, "Antyszczepionkowcy w Częstochowie coraz bardziej aktywni. Demonstrowali na pl. Biegańskiego", *Wyborcza.pl*, 15.01.2022., accessed 30.08.2024, https://czestochowa.wyborcza.pl/czestochowa/7,48725,28005878,antyszczepionkowcy-w-czestochowie-stop-segregacji-i-dyskryminacji.html.

71 Kp, "Protesty w Sieradzu i Zduńskiej Woli", *Nasze.fm*, 15.01.2022, accessed 30.08.2024, https://www.nasze.fm/news,40015.

72 Anna Janiszewska, "Protest antyszczepionkowców na ulicy Piotrkowskiej. Zebrali się przy dawnej »Magdzie« ZDJĘCIA", *Express Ilustrowany*, 16.01.2022, accessed 30.08.2024, https://expressilustrowany.pl/protest-antyszczepionkowcow-na-ulicy-piotrkowskiej-zebrali-sie-przy-dawnej-magdzie-zdjecia/ar/c1-15999603.

73 Klaudia Szynkaruk, "»Stop segregacji sanitarnej«. Przeciwnicy przymusu szczepień spotkali się na placu Łuczkowskiego w Chełmie. Zobacz zdjęcia", *Kurier Lubelski*, 29.01.2022, accessed 30.08.2024, https://kurierlubelski.pl/stop-segregacji-sanitarnej-przeciwnicy-przymusu-szczepien-spotkali-sie-na-placu-luczkowskiego-w-chelmie-zobacz-zdjecia/ar/c14-16021867.

74 Mb, "Gorzów Wlk. Groźby wobec członków Rady Medycznej na wiecu antyszczepionkowców. Sprawą zajmuje się policja", *Gazeta.pl*, 16.01.2022, accessed 30.08.2024, https://wiadomosci.gazeta.pl/wiadomosci/7,173952,28007676,gorzow-wlk-grozby-pod-adresem-rady-medycznej-na-wiecu-antyszczepionkowcow.html.

75 No author, "Protest »Stop segregacji sanitarnej« [ZDJECIA]", *Tarnowska TV*, 15.01.2022, accessed 30.08.2024, https://www.tarnowska.tv/galeria/3381,protest-stop-segregacji-sanitarnej-zdjecia.

76 Walczak, "Kaliski protest antyszczepionkowców".

77 Krzysztof Marczyk, "Protest »Stop segregacji sanitarnej« w Koszalinie. Sprzeciwiają się przymusowi szczepień [ZDJĘCIA]", *GK24*.pl, 15.01.2022, accessed 30.08.2024, https://gk24.pl/protest-stop-segregacji-sanitarnej-w-koszalinie-sprzeciwiaja-sie-przymusowi-szczepien-zdjecia/ar/c1-15998779.

78 KJ, "Protest w Elblągu. »Wyłącz telewizor, włącz myślenie«, »Przestańcie szczepić, zacznijcie leczyć« (film i zdjęcia)", *Info.Elblag.pl*, 24.01.2022, accessed 30.08.2024, https://www.info.elblag.pl/37,68460,Protest-w-Elblagu-Wylacz-telewizor-wlacz-myslenie-Przestancie-szczepic-zacznijcie-leczyc-film-i-zdjecia.html.

79 Sławomir Łopatyński, "Pikieta antyszczepionkowców w Toruniu. Lekarz: – Pod respiratorami umiera 99 proc. niezaszczepionych", *Wyborcza.pl*, 15.01.2022, accessed 30.08.2024, https://torun.wyborcza.pl/torun/7,48723,28004722,rozaniec-pod-pomnikiem-kopernika-w-toruniu-intencja-wstrzymanie.html.

80 Zma, bas, "Bydgoszcz. Groźby śmierci na wiecu antyszczepionkowców. Poseł PO składa zawiadomienie do prokuratury", *Polsat News*, 31.01.2022, accessed 30.08.2024, https://www.polsatnews.pl/wiadomosc/2022-01-31/bydgoszcz-grozby-smierci-na-wiecu-antyszczepionkowcow-posel-po-sklada-zawiadomienie-do-prokuratury.

weaken. Their number decreased, and the number of participants decreased significantly.

It is worth noting that several new organizations and movement leaders emerged during the pandemic, whose main goal was to oppose the lockdown. They mainly brought together entrepreneurs, and, as a result, they differed from the previously existing groups of opponents of vaccinations, which primarily emphasized the alleged health risks that vaccinations were supposed to pose. Entrepreneurs emphasized the second important issue for the movement opposing mandatory vaccination, namely civil liberties especially those related to economics.

In 2020 and 2021, several protests by this social group emerged in Poland. As a result of the protesters' opinion, that the pandemic restrictions were inconsistently and unconstitutionally introduced, as well as an effect of the simple irritation of entrepreneurs with the very necessity of limiting or even closing down some businesses[85] (such as restaurants and bars, transport companies, shops, and other services), strikes broke out in many industries, and street blockades or demonstrations were organized. Farmers and animal breeders, who had already earlier expressed dissatisfaction on the streets, also joined the protests. During this period, organizations such as "the Entrepreneurs' Strike," led by Paweł Tanajno, and "Góralskie Veto!" led by Sebastian Pitoń, were established. Additionally, even before the pandemic, protests by farmers from AgroUnia and breeders from the Polish Fur Industry had been conducted.

---

81 Alina Janusz, "Protest przeciwników szczepień przeciwko COVID-19 w centrum Kielc. Mieli flagi, krzyczeli: »Leczyć, a nie szczepić«", *Wyborcza.pl*, 15.01.2022, accessed 30.08.2024, https://kielce.wyborcza.pl/kielce/7,47262,28005876,protest-przeciwnikow-szczepien-w-centrum-kielc-mieli-flagi.html.

82 Rafał Bolanowski, "Antyszczepionkowcy chcą obejść przepisy. Organizują dwie manifestacje, które dzieli długość płyty rynku", *Wyborcza.pl*, 14.01.2022, accessed 30.08.2024, https://rzeszow.wyborcza.pl/rzeszow/7,34962,27999153,antyszczepionkowcy-chca-obejsc-przepisy-organizuja-dwie-manifestacje.html.

83 Dawid Stefanik, "Antyszczepionkowcy przeciwko obostrzeniom epidemicznym [FOTO, WIDEO]", *TuLegnica.pl*, 23.01.2022, accessed 30.08.2024, https://tulegnica.pl/artykul/marsz-foliarzy-przez/1267167.

84 Joanna Żabska, "Protest przeciwko segregacji sanitarnej w Kłodzku", *24klodzko.pl*, 15.01.2022, accessed 30.08.2024, http://24klodzko.pl/protest-przeciwko-segregacji-sani tarnej-w-klodzku.

85 Dominika Maciejasz, "Przedsiębiorcy i Strajk Kobiet razem wychodzą na ulice. Lempart: Południe Polski zaczęło obywatelskie nieposłuszeństwo", *Wyborcza.pl*, 19.01.2021, accessed 30.08.2024, https://wyborcza.biz/biznes/7,177151,26700795,przedsiebiorcy-i-pioruny-wyjda-na-ulice-marta-lempart-poludnie.html.

Protest groups associating entrepreneurs demanded the lifting of the lockdown and the reduction or lifting of anti-pandemic restrictions.[86] As justification, their representatives used not only arguments related to economic issues but also elements of anti-scientific conspiracy narratives, downplaying the threat of coronavirus infection, negating the effectiveness of medical masks in stopping virus transmission, or casting doubt on the very occurrence of the pandemic.[87] Pitoń demanded the lifting of the lockdown, lowering taxes, and calling on the Highlanders to reopen their businesses but also declared that there is no pandemic and that coronavirus infection itself is "light" and "pleasant".[88] He also said, "The narrative about the pandemic is a socio-technical operation that aims to pull the wool over our eyes and move on to the next phase, which will end with ownership changes".[89] "There is no other way out than to say enough, veto. We count on the opening up of not only Podhale but also Poland as a whole. I think it will be an avalanche effect,"[90] the initiator of Góralski Veto argued at the time. Indeed, in response to the actions of the Podhale people, a similar initiative – Bałtyckie Veto was created in seaside towns. It had the same demands – opening companies closed by the lockdown and lifting the state of the pandemic.[91] Michał Kołodziejczak, the leader of the farmers' protests, also appeared in Zakopane and expressed solidarity with the Góralski Veto fighting the lockdown.[92]

86 Monika Waluś, "Pensjonaty, karczmy, ośrodki narciarskie z Podhala na skraju bankructwa. »Traktują nas jak piąte koło u wozu«", *Wyborcza.pl*, 12.01.2021, accessed 30.08.2024, https://krakow.wyborcza.pl/krakow/7,44425,26678200,pensjonaty-karczmy-osrodki-narciarskie-z-podhala-na-skraju.html.

87 Zdr, ml, "Inicjator Góralskiego Veta Sebastian Pitoń: chorowałem na koronawirusa. To lekka, przyjemna choroba", *Polsat News*, 12.01.2021, accessed 30.08.2024, https://www.polsatnews.pl/wiadomosc/2021-01-12/inicjator-goralskiego-veta-sebastian-piton-chorowalem-na-koronawirusa-to-lekka-przyjemna-choroba.

88 Pmd, "Kim jest Sebastian Pitoń, inicjator »Góralskiego veta«? »Pandemii nie ma, diagnozy lipne«", *Onet*, 12.01.2021, accessed 30.08.2024, https://wiadomosci.onet.pl/krakow/sebastian-piton-goralskie-veto-kim-jest/t4fre4l.

89 Ibid.

90 Szymon Bafia, "Tłumy na konferencji Góralskiego Veta w Zakopanem: jesteśmy strażnikami porządku prawnego", *Onet*, 14.01.2021, accessed 30.08.2024, https://podroze.onet.pl/aktualnosci/tlumy-na-konferencji-goralskiego-veta-w-zakopanem-jestesmy-straznikami-porzadku/3znd3pe.

91 Michał Tokarczyk, "»W portfelu mam 200 zł, a długu już 300 tysięcy«. Bałtyckie Veto, czyli bunt nadmorskich przedsiębiorców", *Wyborcza.pl*, 19.01.2021, accessed 30.08.2024, https://trojmiasto.wyborcza.pl/trojmiasto/7,35612,26698206,rozzaleni-zadluzeni-i-zdesperowani-przedsiebiorcy-znad-morza.html.

92 Monika Waluś, "Górale otwierają 18 stycznia swoje biznesy. »Panie prezydencie, zapraszamy na narty!«", *Wyborcza.pl*, 14.01.2021, accessed 30.08.2024, https://krakow.wy

Additionally, in the context of Russia, the leader of "Góralskie Veto!" Sebastian Pitoń has also repeatedly promoted geopolitical narratives consistent with those disseminated by politicians and institutions of the Russian Federation. Even before the pandemic, in 2017, Pitoń was a panellist at the conference "What Poland in the Age of Asia?" organized by socio-cybernetics from the school of Józef Kossecki, a former ideologist of the national-communist Grunwald Patriotic Union. Pitoń appeared there as a YouTuber in a panel with the monarchist Adam Wielomski, who sympathizes with Russia. The event was opened by Jan Engelgard, editor-in-chief of the weekly "Myśl Polska," which is crucial for promoting Russian geopolitical narratives.[93] In time, Pitoń began writing for this weekly.[94] In subsequent "historical conferences," Pitoń appeared with Grzegorz Braun, who is critical for promoting Russian narratives in Poland, Bartosz Bekier from the pro-Russian terrorist group Falanga, and again with Engelgard and Wielomski. The Bishop Kajetan Sołtyk Education and Development Center published post-conference books with the participants' speeches being run by the pro-Russian activist and former neo-fascist Rafał Mossakowski. Pitoń was also a frequent guest in video materials made by pro-Russian journalists from YouTube channels – Eugeniusz Sendecki from National Television and Piotr Korczarowski from eMisja.tv,[95] as well as Marcin Rola's wRealu24.pl programs.[96] He announced in this programmes that the USA incites conflicts and wars to control the world and that the Americans occupy Poland, whose task is to destabilize the Eurasian continent so that it does not unite. Hitler and Stalin were US agents, and on their orders, they unleashed the war so that the USSR, Germany, Britain, France, and Japan would not unite and the USA could become a world hegemon. On the orders of the USA, PiS is turning Poland into Judeopolonia, and Jews will eventually place Poles in reservations.[97] Russia was forced to intervene in

borcza.pl/krakow/7,44425,26685711,gorale-otwieraja-18-stycznia-swoje-biznesy-panie-prezydencie.html.

93 NAI Warszawa, "Jaka Polska w wieku Azji? – konferencja", *Evenea.pl*, no data, accessed 30.08.2024, https://app.evenea.pl/event/nai?source=youtube.Wsp.

94 Sebastian Pitoń, "Pitoń: Prawybory niedemokratyczne", *Myśl Polska*, no data, accessed 30.08.2024, https://myslpolska.info/2023/03/28/piton-prawybory-niedemokratyczne.

95 EMisja.tv, "Kto tu kogo tak naprawdę opłaca?", *YouTube*, 23.03.2023, accessed 30.08.2024, https://www.youtube.com/watch?v=3GvXfbpS2EI.

96 Ośrodek Monitorowania Zachowań Rasistowskich i Ksenofobicznych, "OTO DLACZEGO WREALU24 ZOSTAŁO ZABLOKOWANE NA YOUTUBE", *Facebook*, 26.09.2022, accessed 30.08.2024, https://www.facebook.com/watch/?v=496862065281497.

97 Jerzy Targalski, "Tworki atakują", *Józef Darski*, no data, accessed 30.08.2024, https://jozefdarski.pl/13612-tworki-atakuja.

Ukraine by the United States, and he would help Putin in "denazification" and build a 10-meter wall on the border with Ukraine.[98] He also considers the Bucha massacre to be a Ukrainian re-enactment, declares that NATO troops should be expelled from Poland, and considers accepting war refugees to be a displacement action.[99] Pitoń also began to engage in active party politics. In 2019, he supported Konfederacja in the European Parliament and Sejm elections. In 2020 he supported Konfederacja leader (Krzysztof Bosak) in his run for the President of Poland office,[100] and in 2021, he supported Braun, in his run for the mayoralty of the city of Rzeszów.[101] In 2023, with another of Braun's associates, the pro-Russian self-proclaimed geopolitician Leszek Sykulski, he began to create the Polish Anti-War Movement.

Protesters against introducing 5G telephony in Poland and mandatory vaccinations didn't limit themselves to legal actions and civil society means. Both movements used criminal threats and acts of vandalism, as well as physical attacks on both objects and people. This radicalization is not a Polish specificity for these movements. In other Western countries, activists involved in both issues took a violent dimension. There were attacks on base stations, which occurred repeatedly in Great Britain. By the end of 2020, over 90 telecommunications masts had been vandalized there[102]), and 39 BT (British Telecom) installers were physically or verbally attacked by people who believed the 5G network caused the coronavirus. There were thirty attempts to set fire to mobile phone masts in the Netherlands and seventeen in France.

98 Nigdy Więcej Szurów, "Sebastian Pitoń: Wybudowałbym mur na granicy z Ukrainą i pomógł Putinowi", *YouTube*, 25.09.2022, accessed 30.08.2024, https://www.youtube.com/watch?v=yGyMTVA_G98.

99 Anna Mierzyńska, "Uwaga na niby-pacyfistyczną narrację! Korzysta na niej Kreml, w Polsce szerzy ją Sykulski", *OKO.Press*, 11.02.2023,accessed 30.08.2024, https://oko.press/uwaga-na-niby-pacyfistyczna-narracje-sykulski.

100 FA, MNIE, "Popiera Konfederację, chce stworzyć Królestwo Podhalańskie. Kim jest Sebastian Pitoń?", *TVP.info*, 12.01.2021, accessed 30.08.2024, https://web.archive.org/web/20210127042215/https://www.tvp.info/51760945/sebastian-piton-kim-jest-popiera-konfederacje-chce-stworzyc-krolestwo-podhalanskie-inicjator-akcji-goralskie-veto-tvp-info.

101 Janusz Jaskółka, "Góralskie Veto i Sebastian Pitoń wspierają Grzegorza Brauna.", *YouTube*, 08.06.2021, za: accessed 30.08.2024, https://www.youtube.com/watch?v=vfXlRNDRqzo.

102 Milo Boyd, "At least 90 phone masts attacked during UK coronavirus lockdown after 5G conspiracy", *Mirror*, 26.05.2020, accessed 30.08.2024, https://www.mirror.co.uk/news/uk-news/least-90-phone-masts-attacked-22086169.

Similar cases also occurred in Ireland, Belgium,[103] Italy,[104] Germany,[105] Cyprus,[106] and Sweden.[107] In the latter, a government official responsible for telecommunications also received death threats.[108] As of May 18, 2020, there had been as many as one hundred and forty-two cases of arson attempted physical destruction, including attacks with saws and axes and dismantling of telecommunications towers and transmitters in Europe. Several telecommunications masts were also destroyed in Bolivia by people who feared that 5G technology could transmit the coronavirus, even though the South American country did not yet have this generation of wireless infrastructure. Four communication towers were destroyed in Yapacani, near Santa Cruz, and others in San Julian and Ichilo.[109]

Attacks on telecommunications infrastructure have also occurred in Poland. In October 2019, one of the leaders of the anti-5G movement, Zbigniew Gelzok, called for physical action, claiming that destroying transmitters would not be a crime but an action "to avert an immediate danger threatening any legally protected good."[110] The first attack took place shortly afterwards. In Kraków, a group of opponents of the new generation network tried to destroy a base station belonging to T-Mobile Polska. The leader of the attackers, Sławomir Dul, was arrested at the time. In May 2020, a telecommunications base station in Łódź was set on fire with Molotov cocktails – two masts of mobile operators, Playa and T-Mobile, suffered damage.[111] In the same month, an at-

---

103 Laurens Cerulus, "5G arsonists turn up in continental Europe", *Politico*, 26.04.2020, accessed 30.08.2024, https://www.politico.com/news/2020/04/26/5g–mast–torchers–turn–up–in–continental–europe–210736.

104 Ibid.

105 Elisabeth Braw, "The Imagined Threats of 5G Conspiracy Theorists Are Causing Real–World Harm", *Foreign Policy,* 29.06.2020, accessed 30.08.2024, https://foreignpolicy.com/2020/06/29/the–imagined–threats–of–5g–conspiracy–theorists–are–causing–real–world–harm.

106 Cerulus, "5G arsonists turn up".

107 Ibid.

108 Braw, "The Imagined Threats of 5G".

109 AFP, "Bolivia Villagers Destroy 5G Masts Over Virus Fears", *Barron's*, 16.06.20, accessed 30.08.2024, https://www.barrons.com/news/bolivia–villagers–destroy–5g–masts–over–virus–fears–01592301304.

110 Mateusz Nowak, "Bojówka >>Anty 5G<< złapana ze szlifierką kątową w ręce. Próbowano uszkodzić krakowską stację bazową", *Spider's Web*, 04.10.2019, accessed 30.08.2024, https://spidersweb.pl/2019/10/anty–5g–niszczenie–nadajnikow.html.

111 He leads the group "Ecological Social Action – 5G – Let the Masts Fall," whose goal is to remove newer-generation telephone towers and base stations from Poland and calls for their destruction. Formerly involved in the activities of the association defending fathers' rights "Brave dad" ("Dzielny tata"), whose leaders were associated with the far-right

tack took place in the Lublin region. In Chełm, police detained a man who had cut metal elements of the structure of a telecommunication tower. Losses were estimated at PLN 45,000.[112]

A similar level of radicalization can be observed in the movement opposing mandatory vaccinations. This process was also not limited to its Polish branch. In December 2021, one of the leaders of the English anti-5G and anti-vaccination movement, Piers Corbyn (and brother of the former Labour Party leader), called for MP s' offices to be set on fire and those who introduce new fascism to be "beaten to death," referring to the anti-pandemic restrictions introduced by the British government.[113] In Germany, the secret services foiled a plot to assassinate Michael Kretschmer, the pro-vaccine governor of Saxony. In Italy, Antonella Viola, an immunologist, had to be placed under police protection after receiving a bullet and a death threat in a letter. In the Netherlands, the home of Deputy Prime Minister Sigrid Kaag was attacked by an anti-vaccination activist armed with a torch. In the UK, an anti-vaccination group, Alpha Men Assemble (AMA), planned attacks on the police, vaccination centres and schools.[114] Anti-vaccination activists were also an essential element of the Citizens of the Reich movement, which planned a coup in Germany. The so-called Querdenker movement, which emerged during the COVID-19 pandemic in Germany and whose key demand was to reject the government's actions against the pandemic, also cooperated closely with the Citizens of the Reich.[115]

Confederation and Sovereign Poland parties, over time he underwent a process of radicalization. He calls for street protests, passive resistance to the police, as well as rebellion against the USA and protested near the US Army barracks in Bolesławiec. Dul claims that the US is using its military to sell us 5G equipment. That is why he organized a protest in front of a military unit in which a dozen or so people took part.

112 Tw, "W Łodzi podpalono maszt telekomunikacyjny. Play: to przestępstwo, będzie zgłaszane organom ścigania", *Wirtualne Media*, 28.05.2020, accessed 30.08.2024, https://www.wirtualnemedia.pl/artykul/w-lodzi-podpalono-maszt-telekomunikacyjny-play-to-przestepstwo-bedzie-zglaszane-organom-scigania.

113 Maya Wolfe-Robinson, "Piers Corbyn arrested on suspicion of calling for MP s' offices to be burned down", *The Guardian*, 19.12.2021, accessed 30.08.2024, https://www.theguardian.com/uk-news/2021/dec/19/piers-corbyn-arrested-on-suspicion-of-calling-for-mps-offices-to-be-burned-down.

114 Ahmed Aboudouh, "»Unparalleled threats«: Anti-vaxxer movement threatens a new wave of extremism", *Independent*, 19.01.2022, accessed 30.08.2024, https://www.independent.co.uk/news/world/europe/antivaxxer-extremism-violence-plots-b1991710.html.

115 Sam Denney, "A Foiled Coup Attempt in Germany and the Danger of Conspiracy Theories", *Lawfare*, 12.12.2022, accessed 30.08.2024, https://www.lawfaremedia.org/article/foiled-coup-attempt-germany-and-danger-conspiracy-theories.

Polish opponents of mandatory vaccinations and government restrictions also demonstrated a similar level of radicalization. Since the outbreak of the pandemic, a whole spectrum of violent behaviour could been observed, starting from aggressive demonstrations and acts of vandalism through attempts at intimidation and the use of punishable threats, including those related to death, to the announcement of plans to organize terrorist attacks.

Anti-vaccination graffiti was made by unknown perpetrators, among others, on the town hall in Sędziszów Małopolski,[116] on the building of the Supreme Medical Chamber in the Beskid Mountains[117] and on the fence of the property of the Civic Coalition MP Dorota Niedziela ("Here lives a traitor to Poland" and "Stop sanitary segregation and vaccinations").[118] There were also more overt actions. In August 2021, a group of a dozen or so opponents of mandatory vaccinations surrounded a bus standing on a boulevard in Gdynia intended for conducting COVID-19 vaccinations. They shouted, among other things: "You are murderers" and "You are children of Dr. Mengele."[119]

Opponents of mandatory vaccinations also published a "blacklist" of doctors and politicians supporting the idea of compulsory vaccination against coronavirus, which caused a wave of hate against such people on the Internet.[120] However, the aggressive criticism did not end with online activity. Protests of opponents of vaccinations also took place in front of the residences of local government officials: the president of Białystok Tadeusz Truskolaski, the mayor of Goniadz Grzegorz Dudkiewicz, the mayor of Grajewo Dariusz Latarowski, and the director of the Augustów school Andrzej Kaszkiel.[121]

---

116 SSŃ, "Antyszczepionkowcy zaatakowali. Sędziszowski ratusz zdewastowany", *O2*, 19.01.2022, accessed 30.08.2024, https://www.o2.pl/informacje/antyszczepionkowcy-zaatakowali-sedziszowski-ratusz-zdewastowany-6728119703563232a.

117 ZKM, WIL, "Zdewastowana elewacja Beskidzkiej Izby Lekarskiej. Prezesi Okręgowych Izb Lekarskich wystosowali wspólne stanowisko", *Wielkopolska Izba Lekarska*, 08.01.2022, accessed 30.08.2024, https://wil.org.pl/zdewastowana-elewacja-beskidzkiej-izby-lekarskiej-prezesi-okregowych-izb-lekarskich-wystosowali-wspolne-stanowisko.

118 Blik, "Kto zniszczył płot u posłanki PO? Napis »tu mieszka zdrajca«", *Rzeczpospolita*, 18.12.2021, accessed 30.08.2024, https://www.rp.pl/przestepczosc/art19210321-kto-zniszczyl-plot-u-poslanki-po-napis-tu-mieszka-zdrajca.

119 MAK, pp, gg, "Antyszczepionkowcy otoczyli mobilny punkt szczepień. Medycy wezwali policję, będzie 15 wniosków do sądu", *TVN24*, 01.08.2021, accessed 30.08.2024, https://tvn24.pl/pomorze/koronawirus-w-polsce-gdynia-antyszczepionkowcy-otoczyli-szczepibus-5163021.

120 Mierzyńska, "Antyszczepionkowcy publikują »czarną listę donosicieli«"

121 Anna Mierzyńska, "Przeciwnicy szczepień grożą, straszą, oskarżają. Są zdeterminowani i eskalują działania", *OKO.Press*, 26.01.2022, accessed 30.08.2024, https://oko.press/przeciwnicy-szczepien-groza-strasza-oskarzaja-sa-zdeterminowani-i-eskaluja-dzialania.

Self-proclaimed "tribunals of justice" were also established (one in Jelenia Góra,[122] another in Podlasie region[123]), which issued "sentences" against doctors and politicians who supported mandatory vaccinations.[124] These included death sentences[125] for "promoting sanitary segregation, exploiting public property, forcing residents to act illegally or participating in an organized armed criminal group."[126]

In Gorzów Wielkopolski in January 2021, during a demonstration one of the speakers spoke of a "bullet in the head" for members of the Medical Council of the Prime Minister.[127] Other opponents of vaccinations threatened with death doctors supporting the idea of protective vaccinations against COVID-19: Krzysztof Simon,[128] Tomasz Karauda,[129] and Bartosz Fiałek.[130] Death threats were also sent to the president of Wałbrzych,[131] the mayor of Mikołów[132] and

122 Koz, "Antyszczepionkowcy mają swój Trybunał Narodowy i rozporządzenie: »Nie ma obowiązku wstrzykiwania do ciała obcego ciała«", *Wyborcza.pl*, 01.02.2022, accessed 30.08.2024, https://wroclaw.wyborcza.pl/wroclaw/7,35771,28058951,antyszczepionkowcy-maja-trybunal-narodowy-i-rozporzadzenie.html.

123 Mierzyńska, "Przeciwnicy szczepień grożą, straszą, oskarżają".

124 Koz, "Antyszczepionkowcy mają swój Trybunał Narodowy".

125 Maciej Chołodowski, "Podlascy antyszczepionkowcy wydawali »wyroki śmierci«. Teraz zostali skazani przez sąd", *Wyborcza.pl*, 20.02.2024, accessed 30.08.2024, https://bialystok.wyborcza.pl/bialystok/7,35241,30718167,podlascy-antyszczepionkowcy-wydawali-wyroki-smierci-teraz.html.

126 Mierzyńska, "Przeciwnicy szczepień grożą, straszą, oskarżają".

127 Mb, "Gorzów Wlk. Groźby wobec członków Rady Medycznej".

128 EW, "Prof. Krzysztof Simon dostał policyjną ochronę po tym, jak antyszczepionkowcy grozili mu śmiercią", *Wyborcza.pl*, 05.10.2021, accessed 30.08.2024, https://wroclaw.wyborcza.pl/wroclaw/7,35771,27650644,prof-krzysztof-simon-dostal-policyjna-ochrone-antyszczepionkowcy.html.

129 Agnieszka Urazińska, "Antyszczepionkowcy nie będą bezkarni? Prokuratura wszczęła śledztwo, atakowany lekarz dostał ochronę", *Wyborcza.pl*, 17.08.2021, accessed 30.08.2024, https://lodz.wyborcza.pl/lodz/7,35136,27460523,antyszczepionkowcy-nie-beda-bezkarni-atakowany-lekarz-ma-ochrone.html.

130 Marta Woźniak, "Prokuratura zajmie się hejterem dr. Bartosza Fiałka. Lekarze chcą, by sprawców ścigano z urzędu", *Wyborcza.pl*, 20.01.2022, accessed 30.08.2024, https://warszawa.wyborcza.pl/warszawa/7,54420,28016999,prokuratura-zajmie-sie-hejterem-dr-bartosza-fialka-lekarze.html.

131 Mateusz Kokoszkiewicz, "Policja przyznała ochronę prezydentowi Wałbrzycha. Po hejcie i groźbach od antyszczepionkowców", *Wyborcza.pl*, 06.05.2021, accessed 30.08.2024, https://walbrzych.wyborcza.pl/walbrzych/7,178336,27056710,policja-przyznala-ochrone-prezydentowi-walbrzycha-po-hejcie.html.

132 RED, "Antyszczepionkowcy grożą śmiercią burmistrzowi Mikołowa. »Rewolucyjna Armia Polaków wydała na ciebie wyrok«", *Gazeta Lubuska*, 14.01.2022, accessed 30.08.2024, https://gazetalubuska.pl/antyszczepionkowcy-groza-smiercia-burmistr

Wągrowiec.[133] An opponent of mandatory vaccinations physically attacked the spokesman for the Ministry of Health, Wojciech Andrusiewicz. The head of the Polish People's Party, Władysław Kosiniak-Kamysz, was attacked on his way to a restaurant in Karpacz, and the attacker threatened the chairman of the Polish People's Party that an attack would be carried out on him.[134] There was also a physical confrontation after opponents of vaccinations entered the office of Grzegorz Cessak, president of the Office for Registration of Medicinal Products, Medical Devices and Biocidal Products.[135] The headquarter of the Supreme Medical Chamber was also invaded by opponents of mandatory vaccinations, who behaved aggressively towards judges of the medical court and destroyed the property of the National Medical Chamber.[136] Another raid took place in July 2021, when a group of protesters broke into a vaccination point in Grodzisk Mazowiecki, trying to prevent there forcibly vaccinations. A fight also broke out there.[137] Opponents of mandatory vaccinations also threatened to blow up the Malta Baths in Poznań[138] and set fire to the sanitary-epidemiological station building and the vaccination point in Zamość.[139]

---

zowi–mikolowa–rewolucyjna–armia–polakow–wydala–na–ciebie–wyrok/ar/c14–159 96143.

133 Redakcja e–grajewo.pl, "Antyszczepionkowcy grożą burmistrzowi Wągrowca", *E–grajewo.pl*, 25.11.2021, accessed 30.08.2024, https://e–grajewo.pl/wiadomosc,antyszcze pionkowcy–groza–burmistrzowi–wagrowca,47621.html.

134 No author, "Prezes ważnego urzędu w szoku. »Poseł Braun i antyszczepionkowcy siłą wtargnęli do mojego gabinetu«", *Fakt*, 03.08.2021, accessed 30.08.2024, https://www.fakt .pl/polityka/grzegorz–braun–i–antyszczepionkowcy–wtargneli–do–grzegorza–cessa–sz efa–urzedu/1djdwpw.

135 No author, "Prezes ważnego urzędu w szoku. »Poseł Braun i antyszczepionkowcy siłą wtargnęli do mojego gabinetu«", *Fakt*, 03.08.2021, accessed 30.08.2024, https://www.fakt .pl/polityka/grzegorz–braun–i–antyszczepionkowcy–wtargneli–do–grzegorza–cessa–sz efa–urzedu/1djdwpw.

136 Redakcja Dentonet, "Wtargnięcie antyszczepionkowców do siedziby Naczelnej Izby Lekarskiej", *Dentonet*, 22.02.2021, accessed 30.08.2024, https://dentonet.pl/wtargniecie –do–siedziby–nil/#gref.

137 Pp, ads, katke, rzw, r, "Szarpanina przed punktem szczepień w Grodzisku Mazowieckim. Dwie osoby zatrzymane przez policję", *TVN24*, 25.07.2021, accessed 30.08.2024, https:// tvn24.pl/polska/grodzisk–mazowiecki–antyszczepionkowcy–wywolali–szarpanine–prz ed–punktem–szczepien–interweniowala–policja–5157297.

138 Tokfm.pl, "Antyszczepionkowcy grozili, że wysadzą Termy Maltańskie. Ewakuowano prawie 2 tysiące osób", *TOK FM*, 09.08.2021, accessed 30.08.2024,https://www.tokfm.pl/ Tokfm/7,103085,27432183,antyszczepionkowcy–grozili–ze–wysadza–termy–maltanskie –ewakuowano.html.

139 PP, adso, "Premier: W Zamościu doszło do bandyckiego aktu. Podpalono sanepid i punkt szczepień", *TVN24*, 02.08.2021, accessed 30.08.2024, https://tvn24.pl/polska/zamosc–pod

A particularly active group, prone to breaking the law in the context of anti-pandemic restrictions or vaccinations, as well as to acts of violence, were supporters of Wojciech Olszański, the so-called Compatriots Comrades. It is worth noting here that this was again a group promoting narratives consistent with Russian ones, whose representatives repeatedly spoke very positively about Russia and Vladimir Putin. The group's leader, actor and director Wojciech Olszański, was initially a marginal participant in the activities of niche nationalist groups. In the 2000s, he underwent a rapid process of radicalization. He became involved in protests against the American search for shale gas in Żurawlów[140] and in the defence of the so called "Smolensk cross"[141] standing in 2010 in front of the Presidential Palace, where he appealed for reconciliation with the Russians. He also protested against the refusal to let a group of nationalist Russian motorcyclists, the Night Wolves, into Poland. His subsequent appearances in nationalist media have featured consistent themes: pro-Russianism, anti-Americanism, a vision of a bloody war in Central European countries allegedly planned by the Americans, and threats of a significant defeat for Poland if it does not change its alliances.[142] Polish political reality is supposedly governed by a Jewish conspiracy, manifested in the behind-the-scenes influence of Zionist plutocrats and attacks by "Jewish communists" on conservative and national values.[143]

Olszański has become known for his numerous vulgar insults and criminal threats against politicians and publicists. He called Marcelina Zawisza, a New Left MP, a "whore."[144] In his opinion, Donald Tusk, Rafał Trzaskowski, and

---

palony–sanepid–i–punkt–szczepien–premier–mateusz–morawiecki–bandycki–akt–st 5163479.

140 EMisja.tv, "Jabłonowski: »Niezwykli Polacy pokonali chciwy koncern«", *YouTube*, 28.05.2016, accessed 30.08.2024, https://www.youtube.com/watch?v=KoPolVoVerM.

141 A wooden cross placed in front of the Presidential Palace in Warsaw by activists of scout organizations on April 15, 2010, 5 days after the Smolensk disaster, to commemorate its 96 victims.

142 EMisja.tv, "»Amerykanie chcą walczyć z Rosją w Polsce«", *YouTube*, 22.04.2016, accessed 30.08.2024, https://www.youtube.com/watch?v=CVjr7yaiVCw.

143 Piotr Żytnicki, "Policja zatrzymała trzy osoby po antysemickim skandalu w Kaliszu. Wśród nich Piotr Rybak", *Wyborcza.pl*, 15.11.2021, accessed 30.08.2024, https://kalisz.wyborcza.pl/kalisz/7,181359,27806550,antysemicki–skandal–w–kaliszu–policja–zatrzymala–trzy–osoby.html.

144 Mariusz Sepioło, Daniel Flis, "Ludwiczek równa krok", *Frontstory*, 13.04.2023, accessed 30.08.2024, https://frontstory.pl/osadowski–radykalizmy–prawica–kamraci.

Szymon Hołownia are "insects," "whores," "traitors," and "pimps".[145] In his opinion, Klaudia Jachira, Greens MP, "has something about her that makes her the perfect specimen for rape."[146] He also wished that "a bomb should have been thrown at the liberal Campus Polska" festival[147] and a civil war front should have been opened in Poland.[148] He also suggested that football hooligans should "take care" of the Ukrainian-Polish journalist Igor Isayev.[149]

Olszański's great sympathy for the authorities of Russia and Belarus is noticeable. Olszański praises the pro-Russian Targowica Confederation[150] and claims that "the Polish national interest is inextricably linked to Russia, which is the guarantor of peace."[151] In his opinion, Crimea "was and always will be Russian,"[152] and in a conversation with the ambassador of the Russian Federation, he praised the services of pro-Soviet nationalist Bolesław Piasecki and NKWD general Ivan Serov for Polish–Russian friendship.[153] He also praised the First Secretary of the Central Committee of the Polish United Workers' Party, Władysław Gomułka, the dictator of the USSR, Joseph Stalin, and the leader of North Korea, Kim Jong-un. In January 2019, he also partici-

---

145 Ośrodek Monitorowania Zachowań Rasistowskich i Ksenofobicznych, "TYLKO WYSZEDŁ Z ARESZTU I JUŻ WZYWA DO MORDOWANIA KOBIET" *Facebook*, 01.09.2022, accessed 30.08.2024, https://www.facebook.com/watch/?v=1998688476993339.

146 Jakub Woroncow, "Prawicowy patostreamer fantazjuje o gwałceniu posłanki. Nie jest wybrykiem, jest problemem", *OKO.Press*, 14.12.2022, accessed 30.08.2024, https://oko.press/prawicowy-patostreamer-fantazjuje-o-gwalceniu-poslanki-nie-jest-wybrykiem-jest-problemem.

147 Ośrodek Monitorowania Zachowań Rasistowskich i Ksenofobicznych, "TYLKO WYSZEDŁ Z ARESZTU I JUŻ WZYWA DO MORDOWANIA KOBIET" *Facebook*, 01.09.2022, accessed 30.08.2024, https://www.facebook.com/watch/?v=1998688476993339.

148 Ibid.

149 Oktawia Kromer, "Rodacy Kamraci grożą śmiercią Igorowi Isajewowi. »Lepszy byłby dla ciebie sznur i gałąź pod ciężarem zgięta«", *Wyborcza.pl*, 04.08.2022, accessed 30.08.2024, https://warszawa.wyborcza.pl/warszawa/7,54420,28754692,olszanski-i-osadowski-namawiaja-do-bicia-ukraincow-i-groza-smiercia.html.

150 Targowica Confederation – a conspiracy of magnates formed on the night of 18 to 19 May 1792 in Targowica (in reality on 27 April 1792 in St. Petersburg), at the request and under the patronage of the Empress of Russia Catherine II, directed against the reforms of the Four-Year Sejm and the Constitution of 3 May, considered a symbol of national treason

151 Grzegorz Rzeczkowski, *Katastrofa posmoleńska. Kto rozbił Polskę* (Warszawa: Tarcza, 2019), 85.

152 Woroncow, "Prawicowy patostreamer fantazjuje o gwałceniu".

153 Suwerenna Myśl, "Wojciech Olszański vel. A. Jabłonowski wyraża wdzięczność ambasadorowi Rosji za ... Iwana Sierowa!", *YouTube*, 02.03.2019, accessed 30.08.2024, https://www.youtube.com/watch?v=SqmFb7Ut5zM&feature=youtu.be.

pated in a protest "in support of Polish–Iranian friendship"[154] and rallied supporting the dictator of Belarus, Alexander Lukashenka.[155] In May 2022, he led a march of eight thousand of his supporters in Warsaw in uniform on the horseback with a Russian banner. He called the Ukrainian flag "Bandera shit" and called for its desecration. He already has four convictions (for criminal threats and inciting hatred on ethnic, racial, and religious grounds), two of which are legally binding. Further trials are ongoing in Warsaw, Kielce, and Bydgoszcz.

An essential element influencing his radicalization was the pandemic. Thanks to it, his political movement, Compatriots Comrades, has been created. It is a network of informal autonomous local nationalist, paramilitary, and anti-vaccination groups whose members became politically active primarily during the COVID-19 pandemic, united by the cult of their leader – Wojciech Olszański. They have the slogan "Death to the Enemies of the Homeland" embroidered on their uniforms. In January 2022, in Bydgoszcz Olszański protested with his people against the introduction of further anti-COVID restrictions and shouted about MP s debating the introduction of mandatory COVID-19 vaccinations: "This is a club for a parliamentary mug. If they vote as they want on February 1 and pass this law, if they do it, they are condemning themselves to death, and we are creating these death lists. We are warning them. If they vote yes, then I, Wojciech Olszański, known on the internet as Aleksander Jabłonowski, want to kill them! And why not? I am filled with joyful hatred, and I want them dead!".[156] He also announced, during one of his broadcasts, that if someone wanted to force him to get vaccinated, he would kill them personally.[157] He also emphasized that outside Warsaw, residents know where MP s live and that during the elections, "we will surprise them".[158] His supporters carried out such raids several times. However, their main target was medical facilities. In July 2021, his uniformed supporters broke into the orphanage in Aleksandrów Kujawski and the "Sanitas" clinic in Bydgoszcz because – in their opinion – the facility was conducting "medical

154 Woroncow, "Prawicowy patostreamer fantazjuje o gwałceniu".

155 Marcin Łuniewski, "Polscy przyjaciele Aleksandra Łukaszenki", *Rzeczpospolita*, 19.09 .2020, accessed 30.08.2024, https://www.rp.pl/Plus-Minus/309119989-Polscy-przyjaciele-Aleksandra-Lukaszenki.html.

156 Dawid Karpiuk, "»Kamraci! To jest kij na poselski ryj!« – krzyczy, wymachując pałką. Tłum zaraz chwyta", *Newsweek*, 03.03.2022, accessed 30.08.2024, https://www.newsweek.pl/polska/spoleczenstwo/antyszczepionkowcy-i-radykalna-prawica-ida-reka-w-reke-dlaczego/23l9v2z.

157 Woroncow, "Prawicowy patostreamer fantazjuje o gwałceniu".

158 Mierzyńska, "Przeciwnicy szczepień grożą, straszą, oskarżają".

experiments on children aged six months to 12 years".[159] When they did not find the management at the facility, they left a written demand to "discontinue the experiment of administering medical preparations of unknown effect to depopulate the Polish Nation".[160]

The activities of radicalized opponents of mandatory vaccinations and anti-pandemic restrictions drew the attention of Polish counterintelligence. In January 2024, Internal Security Agency officers searched the apartments of activists associated with nationalist organizations fighting mandatory vaccinations, such as Compatriots Comrades and "We Defend the Uniform for Future Generations" and the Konfederacja party. The apartment of legal counsel Katarzyna Tarnawa-Gwóźdź, who represented many people associated with the anti-vaccination movement and was an active participant herself, was also searched.[161] According to the resolution of the District Prosecutor's Office in Warsaw, the justification for the search included accusations of preparing for a coup d'état, cooperation with foreign intelligence, inciting people to refuse to follow orders, and inciting murder.[162]

---

159 Mierzyńska, "Rosyjski ślad".

160 Mierzyńska, "Przeciwnicy szczepień grożą, straszą, oskarżają".

161 No author, "Jak można wprowadzić obowiązek szczepień przeciw COVID–19?", *Demagog*, 20.01.2022, accessed 30.08.2024, www.demagog.org.pl/fake_news/jak-mozna-wprowadzic-obowiazek-szczepien-przeciw-covid-19.

162 Paweł Marcinkiewicz, "Nalot ABW na mieszkania Kamratów i antyszczepionkowców. W tle współpraca z obcym wywiadem i podżeganie do zabójstwa", *Wyborcza.pl*, 13.01.2024, accessed 30.08.2024, https://warszawa.wyborcza.pl/warszawa/7,54420,30588856,koniec-bezkarnosci-kamratow-interwencja-abw-w-mieszkaniach.html.

Summary and Conclusions

Anti-scientific movements in Poland are demonstrating a visible dynamic of development. Not only do they organize regular and numerous demonstrations and protests, but they have also established permanently operating organizations and coalitions in the last decade. In the movement opposing mandatory vaccinations, the National Association of Vaccination Awareness "STOP Vaccine Adverse Events" has operated for over a decade, and its leader, Justyna Socha, regularly appears in the mainstream media. Associations of doctors and medical professionals opposing mandatory vaccinations have also been established, such as the Polish Association of Independent Physicians and Scientists, the Ordo Medicus Institute, the Association of Integrated Medicine Physicians, the Association Medics of Hope. There are also devoted to the topic magazines which regularly publish materials created by "experts" and activists of the movement opposing mandatory vaccinations, such as: "HARMONY. Your health. Your responsibility" ("HARMONIA. Twoje zdrowie. Twoja odpowiedzialność"), "Health without medicines" ("Zdrowie bez leków"), as well as the annual conference "What your doctor won't tell you" ("Czego lekarz Ci nie powie"). Activists of the movement were also guests (and often co-organizers) of many festivals of "natural medicine," "new spirituality," or "esotericism," such as: "Harmonia Kosmosu," "Wibracje," "Festiwal Zdrowie," "Pełnia Zdrowia – Festiwal Kultury Zdrowia," "Festiwal Wolności I Zdrowia. Primum Non-Nocere," "Ekologiczne Dożynki," "Festiwal Topinamburu," "Kosmiczny Festiwal" czy "Żywie."

During the festivals, they also met with activists from the movement opposing the introduction of the fifth-generation telephone network to Poland. They appeared together in debates and panels, and their stands were adjacent during these festivals. In the latter movement, although on a smaller scale due to smaller numbers of involved in it people, one can also notice an increasing tendency towards institutionalization. For example, several small organizations formed the Poland Free of 5G Coalition. Also some companies appeared that sell and distribute products that are supposed to protect against the alleged harmful effects of 5G on health.

When we look at the biographies of activists from both movements, we can also see that they often have spent many years being publicly active: fighting GMO s, being active in ecological groups, or alter-globalization organizations. In addition, the movements fighting against mandatory vaccinations and fifth-generation telephony are not separated from each other, and many well-known activists or "experts," such as Jerzy Zięba, Diana Wojtkowiak, or Janusz

 | DOI:10.1163/9789004747852_010

Zagórski, combine their activity in both, often mixing conspiracy narratives related to these issues.

They do not constitute separate narratives but draw from a broad, yet similar, content source. Narratives about vaccines that question the validity of vaccination are based primarily on similar claims known from other countries that vaccines are insufficiently researched and even described as a large-scale medical experiment, harmful, that they change DNA, cause undesirable post-vaccination reactions (including autism), contain supposedly toxic substances (such as aluminium or graphene), and that the vaccines were created based on foetuses from abortion procedures. Opponents of mandatory vaccinations often wove their claims into other conspiracy narratives, such as the plan to depopulate Europe and Poland in particular (allegedly created by the WHO, the "world government", the EU or the Jewish plutocracy), the "Great Replacement" (e.g., Poles to be substituted in Poland by Ukrainians or Jews, or more broadly Europeans by non-European migrants), the "genocide of the white race" (the planned reduction of the white population in Europe), the "Great Reset" (the planned change of the system in Western countries from democracy to totalitarianism), the "New World Order" (the existence of a secret world government), the "Great Lehia" (the existence of a Slavic empire in ancient times), the harmfulness of the fifth-generation telephone network and genetically modified food, the planned spread of the so-called "chemtrails," the possession and use by governments of so-called electromagnetic weapons, the manipulation of the government and the medical community by representatives of medical concerns (the so-called "Big Pharma"), and other anti-Semitic, homophobic, pro-Russian, and anti-Ukrainian narratives.

Particularly noticeable was the linking of conspiratorial narratives regarding COVID-19 with conspiratorial ideas of the "Great Reset," "New World Order," "Big Pharma," "depopulation," and the "Great Replacement." This connection exists in both ordinary adherents of anti-vaccination narratives and "experts," who played an essential role in the movement – doctors, academics, members of parliament, celebrities, artists, and media people. A Polish specificity of anti-vaccination narratives is their linking to the issue of the war in Ukraine and the presence of a large number of Ukrainian refugees in Poland. References to anti-Semitic motifs are also frequent.

Adherents of both studied anti-scientific conspiracy narratives try to present their claims as "science based," similarly to the mainstream science, trying to come up with ideas gathered in a paradigmatic bundle. However, the concepts proclaimed by the aforementioned movements do not find experimental confirmation, are unfalsifiable, and ultimately find their primary support in the concepts of an alleged secret conspiracy of elites wanting to enslave humanity, and are additionally strongly fuelled by anti-Semitic, ufological, and right-wing-populist elements. Therefore, the bundle of concepts attacking vaccines or the fifth-generation telephone network can be considered anti-scientific narrative, which fit into a broader set of anti-Enlightenment movements.

At the same time, activists of the studied movements position themselves as defenders of freedom against the threat posed by large corporations or political elites. As a result, belief in conspiracy narratives regarding 5G, vaccines, or COVID-19 unites people from very different parts of the political spectrum. On the one hand, they attract the right, which sees them as part of *guerrilla* against the assault on individual freedom by the global government. On the other hand, anti-vaccine activists often ally with leftists who distrust corporate "Big Pharma." However, since the main element of their opposition to mandatory vaccination is the regularly raised issue of freedom to decide about one's health or body, it is most accurate to consider the supporters of described in this book conspiracy narrative as a kind of radicalized, populist-minded liberals.

The channels of publicity are also important here – namely, far-right media and right-wing populist politicians. The most important channels for distributing the narrative of opposition to mandatory vaccinations and against using fifth-generation telephone networks in Poland are social media and YouTube in particular. Taking into account the control of fake news or disinformation content introduced on Meta-owned sites (Facebook, Instagram), which results in the rapid removal of pages, posts, and comments of this nature, and the relatively low popularity of Twitter (currently X) and TikTok (additionally used more often by younger users), it can be assumed that YouTube remains the primary source in the Polish info-sphere of propagating anti-vaccination content and conspiratorial narratives.

As a result, the leading media outlets publicizing the movements studied were online television (primarily video channels on YouTube) aimed at supporters of alternative medicine, ufologists, and people associated with broadly understood movements of new spirituality. The conspiratorial nature of anti-vaccination narratives means that their followers most often engage in political movements that describe themselves as anti-establishment. Poland's most active actor in this political segment is the far right. The most important accounts for distributing narratives questioning mandatory vaccinations are YouTube channels and websites combining the presentation of content from the field of alternative medicine and far-right ideology, such as Porozmawiajmy TV, Niezależna Telewizja, Ukryte Terapie, Media Narodowe, wRealu24.pl, Centrum Edukacyjne Polska, Centrum Edukacyjne Powiśle, Centrum Edukacyjne Brynica, Magna Polonia, Telewizja Narodowa, Instytut Spraw Obywatelskich, and eMisja.tv. In these channels, demands against mandatory vaccinations and fifth-generation telephone networks are mixed with content originating from other anti-Enlightenment movements such as integral traditionalism, religious fundamentalism and integralism, new spirituality and new age, monarchism-legitimism, neo-pagan neo-fascism, and Völkism. They also share with them a lack of trust in science and even technophobia, irrationalism, nativism, emotionalism, a solid opposition to progress and modernity, and fantasies of an Arcadian past. In principle, the only different elements are the attitudes towards democracy and liberalism. Therefore, the anti-vaccination and 5G movement can be considered part of a broader wave of anti-Enlightenment movements, and the radicalization of its activists is analogous to other representatives of this attitude.

As a result, in Poland, the topic of the struggle against mandatory vaccinations and 5G, thanks to the aforementioned internet channels, has penetrated mainstream politics, finding its promoters in the mainstream of Polish political life. On the Polish political scene the most active political entities spreading narratives questioning the validity of vaccinations and announcing the harmfulness of fifth-generation telephone networks to human health included the political parties Sovereign Poland, Confederation, and the political movement Kukiz'15, and to a lesser extent the Polish People's Party. Politicians from these parties took part in demonstrations, conferences, and festivals of the movement, publicized its demands, both in the media and in parliament, and also made the electoral lists available to representatives of the movement, putting them forward as candidates in parliamentary, local, and European Parliament elections. The activity of the movements studied was also reflected in the form of parliamentary interpellations and pressure on legislators, both influencing the creation of drafts of new laws consistent with their assumptions and blocking the adoption of regulations contrary to the agenda of the movements studied.

At the same time, the activities of numerous activists of the movement opposing mandatory vaccinations and the introduction of the fifth-generation telephone network to Poland showed many signs of extremism and a significant degree of radicalization. In the last decade, they committed acts of vandalism, illegally entered medical institutions and places with telecommunications infrastructure, destroyed property, created numerous graffiti, and made criminal threats (including death threats) against doctors, politicians, and local government officials. Both in neighbouring countries (such as Germany) and in Poland, activists of the movements (in particular the movement opposing mandatory vaccinations) took part in activities that, according to the security services, had signs of preparing terrorist attacks and striving to overthrow the state system by force.

An essential element of these activities was contact with foreign institutions, primarily from the Russian Federation. This connection was not the only simple contact between activists and institutions or groups from the latter country. The narrative demonizing fifth-generation telephony itself not only comes from Russia but was also publicized and promoted by Russian public media (such as RT television or the Sputnik agency). One cannot speak of a Russian source of its origin in describing the alleged harmfulness of vaccinations. However, numerous scientific studies and journalistic analyses indicate the participation of entities from this country in its promotion, especially in the territories of countries belonging to the EU and NATO.

Additionally, in the case of both movements, an essential role in the dissemination of anti-scientific narratives was played by media channels involved in the promotion of narratives consistent with Russian ones, concerning not only issues related to vaccines and 5G but also the topic of Ukraine, NATO, Jews and their political role, the USA, and Germany. At the same time, politicians representing the movements studied have often operated in international alliances and organizations in close cooperation with regimes and groups in close alliance with Russia. As a result, Russian influence in the movements studied deserves a separate study, not only scientific but also focused on analysing the threats resulting from their activities to the political and systemic stability of the state.

The further development of the movements discussed in this book is essential not only for issues directly related to the main topics of their postulates but also for the political stability of Poland and may significantly affect the further growth of political polarization in this country and constitute a channel of influence on local political life used further by entities of third countries, including, above all, the Russian Federation. Hence, this publication is only a preliminary survey, leaving the task of a more precise definition of these connections to subsequent, more detailed studies.

Bibliography

No author. No title. *Stop 5G*, no data. Accessed 30.08.2024. http://stop5g.com.pl/wp-content/uploads/2020/01/video-1580216127.mp4?_=1.

No author. "[FOTO, WIDEO] »Lutomia mówi NIE, dla wieży 5G!«". *Doba.pl*, 27.04.2019. Accessed 30.08.2024. https://doba.pl/dsw/artykul/-foto-wideo-lutomia-mowi-nie-dla-wiezy-5g-/25194/0.

No author. "»Cud w Sokółce« to bakteria?". *TVN24*, 29.10.2009. Accessed 06.09.2024. https://www.tvn24.pl/polska/cud-w-sokolce-to-bakteria-ra113674-3698513.

No author. "»Plandemia« i szczepionki groźne dla kobiet w ciąży? Fake news!". *Demagog*, 26.07.2022. Accessed 06.09.2024. https://www.demagog.org.pl/fake_news/plandemia-i-szczepionki-grozne-dla-kobiet-w-ciazy-fake-news.

No author. "»To nie kwestia historyczna«. Raport o związkach szczepionek z przemysłem aborcyjnym". *Do Rzeczy*, 12.10.2021. Accessed 06.09.2024. https://www.dorzeczy.pl/religia/210793/powstal-raport-o-zwiazkach-szczepionek-z-przemysle m-aborcyjnym.html.

No author. "15.X.2022 Stowarzyszenie Głos Wiernych zaprasza na wykład dr. n. med. Doroty Jarczewskiej". *Katolickie Stowarzyszenie Lekarzy Polskich. Oddział Mazowiecki bł. Ewy Noiszewskiej*, 15.10.2022. Accessed 06.09.2024. https://www.kslpmazowsze.pl/2022/10/15-x-2022-stowarzyszenie-glos-wiernych-zaprasza-na-wyklad-dr-n-med-doroty-jarczewskiej.

No author. "5G – Pola torsyjne a dewastacja ludzkiego zdrowia – Janusz Zagórski w rozmowie z Dianą Wojtkowiak". *Niezależna Telewizja*, 01.01.2020. Accessed 06.09.2024. https://niezaleznatelewizja.pl/index.php/aiovg_videos/5g-pola-torsyjne-a-dewastacja-ludzkiego-zdrowia-janusz-zagorski-w-rozmowie-z-diana-wojtkowiak .

No author. "Braun: To jest akt barbarzyństwa". *DoRzeczy*, 09.05.2022. Accessed 06 .09.2024. https://dorzeczy.pl/opinie/298624/braun-skomentowal-incydent-z-andriejewem.html.

No author. "Co polecam". *Szaman 21 Wieku*, no data. Accessed 06.09.2024. https://szaman21wieku.com/category/co-polecamy-testy.

No author. "Czy grozi nam wojna? Antyukraińska konferencja w Kielcach", *Demagog*, 03.04.2023. Accessed 06.09.2024. https://www.demagog.org.pl/fake_news/czy-grozi-nam-wojna-antyukrainska-konferencja-w-kielcach.

No author. "Czym jest pole torsyjne i jak z niego korzystać – dr Diana Wojtkowiak". *Porozmawiajmy TV*, 01.09.2020. Accessed 06.09.2024. https://porozmawiajmy.tv/czym-jest-pole-torsyjne-i-jak-z-niego-korzystac-dr-diana-wojtkowiak.

No author. "D.Tusk o ewentualnym obowiązku szczepień: jestem zwolennikiem perswazji". *PolitykaZdrowotna.com*, 09.12.2021. Accessed 5.09.2024. https://politykaz

drowotna.com/artykul/dtusk-o-ewentualnym-obowiazku-szczepien-jestem-zwolennikiem-perswazji/825020.

No author. "Deklaracja warszawska przeciw nowemu totalitaryzmowi". *Norymberga 2.0*, November 2021, Accessed 06.09.2024. https://www.norymberga2.pl/page/view/7/deklaracja-warszawska.html.

No author. "Dementujemy fałszywe informacje w »odkryciach« Grzegorza Płaczka". *Demagog*, 13.10.2021. Accessed 06.09.2024. https://demagog.org.pl/fake_news/dementujemy-falszywe-informacje-w-odkryciach-grzegorza-placzka.

No author. "Dezinformacja na temat szczepień w wywiadzie z dr Sienkiewicz". *Demagog*, 14.02.2022. Accessed 06.09.2024. https://www.demagog.org.pl/fake_news/dezinformacja-na-temat-szczepien-w-wywiadzie-z-dr-sienkiewicz.

No author. "Dokąd zmierzamy jako ludzkość – prof. Ryszard Zajączkowski Lubartów 29.08.2022 r.". *Polskie Stowarzyszenie Niezależnych Lekarzy i Naukowców*, 07.09.2022. Accessed 06.09.2024. https://www.psnlin.pl/video,dokad-zmierzamy-jako-ludzkosc-prof-ryszard-zajaczkowski-lubartow-29-08-2022-r,109.html.

No author. "Dr Anna Martynowska skazana na karę pieniężną przez Naczelny Sąd Lekarski. Ordo Iuris interweniuje". *Instytut na rzecz Kultury Prawnej Ordo Iuris*, no data accessed 06.09.2024, www.ordoiuris.pl/wolnosci-obywatelskie/dr-anna-martynowska-skazana-na-kare-pieniezna-przez-naczelny-sad-lekarski.

No author. "Dr Hubert Czerniak w Gdańsku, 17.02.2018, część 1". *Porozmawiajmy TV*, 17.03.2018. Accessed 06.09.2024. https://porozmawiajmy.tv/dr-hubert-czerniak-w-gdansku-17-02-2018-czesc-1.

No author. "Ekolog mówi prawdę o 5G! | Olaf Swolkień". *CDA*, no data. Accessed 06.09.2024. https://ebd.cda.pl/620x368/5263092ac.

No author, "EEAS SPECIAL REPORT UPDATE: Short Assessment of Narratives and Disinformation Around the COVID-19 Pandemic (UPDATE DECEMBER 2020 – APRIL 2021), *European External Action Service*, 28.04.2021, https://euvsdisinfo.eu/uploads/2021/04/EEAS-Special-Report-Covid-19-vaccine-related-disinformation-6.pdf.

No author. "Ekranowanie pól elektromagnetycznych". *EMF Busters*, no data. Accessed 06.09.2024. https://emfbusters.pl.

No author. "GLOBALNY PROTEST STOP 5G – Jadwiga Łopata i Julian Rose". *Niezależna Telewizja*, 07.04.2020. Accessed 06.09.2024. https://niezaleznatelewizja.pl/index.php/aiovg_videos/globalny-protest-stop-5g-jadwiga-lopata-i-julian-rose.

No author. "Guru antyszczepionkowców »fotografuje młode dziewice«. »To był kiedyś niezły lekarz«". *Medonet*, 12.01.2022. Accessed 06.09.2024. https://www.medonet.pl/zdrowie/wiadomosci,guru-antyszczepionkowcow-fotografuje-dziewice--to-byl-kiedys-niezly-lekarz,artykul,30960450.html.

No author. "Handel dziećmi, gwiazdy Dawida i znaczenie sądownictwa – Janusz Korwin-Mikke, z komentarzem Leszka Bubla". *Porozmawiajmy TV*, 17.03.2020.

Accessed 06.09.2024. https://porozmawiajmy.tv/handel-dziecmi-gwiazdy-dawida-i-znaczenie-sadownictwa-janusz-korwin-mikke-z-komentarzem-leszka-bubla.

No author. "How Russia Influences Vaccination". *Center for Strategic Communication*, 12.11.2021. Accessed: 30.08.2024. http://www.spravdi.gov.ua/en/how-russia-influences-vaccination.

No author. "Hubert Czerniak – fake newsy o 5G i szczepieniach przeciw HPV". *Demagog*, 09.12.2021. Accessed 06.09.2024. https://demagog.org.pl/fake_news/hubert-czerniak-fake-newsy-o-5g-i-szczepieniach-przeciw-hpv.

No author. "Jak można wprowadzić obowiązek szczepień przeciw COVID-19?". *Demagog*, 20.01.2022. Accessed 30.08.2024. www.demagog.org.pl/fake_news/jak-mozna-wprowadzic-obowiazek-szczepien-przeciw-covid-19.

No author. "Jaworzno. Konferencja niezależnych naukowców i lekarzy". *Tarnogorski.info*, 07.10.2021. Accessed 06.09.2024. https://www.tarnogorski.info/jaworzno-w-niedziele-17-pa.

No author. "Jerzy Jaśkowski kontra lekarze? W odpowiedzi podaje fake newsy!". *Demagog*, 09.12.2022. Accessed 06.09.2024. https://www.demagog.org.pl/fake_news/jerzy-jaskowski-kontra-lekarze-w-odpowiedzi-podaje-fake-newsy.

No author. "Katarzyna Ratkowska ujawnia tajne dokumenty Pfizera? Fake news!". *Razem Przeciw Dezinformacji*, 23.06.2022. Accessed 06.09.2024. www.razemprzeciwdezinformacji.pl/katarzyna-ratkowska-ujawnia-tajne-dokumenty-pfizera-fake-news-2.

No author. "Katarzyna Ratkowska w Polsat News. Gdzie minęła się z prawdą?". *Demagog*, 02.07.2021. Accessed 06.09.2024. https://www.demagog.org.pl/fake_news/katarzyna-ratkowska-w-polsat-news-gdzie-minela-sie-z-prawda.

No author. "Każda choroba przewlekła jest do wyleczenia? To nie jest prawda", *Demagog*, 23.08.2023. Accessed 06.09.2024. https://www.demagog.org.pl/fake_news/kazda-choroba-przewlekla-jest-do-wyleczenia-to-nie-jest-prawda.

No author. "Konferencja w Krakowie organizowana przez J. Rose i J. Łopata Wystąpienie Janusza Zagorskiego". *Niezależna Telewizja*, 01.04.23. Accessed 06.09.2024. https://niezaleznatelewizja.pl/index.php/aiovg_videos/konferencja-w-krakowie-organizowana-przez-j-rose-i-j-lopata-wystapienie-janusza-zagorskiego.

No author. "Koronasceptycy próbowali wtargnąć do Reichstagu. W proteście w Berlinie uczestniczył Grzegorz Braun". *Wprost*, 31.08.2020. Accessed 06.09.2024. https://www.wprost.pl/swiat/10358116/berlin-protest-koronasceptykow-i-proba-wtargniecia-do-reichstagu-posel-grzegorz-braun-uczestniczyl-w-demonstracji.html.

No author. "KTO NAM ZMIENIŁ MEDYCYNĘ? Prof. Krystyna Lisiecka-Opalko PSNLiN", *Polskie Stowarzyszenie Niezależnych Lekarzy i Naukowców*, 17.05.2022. Accessed 06.09.2024. https://www.psnlin.pl/video,kto-nam-zmienil-medycyne-prof-krystyna-lisiecka-opalko-psnlin,93.html.

No author. "Lewica składa projekt ustawy o obowiązkowych szczepieniach". *Dziennik. Gazeta Prawna*, 06.12.2021. Accessed 5.09.2024. https://www.gazetaprawna.pl/wiadomosci/kraj/artykuly/8307534,lewica-sklada-projekt-ustawy-obowiazkowe-szczepienia.html.

No author. "Lockdowny nie działały, a szczepienia są nieskuteczne? Fałsz!". *Demagog*, 28.12.2022. Accessed 06.09.2024. https://www.demagog.org.pl/fake_news/lockdowny-nie-dzialaly-a-szczepienia-sa-nieskuteczne-falsz.

No author. "Ludobójstwo Polaków w XXI-wieku – poseł Majka i grupa anty5G". *Porozmawiajmy TV* 09.06.2019. Accessed 06.09.2024. https://porozmawiajmy.tv/ludobojstwo-polakow-w-xxi-wieku-posel-majka-i-grupa-anty5g.

No author. "Measels". *World Health Organization*, 31.05.2023. Accessed: 30.08.2024. https://www.who.int/news-room/fact-sheets/detail/measles.

No author. "Mentzen tłumaczy się z szurów na listach Konfederacji i nie przeszkadzałby mu poseł wierzący w planetę Nibiru". *Donald*, 11.10.2023. Accessed 06.09.2024. https://www.donald.pl/artykuly/wwMn66Y2/mentzen-tlumaczy-sie-z-szurow-na-listach-konfederacji-i-nie-przeszkadzalby-mu-posel-wierzacy-w-planete-nibiru.

No author. "Mocny głos ekspertów Ordo Iuris w sprawie zarządzania kryzysem COVID-19. Wystąpili: Chazan, Rieske, Basiukiewicz, Witczak"/ *Stowarzyszenie Polonia Christiana*, 13.01.2022. Accessed 06.09.2024. https://www.poloniachristiana.org/mocny-glos-ekspertow-ordo-iuris-w-sprawie-zarzadzania-kryzysem-covid-19-wystapili-chazan-rieske-basiukiewicz-witczak.

No author. "O nas". *OrdoMedicus.org*, no data. Accessed 06.09.2024.https://www.ordomedicus.org/o-nas.

No author. "Ordo Medicus o skuteczności szczepionki przeciw Covid-19 i szczepieniach dzieci". *Stowarzyszenie Polonia Christiana*, 20.08.2021. Accessed 06.09.2024. https://www.poloniachristiana.org/ordo-medicus-o-skutecznosci-szczepionki-przeciw-covid-19-i-szczepieniach-dzieci.

No author. "Petycja »Stop zagrożeniu zdrowia i życia«". *Urząd Gminy Stryszów*, 23.03.2020. Accessed 30.08.2024. https://bip.malopolska.pl/ugstryszow,a,1752178,petycja-stop-zagrozeniu-zdrowia-i-zycia.html.

No author. "Polscy lekarze bronią swoich praw – Konferencja prasowa w Sejmie RP", *Polskie Stowarzyszenie Niezależnych Lekarzy i Naukowców*. 12.04.2022. Accessed 06.09.2024. https://www.psnlin.pl/video,polscy-lekarze-bronia-swoich-praw-konferencja-prasowa-w-sejmie-rp,80.html.

No author. "Pomoc specjalistów. lek. stom. Daniel Mazur" *Polskie Stowarzyszenie Niezależnych Lekarzy i Naukowców*, no data. Accessed 06.09.2024.https://www.psnlin.pl/pomoc-specjalistow,17.html.

No author. "Pomoc specjalistów. lekarz Izabela Kaczmarek". *Polskie Stowarzyszenie Niezależnych Lekarzy i Naukowców*, no data. Accessed 06.09.2024. https://www.psnlin.pl/pomoc-specjalistow,45.html.

No author. "POWSTAŁ RUCH POLSKA JEST JEDNA SPRZECIWIAJĄCY SIĘ SEGREGACJI SANITARNEJ. APELUJE DO PREZYDENTA O »WETOWANIE AKTÓW PRAWNYCH, KTÓRE GODZĄ W PRAWA I WOLNOŚCI OBYWATELSKIE«". *Radio Maryja*, 26.09.2021. Accessed 5.09.2024. https://www.radiomaryja.pl/informacje/powstal-ruch-polska-jest-jedna-sprzeciwiajacy-sie-segregacji-sanitarnej-apeluje-do-prezydenta-o-wetowanie-aktow-prawnych-ktore-godza-w-prawa-i-wolno sci-obywatelskie.

No author. "Prezes ważnego urzędu w szoku. »Poseł Braun i antyszczepionkowcy siłą wtargnęli do mojego gabinetu«"/ *Fakt*, 03.08.2021. Accessed 30.08.2024. https://www.fakt.pl/polityka/grzegorz-braun-i-antyszczepionkowcy-wtargneli-do-grzegorza-cessa-szefa-urzedu/1djdwpw.

Noawww.zdrowiebezlekow.pl/blog/andrzej-au96uthor. "Prof. Andrzej Frydrychowski". *Zdrowie bez Leków*, no data. Accessed 06.09.2024..

No author. "Prof. Andrzej Frydrychowski: izby lekarskie to organizacje przestępcze". *Pl1.tv*, no data. Accessed 06.09.2024. https://www.pl1.tv/title/prof-andrzej-frydry chowski-izby-to-organizacja-przestepcza.

No author, "Protest »Stop segregacji sanitarnej« [ZDJECIA]", *Tarnowska TV*, 15 .01.2022, accessed 30.08.2024, https://www.tarnowska.tv/galeria/3381,protest-stop-segregacji-sanitarnej-zdjecia.

No author. "Protest antyszczepionkowców przed Sejmem. »Koronawirus mutuje w faszystowską dyktaturę«". *Dziennik.pl*, 05.01.2022. Accessed 30.08.2024. https://gos podarka.dziennik.pl/news/galeria/8327958,protest-antyszczepionkowcy-sejmem .html.

No author. "Stop smog 5g". *Obywatele Decydują*, no data. Accessed 06.09.2024. https://www.obywateledecyduja.pl/kampania/stop-smog-5g.

No author. "Radomszczanie nie chcą masztu z nadajnikiem sieci telefonii komórkowej". *TVP3 Łódź*, 13.07.2020. Accessed 30.08.2024. https://lodz.tvp.pl/48949780/ra domszczanie-nie-chca-masztu-z-nadajnikiem-sieci-telefonii-komorkowej.

No author. "REGULAMIN Parlamentarnego Zespołu ds. Bezpieczeństwa Programu Szczepień Ochronnych Dzieci i Dorosłych". *Sejm Rzeczypospolitej Polskiej*, no data. Accessed 06.09.2024. https://orka.sejm.gov.pl/opinie8.nsf/nazwa/zesp_szcze pienochronnych/$file/zesp_szczepienochronnych.pdf .

No author. "STANOWISKO KONFEDERACJI KORONY POLSKIEJ W SPRAWIE OBOWIĄZKOWYCH SZCZEPIEŃ DZIECI". *Konfederacja Korony Polskiej*, no data. Accessed 06.09.2024. https://www.konfederacjakoronypolskiej.pl/stanowisko-konfe deracji-korony polskiej-w-sprawie-obowiazkowych-szczepien-dzieci.

No author. "Szczepienie na COVID-19 zagraża dzieciom? Fałszywe informacje!". *Demagog*, 10.11.2022. Accessed 06.09.2024. https://demagog.org.pl/fake_news/szczepie nie-na-covid-19-zagraza-dzieciom-falszywe-informacje.

No author. "Szczepionki nie powstrzymują koronawirusa? Fake news!". *Demagog*, 12.10.2021. Accessed 06.09.2024. https://www.demagog.org.pl/fake_news/szczepionki-nie-powstrzymuja-koronawirusa-fake-news.

No author. "Szczepionki to niepotrzebne i nieprzebadane zagrożenie? Fake news!". *Demagog*, 04.08.2021. Accessed 06.09.2024. https://www.demagog.org.pl/fake_news/szczepionki-to-niepotrzebne-i-nieprzebadane-zagrozenie-fake-news.

No author, "Tackling COVID-19 Disinformation – Getting the Facts Right", *European Commission and European External Action Service*, 10.06.2020, accessed 30.03.2025 https://commission.europa.eu/strategy-and-policy/coronavirus-response/fighting-disinformation/tackling-coronavirus-disinformation_en.

No author. "Technologia 5G jako broń technotroniczna – Ewa Pawela". *Porozmawiajmy TV*, 28.01.2019, accessed 06.09.2024 https://porozmawiajmy.tv/technologia-5g-jako-bron-technotroniczna-ewa-pawela.

No author. "The mystery of mass bird deaths". *BBC*, 07.01.2011. Accessed: 30.08.2024. https://www.bbc.com/news/world-us-canada-12135380.

No author. "Teoria Anny Martynowskiej o pandemicznych oszustwach jest fałszywa". *Demagog*, 23.02.2021. Accessed 06.09.2024. www.demagog.org.pl/fake_news/teoria-dr-anny-martynowskiej-o-pandemicznych-oszustwach-jest-falszywa.

No author. "Traktat antypandemiczny WHO – Wysłuchanie publiczne – dr Emilia Sykut-Domańska z PSNLiN". *Polskie Stowarzyszenie Niezależnych Lekarzy i Naukowców*, 12.04.2022. Accessed 06.09.2024. https://www.psnlin.pl/news-psnlin,traktat-antypandemiczny-who-wysluchanie-publiczne-dr-emilia-sykut-domanska-z-psnlin,56.html.

No author. "Trumna dla Ministra Zdrowia – VI Międzynarodowy Protest Przeciw Przymusowi Szczepień – cz.2". *Polskie Stowarzyszenie Niezależnych Lekarzy i Naukowców*, 08.06.2022. Accessed 06.09.2024. https://www.psnlin.pl/video,trumna-dla-ministra-zdrowia-vi-miedzynarodowy-protest-przeciw-przymusowi-szczepien-cz-2,99.html.

No author. "Uczyńmy Polskę wolną od 5G – Jadwiga Łopata". *Porozmawiajmy TV*, 07.03.19. Accessed 06.09.2024. https://porozmawiajmy.tv/uczynmy-polske-wolna-od-5g-jadwiga-lopata.

No author. "Ufolog u Prokopa. Janusz Zagórski wierzy, że obcy są wśród nas". *WP*, no data. Accessed 06.09.2024. https://web.archive.org/web/20191120185150/https://ztymicosieznaja.wp.pl/artykul/ufolog-u-prokopa-janusz-zagorski-wierzy-ze-obcy-sa-wsrod-nas.

No author. "W Polsce stopniowo wprowadza się faszyzm – Benjamin Fulford, Fabiola Tsugami-Shaba i Hanna Kazahari". *Porozmawiajmy TV*, 29.04.2020. Accessed 06.09.2024. https://porozmawiajmy.tv/w-polsce-stopniowo-wprowadza-sie-faszyzm-benjamin-fulford-fabiola-tsugami-shaba-i-hanna-kazahari.

No author. "Wiejska lekarka jak szamanka. Kazała pacjentom przytulać się do drzew". *WP.pl*, 20.11.2015. Accessed 06.09.2024. https://www.kobieta.wp.pl/wiejska-le

karka-jak-szamanka-kazala-pacjentom-przytulac-sie-do-drzew-5982777517777537a.

No author. "Witamina C na ciężki COVID-19? Fake newsy Jerzego Zięby!". *Demagog*, 17.02.2022. Accessed 06.09.2024. https://demagog.org.pl/fake_news/witamina-c-na-ciezki-covid-19-fake-newsy-jerzego-zieby.

No author. "Współczesne niewolnictwo, plany Żydów przeciw Żydom i palenie zielarek – Grzegorz Braun". *Porozmawiajmy TV*, 18.08.2016. Accessed 06.09.2024. https://porozmawiajmy.tv/wspolczesne-niewolnictwo-plany-zydow-przeciw-zydom-i-palenie-zielarek-grzegorz-braun.

No author. "Wszystko co musisz wiedzieć o tej epidemii – Jerzy Zięba". *Porozmawiajmy TV*, 19.04.2020. Accessed 06.09.2024. https://porozmawiajmy.tv/wszystko.

No author. "Wybory 2011: Komitet Wyborczy Nowa Prawica – Janusza Korwin-Mikke". *Państwowa Komisja Wyborcza*, no data. Accessed 06.09.2024. https://wybory2011.pkw.gov.pl/rfl/pl/393d3249f302f85c28fdca91340862b9.html.

No author. "Wyniki głosowania: Komitet Wyborczy Unia Polityki Realnej". *Państwowa Komisja Wyborcza*, no data. Accessed 06.09.2024. https://pe2009.pkw.gov.pl/PUE/PL/WYN/W110/2.htm.

No author. "Wywiad z Romanem Zielińskim zawiera nieprawdziwe informacje". *Demagog*, 15.01.2021. Accessed 06.09.2024. https://www.demagog.org.pl/fake_news/wywiad-z-romanem-zielinskim-zawiera-nieprawdziwe-informacje.

No author. "Zakazane krojenie marchewki,
rou-co-musisz-wiedziec-o-tej-epidemii-jerzy-ziebandup na deser i referendum bez mocy – dr Jarosław Sachajko". *Porozmawiajmy TV*, 09.06.2016. Accessed 06.09.2024. https://porozmawiajmy.tv/zakazane-krojenie-marchewki-roundup-na-deser-i-referendum-bez-mocy-dr-jaroslaw-sachajko.

No author "ZAPOWIEDZI. V Festiwal Ekologii Neohumanistycznej. 20.–24.7.95 Głębock". *Zielone Brygady*, no data. Accessed 06.09.2024. http://www.zb.eco.pl/zb/72/zapowie1.htm.

No author. "Zbigniew Hałat o szczepionkach? Kolejny dezinformujący materiał!". *Demagog*, 11.01.2022. Accessed 06.09.2024. https://www.demagog.org.pl/fake_news/zbigniew-halat-o-szczepionkach-kolejny-dezinformujacy-material.

No author. "Zdrowie i Pro Life to tematy dwóch ostatnich dni Tygodnia Życia i Wolności". *Stowarzyszenie im. Piotra Skargi*, 11.06.2021. Accessed 06.09.2024. https://www.piotrskarga.pl/zdrowie-i-pro-life-to-tematy-dwoch-ostatnich-dni-tygodnia-zycia-i-wolnosci,15568,i.html.

No author. "Zdrowie to wolność – dr Ewelina Gierszewska PSNLiN". *Polskie Stowarzyszenie Niezależnych Lekarzy i Naukowców*, 18.03.2022. Accessed 06.09.2024. https://www.psnlin.pl/video,zdrowie-to-wolnosc-dr-ewelina-gierszewska-psnlin,72.html.

No author. "Zrób to sam! Przewodnik jak mierzyć i oceniać promieniowanie elektromagnetyczne", *Tygodnik Spraw Obywatelskich*,. 21.07.2020. Accessed 06.09.2024.

https://instytutsprawobywatelskich.pl/zrob-to-sam-przewodnik-swiadomego-obywatela-pomiar-i-ocena-promieniowania-elektromagnetycznego.

Aboudouh, Ahmed. "»Unparalleled threats«: Anti-vaxxer movement threatens a new wave of extremism". *Independent*, 19.01.2022. Accessed 30.08.2024. https://www.independent.co.uk/news/world/europe/antivaxxer-extremism-violence-plots-b1991710.html.

Adolphs, Ralph. "The biology of fear", *Current Biology* 23, no. 2 (2013):79–93.

Adm. "Grzegorz Braun o Robercie Biedroniu: Powinien trafić do więzienia",=. *Rzeczpospolita*, 21.03.2019. Accessed 06.09.2024. https://www.rp.pl/Polityka/190329834-Grzegorz-Braun-o-Robercie-Biedroniu-Powinien-trafic-do-wiezienia.html.

Admin. "Warto przeczytać: »Polityka Polska« – nowy miesięcznik." *Geopolityka.net*, 29.05.2015. Accessed 06.09.2024. https://geopolityka.net/polityka-polska-nowy-miesiecznik.`

Administrator. "Czy 5G to wojskowa broń światowych elit do selektywnego zabijania ludzi?! Na to wygląda!". *Warszawska Gazeta*, 02.03.2019. Accessed 06.09.2024. https://gazetawarszawska.com/index.php/pugnae/3151-czy-5g-to-wojskowa-bron-swiatowych-elit-do-selektywnego-zabijania-ludzi-na-to-wyglada.

ADMINIWJKK. "Kolejna konferencja zorganizowana przez Wolne Wybory! Lekarze w Jaśle". *Wolne Wybory*, 06.11.2021. Accessed 06.09.2024. https://www.wolnewybory.org/kolejna-konferencja-zorganizowana-przez-wolne-wybory-niezalezni-lekarze-w-jasle.

AFP. "Bolivia Villagers Destroy 5G Masts Over Virus Fears". *Barron's*, 16.06.20. Accessed 30.08.2024. https://www.barrons.com/news/bolivia-villagers-destroy-5g-masts-over-virus-fears-01592301304.

Ajir, Media; Vailliant, Bethany. "Russian Information Warfare: Implications for Deterrence Theory". *Strategic Studies Quarterly* 12, no. 3 (2018):70–89.

Akimov, Anatoly E.; Shipov, Gennady I. "Torsion fields and experimental manifestation".? *New Energy News* (2017): 11–14.

akw, tmw, rzw. "Jaki dwa razy o szczepieniach. Rok temu przeciw »przymuszaniu«, teraz obiecuje »promocję«, *TVN24*, 03.10.2018. Accessed 5.09.2024. https://www.tvn24.pl/wiadomosci-z-kraju,3/patryk-jaki-o-obowiazkowych-szczepieniach-w-sejmie-projekt,873150.html.

Albarracin, Dolores; Albarracin, Julia; Chan, Man-pui Sally and Hall Jamieson, Kathleen. *Creating Conspiracy Beliefs: How Our Thoughts Are Shaped.* Cambridge: Cambridge University Press, 2022.

Althusser, Louis. "Ideologia i aparaty ideologiczne państwa". *Recykling Idei*, 14.06.2006. Accessed 07.09.2024. http://www.web.archive.org/web/20180722174007/http://recyklingidei.pl/althusser-ideologie-aparaty-ideologiczne-panstwa.

AM. "Uniwersytet Medyczny odcina się od opinii »ekspertki« antyszczepionkowców". *Polityka Zdrowotna*, 14.11.2018. Accessed 06.09.2024. https://politykazdrowotna.

com/artykul/uniwersytet-medyczny-odcina-sie-od-opinii-ekspertki-antyszczepion kowcow-n835455.

Amk, "Godek na komisji zdrowia: Wszystkie szczepionki korzystają ze zbrodni aborcji". *Rzeczpospolita*, 05.01.2022. Accessed 06.09.2024. www.rp.pl/polityka/art19258431–godek–na–komisji–zdrowia–wszystkie–szczepionki–korzystaja–ze–zbrodni–aborcji.

Aml. "Warszawa. Protest antyszczepionkowców przed Sejmem. Przynieśli transparent nawiązujący do Auschwitz". *Polsat News*, 15.12.2021. Accessed 30.08.2024. https://www.polsatnews.pl/wiadomosc/2021–12–15/warszawa–protest–antyszczepion kowcow–przed–sejmem–przyniesiono–transparent–nawiazujacy–do–obozow.

Antosiewicz, Marcin. "Straszyła LGBT+ i imigrantami. Wspierała antyszczepionkowców i płaskoziemców. Tak przez lata działała rosyjska dezinformacja w Polsce". *Newsweek*, 15.12.2019. Accessed 30.08.2024. https://www.newsweek.pl/swiat/rosyj ska–dezinformacja–w–polsce–straszyla–lgbt–wspierala–antyszczepionkowcow/p87g1w5.

Astromaria. "Świadectwa rodziców, których dzieci doświadczyły problemów zdrowotnych po szczepieniach". *Jestem za, a nawet przeciw* (blog), 30.09.2014. Accessed: 30.08.2024. https://www.astromaria.wordpress.com/2014/09/30/swia dectwa–rodzicow–ktorych–dzieci–doswiadczyly–problemow–zdrowotnych–po–szczepieniach.

Backhaus, Insa; Hoven, Hanno, Kawachi, Ichiro. "Far–right political ideology and COVID–19 vaccine hesitancy: Multilevel analysis of 21 European countries". *Social Science & Medicine*, 335 (2023):116–227.

Bafia, Szymon. "Tłumy na konferencji Góralskiego Veta w Zakopanem: jesteśmy strażnikami porządku prawnego". *Onet*, 14.01.2021. Accessed 30.08.2024. https://po droze.onet.pl/aktualnosci/tlumy–na–konferencji–goralskiego–veta–w–zakopane m–jestesmy–straznikami–porzadku/3znd3pe.

Bajtek, Mateusz. "Na wojnie z kłamstwem". *Interia.pl*, 15.01.2018. Accessed 06.09.2024. https://wydarzenia.interia.pl/zagranica/news–na–wojnie–z–klamstwem,nId,2508 561#utm.

Baker Stephanie Alice and Walsh, Michael James. "A mother's intuition: it's real and we have to believe in it': how the maternal is used to promote vaccine refusal on Instagram". *Information, Communication & Society* 26, no. 8 (2023):1675–1692.

Bakselerowicz, Piotr. "Antyukraińskie spotkanie Grzegorza Brauna z wyborcami. Pomoc uchodźcom porównuje do grzechu". *Wyborcza.pl*, 19.03.2022. Accessed 06.09.2024. https://zielonagora.wyborcza.pl/zielonagora/7,35182,28238667,spotka nie–grzegorza–brauna–z–wyborcami–pomaganie–uchodzcom.html.

Bałuka, Mateusz. "Narodowcy z rządowym wsparciem. Ich media promują antyszczepionkowe treści". *Onet*, 21.09.2021. Accessed 06.09.2024. https://www.wiadomosci .onet.pl/kraj/koronawirus–narodowcy–z–rzadowym–wsparciem–promuja–antys zczepionkowe–tresci/wold46f.

Bar–On, Tamir. *Rethinking the French New Right. Alternatives to Modernity*. London, Routledge, 2013.

Bar–On, Tamir. *Where Have All The Fascists Gone?* London: Routledge, 2019.

Barkun, Michael. *A Culture of Conspiracy: Apocalyptic Visions in Contemporary America*. Berkeley: University of California Press, 2013.

Barlinek 5G, "Eksperci ostrzegają przed zgubnymi skutkami 5G Zbigniew Gelzok i Rafik Ennaoui!". *YouTube*, 24.11.2019. Accessed 06.09.2024. https://www.youtube.com/watch?v=zmei_MF_FUQ.

Barthes, Roland. *Mythologies*. New York: Hill and Wang, 1972.

Bartyzel, Jacek. *Umierać, ale powoli. O monarchistycznej i katolickiej kontrrewolucji w krajach romańskich 1815–2000*. Krakow: ARCANA, 2006.

Basha, Saddiq. "»Death to the Grid«: Ideological Narratives and Online Community Dynamics in Encouraging Far–Right Extremist Attacks on Critical Infrastructure". *Counter Terrorist Trends & Analysis* 15, no. 4 (2023):17–24.

Bauman, Zygmunt. *Nowoczesność i Zagłada*. Krakow, Wydawnictwo Literackie, 2009.

Bazzell, Michael. *Open Source Intelligence Techniques: Resources For Searching And Analyzing Online Information*. Charleston: Createspace Independent Publishing Platform, 2018.

Beck, Ulrich. *Społeczeństwo ryzyka. W drodze do innej nowoczesności*. Warsaw: Wydawnictwo Scholar, 2002.

Bednarczuk, Bartosz. "Śląskie: Bunt przeciwko sieci 5G! Nadajnika nie chcą u siebie mieszkańcy Bierunia. Głos zabrał burmistrz!". *TVS*, 12.06.2020. Accessed 30.08.2024. https://tvs.pl/informacje/slaskie–bunt–przeciwko–sieci–5g–nadajnika–nie–chca–u–siebie–mieszkancy–bierunia–glos–zabral–burmistrz.

Bednarek Adam. "Mapa wstydu. W tych polskich miastach protestowano przeciwko 5G". *Dobre Programy*, 24.09.2020. Accessed 30.08.2024. https://www.dobreprogramy.pl/mapa–wstydu–w–tych–polskich–miastach–protestowano–przeciwko–5g,6628692371286145a.

Bednarek Adam. "Maszt jak sklep monopolowy. To już poszło za daleko". *Spider's Web*, 28.11.2023. Accessed 30.08.2024. https://spidersweb.pl/2023/11/radom–maszt–protest–odleglosc–od–dzialek.html.

Berman, Jonathan M. *Anti–vaxxers. How to challenge a misinformed movement*. Cambridge: The MIT Press, 2000.

Bernaciński, Łukasz and Kwaśniewski, Jerzy (editorpolitykazdrowotna.com/artykul/uniwersytet–medyczny–odcina–sie–od–opinii–ekspertki–antyszczepionkowcow/835455s) *Ograniczenia praw i wolności wprowadzone w Polsce w zw. z epidemią Covid–19. Aspekty prawne, medyczne i ekonomiczne*. Warszawa, Wydawnictwo Naukowe Instytutu na rzecz Kultury Prawnej Ordo Iuris, 2022.

Benkler, Yochai; Faris, Robert and Roberts, Hal. *Network Propaganda: Manipulation, Disinformation, and Radicalization in American Politics*. New York: Oxford Academic, 2018. Kindle.

Bielsko TV, "Protest antyszczepionkowców w obronie lekarki". *YouTube*, 17.11.2021. Accessed 30.08.2024. https://www.youtube.com/watch?v=ztU3WxpYgwQ.

Bieniek, Martyna. "Ramowanie a metafora w ujęciu kognitywistycznym – czy mają ze sobą coś wspólnego?". *Philosophical Discourses Prace Naukowe Uniwersytetu Humanistyczno-Przyrodniczego im. Jana Długosza w Częstochowie* 2 (2020):149–164.

Bittman, Ladislav. *The deception game: Czechoslovak intelligence in Soviet political warfare*. Syracuse: Syracuse University Research Corp, 1972.

Biuro Prasowe OWP, "Warszawa: Miesięcznica Wołyńska". *Obóz Wielkiej Polski*, no data. Accessed 06.09.2024. https://www.owp.org.pl/index.php/703-warszawa-miesiecznica-wolynska-2.

BJS, AGIE, SC. "Marsz przeciwko restrykcjom. Wśród uczestników poseł Braun [WIDEO]". *TVP Info*, 20.03.2021. Accessed 06.09.2024. https://www.tvp.info/52885484/pandemia-koronawirusa-polska-warszawa-marsz-przeciwko-restrykcjom-epidemicznym-wsrod-uczestnikow-posel-konfederacji-grzegorz-braun.

Blaskiewicz, Robert. "The Big Pharma conspiracy theory". *Medical Writing* 22, no. 4 (2013):259–261.

Błaszkiewicz, Paulina. "Koronasceptycy w szkole Rydzyka: Lockdown to katastrofa, maseczki to absurd, a pandemia generuje hajs w mediach". *Wyborcza.pl*, 29.01.2021. Accessed 06.09.2024. https://torun.wyborcza.pl/torun/7,48723,28052653,koronasceptycy-w-szkole-rydzyka-lockdown-to-katastrofa-maseczki.html.

Blik. "Kto zniszczył płot u posłanki PO? Napis »tu mieszka zdrajca«". *Rzeczpospolita*, 18.12.2021. Accessed 30.08.2024. https://www.rp.pl/przestepczosc/art19210321-kto-zniszczyl-plot-u-poslanki-po-napis-tu-mieszka-zdrajca.

Błochowiak, Mariusz. "Biała księga pandemii koronawirusa – fakty i dane ukrywane przed opinią publiczną", *Polskie Stowarzyszenie Niezależnych Lekarzy i Naukowców*, 05.09.2022. Accessed 06.09.2024. https://www.psnlin.pl/artykuly,biala-ksiega-pandemii-koronawirusa-fakty-i-dane-ukrywane-przed-opinia-publiczna,25,254.html.

Błochowiak, Mariusz (editor). *Fałszywa pandemia. Krytyka naukowców i lekarzy*. Warszawa: Fundacja Osuchowa, 2020.

Błochowiak, Mariusz (editor). *Fałszywa pandemia. Krytyka naukowców i lekarzy 2*. Warszawa: Fundacja Osuchowa, 2020.

Błochowiak, Mariusz (editor). *Fałszywa pandemia. Krytyka naukowców i lekarzy 3*. Warszawa: Fundacja Osuchowa, 2021.

Błochowiak, Mariusz (editor). *Fałszywa pandemia. Krytyka naukowców i lekarzy 4*. Warszawa: Fundacja Osuchowa, 2021.

Błochowiak, Mariusz. "SKUTKI RESTRYKCJI RZĄDOWYCH DLA FUNKCJONOWANIA SŁUŻBY ZDROWIA W 2020 I 2021 ROKU". *Fundacja Ordo Medicus*, 01.03.2022. Accessed 06.09.2024. https://www.ordomedicus.org/skutki-restrykcji-rzadowych-dla-funkcjonowania-sluzby-zdrowia-w-2020-i-2021-roku.

Błochowiak, Mariusz. "XVI POSIEDZENIE KOMISJI ŚLEDCZEJ – LEK. PAWEŁ BASIUKIEWICZ". *Fundacja Ordo Medicus*, 09.05.2022. Accessed 06.09.2024. https://www.ordomedicus.org/xvi-posiedzenie-komisji-sledczej-lek-pawel-basiukiewicz.

Bodalska, Barbara. "Polska: Koniec bezradnej Rady Medycznej przy premierze". *Euractiv.pl*, 17.01.2022. Accessed 5.09.2024. https://www.euractiv.pl/section/zdrowie/news/polska-pandemia-koronawirus-zdrowie-rada-medyczna-rezygnacja-premier-morawiecki-niedzielski.

Bodakowski, Jan. "Za duża jest liczba zgonów po wstrzyknięciu preparatu zwanego szczepionką na covid [OPINIA]". *Media Narodowe*, 19.08.2021. Accessed 06.09.2024. https://www.medianarodowe.com/2021/08/19/za-duza-jest-liczba-zgonow-po-wstrzyknieciu-preparatu-zwanego-szczepionka-na-covid-opinia.

Bodner, John; Welch, Wendy; Brodie, Ian. *COVID-19 Conspiracy Theories: Qanon, 5G, the New World Order and Other Viral Ideas*. Jefferson: McFarland & Company, 2020. Kindle.

Boellstorff, Tom; Nardi, Bonnie; Pearce, Celia and Taylor, T. L. *Ethnography and Virtual Worlds: A Handbook of Method*. Princeton, NJ: Princeton University Press, 2012.

Bolanowski, Rafał. "Antyszczepionkowcy chcą obejść przepisy. Organizują dwie manifestacje, które dzieli długość płyty rynku". *Wyborcza.pl*, 14.01.2022. Accessed 30.08.2024. https://rzeszow.wyborcza.pl/rzeszow/7,34962,27999153,antyszczepionkowcy-chca-obejsc-przepisy-organizuja-dwie-manifestacje.html.

Bolczyk, Bartłomiej. "W Koninie odbędzie się marsz »Z miłości do wolności«. Będą protestować przeciwko obostrzeniom antyepidemicznym". *Konin. Nasze Miasto*. 21.03.2021. Accessed 30.08.2024. https://konin.naszemiasto.pl/w-koninie-odbedzie-sie-marsz-z-milosci-do-wolnosci-beda/ar/c1-8160855.

Bomba Prawdy Kontynuacja 3. "Obecne Państwo ukraińskie które nie zasługuje na pomoc Polski Iwan Komarenko". *YouTube*, 24.02.2022. Accessed 06.09.2024. https://www.youtube.com/watch?v=prQQahOOUdM.

Bomba Prawdy Kontynuacja 3. "Rosja na Ukrainie to gwarant normalizacji zgniłej i upadającej tkanki tego kraju Henryk Pająk". *YouTube*, 20.02.2022. Accessed 06.09.2024. https://www.youtube.com/watch?v=WpACcYDGgEA.

Borakiewicz, Wojciech. "Konfederacja: Państwo ma Ukraińcom pomagać w drugiej kolejności. Ks. Kneblewski: Słusznie". *Wyborcza.pl*, 09.03.2022. Accessed 06.09.2024. https://bydgoszcz.wyborcza.pl/bydgoszcz/7,48722,28200951,konfederacja-panstwo-ma-ukraincom-pomagac-w-drugiej-kolejnosci.html.

Borenstein, Seth. "Technology to blame for animal die-off panic". *NBC News*, 07.01.2011. Accessed 30.08.2024. https://www.nbcnews.com/id/wbna40961721.

Boyd, Milo. "At least 90 phone masts attacked during UK coronavirus lockdown after 5G conspiracy". *Mirror*, 26.05.2020. Accessed 30.08.2024. https://www.mirror.co.uk/news/uk-news/least-90-phone-masts-attacked-22086169.

Bp. "Agata Osiniak z Kolbuszowej: Ponad 50 procent naszego społeczeństwa została oszukana? Lekarz ostro o szczepionkach [WIDEO]". *Korso Kolbuszowskie*, 20.01.22. Accessed 06.09.2024. https://korsokolbuszowskie.pl/strefa-covid-19/dr-agata-osiniak-ponad-50-procent-naszego-spoleczenstwa-zostala-oszukana-lekarz-ostro-o-szczepionkach-wideo/tGlgcEuYwybpu2P9mqrH.

Bp. "Manifestowano przeciwko przymusowi szczepień". *DKL24*.pl, 15.01.2022. Accessed 06.09.2024. https://www.dkl24.pl/pl/a/12046/manifestowano-przeciwko-przymusowi-szczepien.html.

Braun, Grzegorz. "Przymus szczepień? Owszem – przeciw totalniackiej wściekliźnie". *Grzegorz Braun*, 16.06.2017. Accessed 06.09.2024. https://www.grzegorzbraun.pl/2017/06/16/przymus-szczepien-owszem-przeciw-totalniackiej-wsciekliznie.

Braun, Grzegorz. "TYLKO U NAS! Grzegorz Braun: Sojusznicza fatamorgana". *Polska Niepodległa*, 07.10.2018. Accessed 06.09.2024. https://polskaniepodlegla.pl/opinie/item/17598-tylko-u-nas-grzegorz-braun-sojusznicza-fatamorgana.

Braw, Elisabeth. "The Imagined Threats of 5G Conspiracy Theorists Are Causing Real-World Harm". *Foreign Policy*, 29.06.2020. Accessed 30.08.2024. https://foreignpolicy.com/2020/06/29/the-imagined-threats-of-5g-conspiracy-theorists-are-causing-real-world-harm.

Brezar, Aleksandar. "From anti-vax to pro-Putin, how conspiracy theorists are backing Russia's war in Ukraine". *Euronews*, 13.05.22. Accessed 30.08.2024. https://www.euronews.com/my-europe/2022/05/13/from-anti-vax-to-pro-putin-how-conspiracy-theorists-are-backing-russia-s-war-in-ukraine.

Broad, William J. "Your 5G Phone Won't Hurt You. But Russia Wants You to Think Otherwise". *The New York Times*, 12.05.19. Accessed 07.09.2024. http://www.nytimes.com/2019/05/12/science/5g-phone-safety-health-russia.html.

Broderick, Ryan. "A Conspiracy Theory That 5G Is Causing The Coronavirus Is Spreading Alongside The Pandemic". *Buzzfeed News*, 03.04.2020. Accessed 30.08.2024. https://www.buzzfeednews.com/article/ryanhatesthis/conspiracy-theory-5g-coronavirus-qanon.

Broniatowski, David A; Jamison, Amelia M; Qi, SiHua; AlKulaib, Lulwah; Chen, Tao; Benton, Adrian; Quinn, Sandra C; Dredze, Mark. "Weaponized Health Communication: Twitter Bots and Russian Trolls Amplify the Vaccine Debate". *American Journal of Public Health* 108 (2018): 1378–1384.

Bruns, Alex; Hurcombe, Edward; Harrington, Stephen. "»Corona? 5G? or both?«: the dynamics of COVID-19/5G conspiracy theories on Facebook". *Media International Australia* 177, no. 1 (2020):12–29.

Bruns, Alex; Hurcombe, Edward; Harrington, Stephen. "Covering Conspiracy: Approaches to Reporting the COVID/5G Conspiracy Theory". *Digital Journalism* 10, no. 6 (2022):930–951.

Brzeska, Anna. "Medycy – antyszczepionkowcy. Co z nimi zrobić?",. *Onet*, 01.08.2021. Accessed 06.09.2024. https://wiadomosci.onet.pl/kraj/koronawirus–i–medycy–antyszczepionkowcy–co–z–nimi–zrobic/p25bnjx.

Bs. "Antyszczepionkowcy w Częstochowie coraz bardziej aktywni. Demonstrowali na pl. Biegańskiego". *Wyborcza.pl*, 15.01.2022. Accessed 30.08.2024. https://czestochowa.wyborcza.pl/czestochowa/7,48725,28005878,antyszczepionkowcy–w–czestochowie–stop–segregacji–i–dyskryminacji.html.

BS. "TV Republika "amerykańską i żydowską agenturą", czyli Grzegorz Braun w formie". *Strajk.eu*, 09.05.2016. Accessed 06.09.2024. https://strajk.eu/tv–republika–amerykanska–i–zydowska–agentura–czyli–grzegorz–braun–w–formie.

Buarque, Beatriz. "Is It Conspiracy or »Truth«? Examining the Legitimation of the 5G Conspiracy Theory during the Covid–19 Pandemic". *Social Epistemology* 36, no. 3 (2023):317–328.

Burda, Katarzyna. "Szczepionkowy i nowotworowy szarlatan oraz jego wyznawcy", *Newsweek*, 16.04.2016, www.archive.is/20170928113638/http://www.newsweek.pl/wiedza/nauka/szczepionkowy–szarlatan–i–jego–wyznawcy,artykuly,384030,1.html.

Burdzy, Barbara. "Na kawie z dyktatorem Syrii. Niebezpieczne gry posła Kukiz'15". *Fakt*, 10.12.2018. Accessed 06.09.2024. https://www.fakt.pl/wydarzenia/polityka/pawel–skutecki–kto–stoi–za–poslem–kukiz15/bxzrmq2.

Burgess, Adam. *Cellular Phones, Public Fears, and a Culture of Precaution*. Nowy Jork: Cambridge University Press, 2003.

Butler, Josh; Martin, Sarah. "Australian online anti–vaccine groups switch to Putin praise and Ukraine conspiracies". *The Guardian*, 01.03.2022. Accessed 30.08.2024. https://www.theguardian.com/australia–news/2022/mar/02/australias–anti–vaccine–groups–switch–focus–to–putin–praise–and–ukraine–conspiracies.

Butter, Michael; Knight, Peter. *Routledge Handbook Of Conspiracy Theories*. London: Routledge, 2020.

Byford, Jovan. *Conspiracy theories: a critical introduction*. Houndmills, Basingstoke, Hampshire: Palgrave Macmillan, 2011), 32–37.

Bż. "Piszą skargi na lekarkę, bo »kazała się przytulać do drzew«. »Nawet nie wiem, jak się bronić«". *TVN24*, 19.11.2015. Accessed 06.09.2024. https://www.tvn24.pl/lodz/wiejska–lekarka–jak–szamanka–kazala–nam–przytulac–sie–do–drzew–ra595889–3318448.

Callahan, Gene and McIntyre, Kenneth B. *Critics of Enlightenment Rationalism*. London, Palgrave, 2020.

Canadian Security Intelligence Service, *Who Said What? The Security Challenges of Modern Disinformation*, (Ottawa: CSIS, 2017), accessed 13.04.2025 https://www.canada.ca/content/dam/csis-scrs/documents/publications/disinformation_post-report_eng.pdf.

Candlin, Christopher N. "General editor's preface". In: *The Construction of Professional Discourse* edited by Gunnarsson, Britt-Luoise; Linell, Per; Nordberg, Bengt. New York: Routledge, 2014.

CEPowiśle. "Etyka informacyjna: Justyna Socha, Grzegorz Płaczek, Rafał Mossakowski, Piotr Szlachtowicz – panel". *CDA*, 23.09.21. Accessed 06.09.2024. https://www.cda.pl/video/84786661a/vfilm.

CEPowiśle. "Hanna Kazahari (Tokio): Plandemia – zaplanowana przez globalistów i syjonistów fałszywa flaga wirusa". *BitChute*, 08.04.2020. Accessed 06.09.2024. https://www.bitchute.com/video/ccEa9MRbElOQ.

Cerulus, Laurens. "5G arsonists turn up in continental Europe". *Politico*, 26.04.2020. Accessed 30.08.2024. https://www.politico.com/news/2020/04/26/5g-mast-torchers-turn-up-in-continental-europe-210736.

Charteris-Black, Jonathan. *Politicians and Rhetoric: The Persuasive Power of Metaphor*. Basingstoke: Palgrave Macmillan, 2011.

Chaunu, Pierre. *Cywilizacja wieku oświecenia*. Warsaw, PIW, 1993.

Chen, Sibo. "Far-Right Political Extremism and the Radicalisation of the Anti-vaccine Movement in Canada". In: *Communicating COVID-19*, edited by Lewis, Monique; Govender, Eliza; Holland, Kate. 303–323. Cham: Palgrave Macmillan, 2024.

Chojnowska, Ada. "Kontrola umysłu i eksperymenty na ludziach ... Koalicja przeciw 5G obradowała w Krakowie". *Wyborcza.pl*, 28.02.2019. Accessed 06.09.2024. https://krakow.wyborcza.pl/krakow/7,44425,24504806,kontrola-umyslu-i-eksperymenty-na-ludziach-koalicja-przeciw.html.

Chmielewska-Szlajfer, Helena; Staniszewska, Zuzanna; Sobol, Emilia; Kalinowska-Balcerzak, Anna. "Badania jakościowe na temat postaw rodziców mających wątpliwości czy szczepić swoje dzieci oraz postaw antyszczepionkowych w Internecie – raport z badań ALK". *Reaserchgate*, 09.12.2022. Accessed 07.09.2024. https://www.researchgate.net/publication/366155169_Badania_jakosciowe_na_temat_postaw_rodzicow_majacych_watpliwosci_czy_szczepic_swoje_dzieci_oraz_postaw_antyszczepionkowych_w_Internecie_-_raport_z_badan_ALK.

Chołodowski, Maciej. "Białystok. Pediatra przeciwna szczepieniom dzieci stanęła przed sądem lekarskim. Bronią jej antyszczepionkowcy". *Wyborcza.pl*, 09.02.2022. Accessed 06.09.2024. https://www.bialystok.wyborcza.pl/bialystok/7,35241,28090535,bialystok-pediatra-przeciwna-szczepieniom-dzieci-przed-sadem.html.

Chołodowski, Maciej. "Koronawirus. Prawicowi lekarze i naukowcy przeciwni szczepieniom. Prof. Flisiak: Ten apel to nieodpowiedzialność". *Wyborcza.pl*, 02.12.2020. Accessed 06.09.2024. https://www.bialystok.wyborcza.pl/bialystok/7,35241,26567278,koronawirus-prawicowi-lekarze-i-naukowcy-sprzeciwprzeciwni.html.

Chołodowski, Maciej. "Podlascy antyszczepionkowcy wydawali »wyroki śmierci«. Teraz zostali skazani przez sąd". *Wyborcza.pl*, 20.02.2024. Accessed 30.08.2024. https://bialystok.wyborcza.pl/bialystok/7,35241,30718167,podlascy-antyszczepionkowcy-wydawali-wyroki-smierci-teraz.html.

Chrobak, Barbara. "Interpelacja nr 23562 do ministra zdrowia, ministra środowiska w sprawie rozpylania w atmosferze szkodliwych substancji". *Sejm Rzeczpospolitej Polskiej*, 29.06.2018. Accessed 06.09.2024. https://www.sejm.gov.pl/Sejm8.nsf/InterpelacjaTresc.xsp?key=3323CEFA.

Chrobry Szlak. "Polska w cieniu p(l)andemii – Rafał Mossakowski". *YouTube*, 20.04.2020. Accessed 06.09.2024. https://www.youtube.com/watch?v=-RiXUVcM1DY.

Chrobry Szlak. "Wojna na Ukrainie i obłęd »patriotów« – Kamil Klimczak, Rafał Mossakowski". *YouTube*, 24.02.2022. Accessed 06.09.2024. https://youtu.be/JqPx791CoTM.

Cieślik, Kamila. "Sejm miał uznać Rosję za państwo sponsorujące terroryzm. "Braun zgłosił sprzeciw", obrady przerwano". *Gazeta.pl*, 01.12.2022. Accessed 06.09.2024. https://wiadomosci.gazeta.pl/wiadomosci/7,114884,29207224,sejm-mial-zajac-sie-ustawa-uznajaca-rosje-za-panstwo-sponsorujace.html .

Cimek, Gracjan. "Czas Globalczyka (1 i 2)". *Tygodnik Spraw Obywatelskich*, 07.03.2023. Accessed 06.09.2024. https://instytutsprawobywatelskich.pl/czas-globalczyka-1-i-2.

Cios, Janek. "Jarosław Sachajko (Kukiz'15) – przeciw GMO". *Życie Zamościa*, 30.11.2016. Accessed 06.09.2024. https://zyciezamoscia.pl/informacje/jaroslaw-sachajko-kukiz15-przeciw-gmo.

Claassen, Liesbeth; Smid, Tjabe; Woudenberg, Fred; Timmermans, Danielle R.M. "Media Coverage on Electromagnetic Fields and Health: Content Analysis of Dutch Newspaper Articles and Websites: Health Risks in the Media". *Health, Risk & Society* 14 no. 7–8 (2012): 681–696.

Co tam w Sejmie. "2019.02.19 Parlamentarny Zespół ds. Bezpieczeństwa Programu Szczepień Ochronnych Dzieci i Dorosłych". *YouTube*, 19.02.2019. Accessed 06.09.2024. https://www.youtube.com/watch?v=Q4UrZntXAaI.

Cockerell, Isobel. "Why anti-vaccine groups are running with Putin's narrative". *Coda*, 01.04.2022. Accessed 30.08.2022, https://www.codastory.com/newsletters/putin-anti-vaxxers.

Cohen, Stanley. *Folk Devils and Moral Panics: The Creation of the Mods and Rockers*. London and New York: Routledge, 2002.

Confessore, Nicholas; Dance, Gariel J. X.; Harris, Richard; Hessen, Mark. "The Follower Factory". *The New York Times*, 27.01.2018. Accessed 01.09.2024. https://www.nytimes.com/interactive/2018/01/27/technology/social-media-bots.html.

Conti, Nicolò. "National political elites, the EU, and the populist challenge". *Politics* 38, no. 3 (2018):361–377.

Contreras, Brian; Lee, Wendy. "Putin targets lots of Americans with disinformation. One example? Anti-vaccine groups". *Los Angeles Times*, 25.02.2022. Accessed 30.08.2024. https://www.latimes.com/business/technology/story/2022-02-25/heres-what-putins-disinformation-war-looks-like-on-the-internet.

Crossley, Nick. *Towards Relational Sociology*. London: Routledge, 2011.

Cuda, Heidi. "How Russia Joined With Anti–Vaxxers to Wage Biological Warfare by Proxy Against the West". *Byline Supplement*, 18.03.2023. Accessed 30.08.2024. https://www.bylinesupplement.com/p/how–russia–joined–with–anti–vaxxers.

Cybersecurity and Infrastructure Security Agency, *Malicious Activity Targeting COVID -19 Research, Vaccine Development*, 16.07.2020, accessed 30.03.2025 https://www .cisa.gov/news-events/alerts/2020/07/16/malicious-activity-targeting-covid-19-rese arch-vaccine-development.

Cylka, Tomasz. "Podejrzane 470 tys. zł Justyny Sochy. Prokuratura w Poznaniu zajęła się liderką antyszczepionkowców". *Wyborcza.pl*, 06.02.2022. Accessed 06.09.2024. https://poznan.wyborcza.pl/poznan/7,36001,28078738,podejrzane–470–tys–zl–jus tyny–sochy–prokuratura–w–poznaniu.html#S.embed_article–K.C–B.1–L.2.zw.

Cywiński, Paweł; Katner, Filip; Ziółkowski, Jarosław. *Zarządzanie strachem. Jak prawica wygrywa debatę publiczną w Polsce*. Warsaw: Fundacja im. Stefana Batore go, 2019.

Czajkowski, Paweł. "Boisz się 5G? Kup do ochrony pendrive z »kwantową technologią« za 346 USD". *IT Hardware*, 29.05.2020. Accessed 06.09.2024. https://ithar dware.pl/aktualnosci/boisz_sie_5g_kup_do_ochrony_pendrive_z_quot_kwantowa _technologia_quot_za_346_usd–12352.html.

Czarczyńska, Kira. "Nadajnik Play stanie mimo protestów mieszkańców". *Komórkomania*, 05.04.2010. Accessed 30.08.2024. https://komorkomania.pl/18372,nadaj nik–play–stanie–mimo–protestow–mieszkancow.

Czarniawska, Barbara. *Narratives in Social Science Research*. London, SAGE, 2004.

Czech, Franciszek; Ścigaj, Paweł. "Popularność narracji spiskowych w Polsce czasu pandemii COVID–19. Raport z reprezentatywnych badań sondażowych". *Instytut Studiów Międzynarodowych Uniwersytetu Jagiellońskiego*, 2020. Accessed 07.09.2024. http://www.ism.uj.edu.pl/wp–content/uploads/2020/07/covid–narrac je–spiskowe.pdf.

Czubkowska, Sylwia. "Komu zależy, by płonęły maszty 5G? Ponad 140 ataków w całej Europie!". *Spider's Web*, 04.06.2020. Accessed 30.08.2024. https://spidersweb.pl/ plus/2020/06/5g-potesty-covid-ataki-teorie-spiskowe.

Czubkowska, Sylwia. "Strach przed elektrosmogiem już blokuje inwestycje telekomom. Czy może opóźnić wprowadzenie 5G?". *Wyborcza.pl*, 12.04.2019. Accessed 5.09.2024. https://wyborcza.biz/biznes/7,177150,24642674,strach–przed–elektros mogiem–juz–blokuje–inwestycje–telekomom.html.

Czermiński Jakub and Mikulski, Jakub. "Ojciec Rydzyk bierze się za covid. W Toruniu zorganizuje sympozjum", *Rzeczpospolita*. 24.01.2022. Accessed 06.09.2024. https:// www.rp.pl/ochrona–zdrowia/art19320451–ojciec–rydzyk–bierze–sie–za–covid –w–toruniu–zorganizuje–sympozjum.

Diani, Mario. "The Concept of Social Movement." *The Sociological Review* 40, no. 1 (1992): 1–25.

Danielewicz, Marta. "Maszt 5G będzie górował nad Puszczą Zielonką? Mieszkańcy Dębogóry nie chcą takiej inwestycji pod swoimi oknami przy wejściu do lasu". *Gazeta Pomorska*", 19.06.2020. Accessed 30.08.2024. https://plus.pomorska.pl/maszt-5g-bedzie-gorowal-nad-puszcza-zielonka-mieszkancy-debogory-nie-chca-takiej-inwestycji-pod-swoimi-oknami-przy-wejsciu-do/ar/c1-15031340.

Danielewski, Michał. "Oto pożytek z koronawirusa w Polsce: PiS docenił standardy WHO. Już nie są antypolskim spiskiem". *OKO.Press*, 11.03.2020. Accessed 5.09.2024. https://www.oko.press/oto-pozytek-z-koronawirusa-w-polsce.

Dauksza, Julia and Gielewska, Anna. "Ukrainizacja w kolorze braun", *Frontstory*, 05.08.2022. Accessed 06.09.2024. https://frontstory.pl/ukrainizacja-grzegorz-braun-propaganda-rosja-ukraina-dezinformacja.

Dauksza Julia and Sepioło, Mariusz. "Orkiestra Putina: Dyrygenci i chórek". *Frontstory*, 31.03.2022. Accessed 07.09.2024. https://frontstory.pl/dezinformacja-antyszczepionkowcy-wojna-rosja-ukraina-konfederacja-braun

Davies, Philip and Rosemont. Hugo. *Russia and Disinformation: Maskirovka*. Centre for Research and Evidence on Security Threats, 2019. Accessed 13.04.2025. https://www.crestresearch.ac.uk/resources/russia-and-disinformation-maskirovka-full-report.

De Keersmaecker, Jonas and Roets, Arne. "»Fake news«: Incorrect, but hard to correct. The role of cognitive ability on the impact of false information on social impressions", *Intelligence* 65 (2017):107–110.

Deer, Brian. *The Doctor Who Fooled the World: Science, Deception, and the War on Vaccines*. Baltimore: Johns Hopkins University Press, 2020.

Demczuk, Agnieszka. "Fenomen ruchu antyszczepionkowego w cyberprzestrzeni, czyli fake news i postprawda na usługach hipotezy Andrew Wakefielda". *Studia de Cultura* 4 (2018):92–113.

Demczuk, Agnieszka. "Ruch antyszczepionkowy w Polsce i jego kabiny pogłosowe w alt-internet w latach 2020–2022". *Rocznik Instytutu Europy Środkowo-Wschodniej* 3 (2022):9–35.

Demczuk, Agnieszka. *SARS-CoV-2 i COVID-19. Plotki, dezinformacje i narracje spiskowe w polskim dyskursie publicznym*. Lublin, Wydawnictwo UMCS, 2021.

Denney, Sam. "A Foiled Coup Attempt in Germany and the Danger of Conspiracy Theories". *Lawfare*, 12.12.2022. Accessed 30.08.2024. https://www.lawfaremedia.org/article/foiled-coup-attempt-germany-and-danger-conspiracy-theories.

Dentith, Matthew R.X. *The Philosophy of Conspiracy Theories*. Londyn: Palgrave Macmillan, 2015.

Docent, Vincenty. "»Jeżeli to co zrobię pomoże chociaż jednej osobie, to warto to zrobić«"/ *Patronite.pl*, no data. Accessed 06.09.2024. https://patronite.pl/Vincenty.

Docent, Vincenty. "MISTYK W ROLLS ROYCE ✸ MOJA HISTORIA – Sport, Plastyka, Psylocybina, Przebudzenie, Port Kosmiczny ✸". *YouTube*, 20.09.2023. Accessed 06.09.2024. https://youtu.be/SCDkttOojGs.

Domański, Tomasz. "Poseł pyta o biologiczny wpływ 5G na populację Polaków. Spider's Web odpowiada". *Spider's Web*, 25.01.2019. Accessed 06.09.2024. https://spidersweb.pl/2019/01/posel-robert-majka-5g-teorie-spiskowe.html.

Domański, Tomasz. "Rozdają ulotki, na których straszą 5G. Za pieniądze zabezpieczą twoje mieszkanie". *Spider's Web*, 08.05.2020. Accessed 06.09.2024. https://spidersweb.pl/2020/05/alert-5g-promieniowanie-badania-skutki.html.

Donovan, Joan; Dreyfuss, Emily and Friedberg, Brian. *Meme Wars: The Untold Story of the Online Battles Upending Democracy in America*. New York: Bloomsbury Publishing, 2022.

Dorius, Shawn F; Lawrence-Dill, Carolyn J. "Sowing the seeds of skepticism: Russian state news and anti-GMO sentiment". *GM crops & food* 9, no. 2 (2018):53–58.

DoRzeczy.pl. "Dr Sobolewski: Nie ma dowodów skuteczności środków niefarmaceutycznych". *DoRzeczy*, 12.02.2022. Accessed 06.09.2024. https://www.dorzeczy.pl/opinie/262321/dr-sobolewski-nie-ma-dowodow-skutecznosci-srodkow-niefarmaceutycznych.html.

Douglas, Karen M.; Sutton, Robbie M. and Cichocka, Aleksandra. *The Psychology of Conspiracy Theories*. London: Routledge, 2024.

DSZ. "»A piersi rosną i rosną«. Eksperci ostrzegają przed jego teoriami, on nadal chce »leczyć« ludzi". *Onet*, 13.12.2021. Accessed 06.09.2024. https://wiadomosci.onet.pl/kraj/kim-jest-jerzy-zieba-i-czemu-lekarze-go-nie-uznaja-tresci-zagrazajace-zyciu/8q6nb4b.

Dubow, Ben; Lucas, Edward; and Morris, Jake. *Jabbed in the Back: Mapping Russian and Chinese Information Operations During the COVID-19 Pandemic*. Center for European Policy Analysis, December 2, 2021. Accessed March 30, 2025. https://cepa.org/comprehensive-reports/jabbed-in-the-back-mapping-russian-and-chinese-information-operations-during-the-covid-19-pandemic.

Dudzik, Iwona. "Mobilizacja wiedzy kontra demonstracja". *Podyplomie.pl*, 08.07.2017. Accessed 06.09.2024. https://podyplomie.pl/medical-tribune/27896,mobilizacja-wiedzy-kontra-demonstracja.

Dugin, Aleksander. "Post modernizm?", *Obywatel* 1, no. 9 (2003).

Dura, Maksymilian. "Ataki elektromagnetyczne na polskich obywateli? »Nieporozumienie«". *Defence24*, 16.03.2016. Accessed 5.09.2024. .https://defence24.pl/geopolityka/ataki-elektromagnetyczne-na-polskich-obywateli-nieporozumienie

Durbach, Nadja."They might as well brand us: Working class resistance to compulsory vaccination in Victorian England". *The Society for the Social History of Medicine* 13 (2000):45–62.

Durmaz, Nihal, Hengirmen, Engin."The dramatic increase in anti-vaccine discourses during the COVID-19 pandemic: a social network analysis of Twitter". *Human Vaccines & Immunotherapeutics* 18, no. 1 (December 2022): n.pag.

Dziermański, Arkadiusz. "Działalność krakowskich ekspertów ds PEM kosztowała miasto ponad milion złotych. W planach było dużo więcej". *What Next*, 31.05.2020.

Accessed 06.09.2024. https://whatnext.pl/krakow–przeciwnicy–sieci–komorkowych–wydatki–elektrosmog.

Dziermański, Arkadiusz. "Intel, Telia i Ericsson uruchamiają pierwszą sieć 5G w Europie". *Antyweb*, 03.10.2017. Accessed: 30.08.2024. https://www.telepolis.pl/wiadomosci/wydarzenia/intel–telia–i–ericsson–uruchamiaja–pierwsza–siec–5g–w–europie.

Dziermański, Arkadiusz. "Krzyczą, że 5G zabija i twierdzą, że mają dowody. Podsumowanie krakowskiego zlotu przeciwników 5G". *Telepolis*, 01.03.2019. Accessed: 30.08.2024. https://www.telepolis.pl/artykuly/wydarzenia/siec–5g–w–polsce–szkodliwa–technologia–dowody.

Dziermański, Arkadiusz. "Pod przykrywką cyberbezpieczeństwa dzieci, sejm odwiedzili przeciwnicy łączności komórkowej", *Telepolis*, 20.12.2019. Accessed 06.09.2024. https://www.telepolis.pl/artykuly/felietony/zmiany–norm–pem–5g–sejm–sachajko–braun.

Dziermański, Arkadiusz. "Protest przeciwko sieci 5G w Polsce był smutny. Teorie o szkodliwym PEM wybrzmiały w sejmie", *Telepolis*, 01.05.2019., https://www.telepolis.pl/artykuly/wydarzenia/siec–5g–w–polsce–protest–szkodliwe–pem–sejm.

Dzierżawski, Mariusz. "Przymus szczepienia na COVID w Watykanie". *Do Rzeczy*, 30.09.2021. Accessed 06.09.2024. www.dorzeczy.pl/tylko–na–dorzeczy/205807/watykan–przymus–szczepienia–na–covid–pracownikow–stolicy–apostolskiej.html.

Ebeling, Régis; Córdova Saenz, Carlos Abel; Campos Nobre, Jéferson; Becker, Karin. "Analysis of the Influence of Political Polarization in the Vaccination Stance: The Brazilian COVID–19 Scenario". *Proceedings of the International AAAI Conference on Web and Social Media* 16, no. 1 (2022):159–170.

Eco, Umberto. *A Theory of Semiotics*. Bloomington: Indiana University Press, 1976.

Eliade, Mircea . *Sacrum, mit, historia. Wybór esejów*. Warsaw:PIW, 1974.

Eliassen, Ingeborg; Pena, Paulo. "Real 5G issues overshadowed by Covid–19 conspiracy theories". *Investigate Europe*, 12.06.2020. Accessed 30.08.2024. https://www.investigate–europe.eu/posts/5g–covid–conspiracy.

Elzanaty, Ahmed; Chiaraviglio, Luca; Slim-Alouini, Mohamed. "5G and EMF Exposure: Misinformation, Open Questions, and Potential Solutions". *Frontiers in Communications and Networks* 2 (2021): 1–11.

EMisja.tv. "»Amerykanie chcą walczyć z Rosją w Polsce«". *YouTube*, 22.04.2016. Accessed 30.08.2024. https://www.youtube.com/watch?v=CVjr7yaiVCw.

EMisja.tv. "dr Czerniak: »Zaczyna się holocaust Polaków«". *YouTube*, 25.09.2018. Accessed 06.09.2024. https://www.youtube.com/watch?v=d3ykgqBbCk4.

EMisja.tv. "Jabłonowski: »Niezwykli Polacy pokonali chciwy koncern«". *YouTube*, 28.05.2016. Accessed 30.08.2024. https://www.youtube.com/watch?v=KoPolVoVerM.

EMisja.tv. "Kto tu kogo tak naprawdę opłaca?". *YouTube*, 23.03.2023. Accessed 30.08.2024. https://www.youtube.com/watch?v=3GvXfbpS2FI.

European External Action Service (EEAS), "EEAS Special Report Update: Short Assessment of Narratives and Disinformation around the COVID-19 Pandemic", *EEAS* June 2020, accessed 13.04.2025, https://www.eeas.europa.eu/delegations/georgia/eeas-special-report-update-short-assessment-narratives-and-disinformation-around-covid-19-pandemic_en.

EUvsDisinfo, "Figure of the Week: 5G," *EUvsDisinfo*, 12.04.2021, accessed 13.04.2025 https://euvsdisinfo.eu/figure-of-the-week-5g.

EugeniuszSendecki71. "ALEKSANDER JABŁONOWSKI I EUGENIUSZ SENDECKI NA PROTEŚCIE PRZECIWKO PRZYMUSOWI SZCZEPIEŃ 03–06–2017". *YouTube*, 03.06.2017. Accessed 06.09.2024. https://www.youtube.com/watch?v=MsAEu3SjO64.

EugeniuszSendecki71. "IWAN KOMARENKO I MACIEJ PORĘBA ROZMAWIAJĄ W TELEWIZJI NARODOWEJ". *YouTube*, 06.11.2021. Accessed 06.09.2024. https://www.youtube.com/watch?v=4FKydSpDLtA.

EugeniuszSendecki71. "OPÓR POLSKI PRZECIWKO PRZYMUSOWI SZCZEPIEŃ I UBECKI ZARZĄD NAD POLSKĄ SŁUŻBĄ ZDROWIA". *YouTube*, 03.06.2017. Accessed 06.09.2024. https://www.youtube.com/watch?v=dJ3ZOb_JPZs&t=91s.

EugeniuszSendecki71. "PROTEST PRZECIWKO NADUŻYCIOM W SYSTEMIE SZCZEPIEŃ W POLSCE NOP (6)". *YouTube* 04.06.2017. Accessed 06.09.2024. https://www.youtube.com/watch?v=24fBHI2sZ90.

EW. "Prof. Krzysztof Simon dostał policyjną ochronę po tym, jak antyszczepionkowcy grozili mu śmiercią". *Wyborcza.pl*, 05.10.2021. Accessed 30.08.2024. https://wroclaw.wyborcza.pl/wroclaw/7,35771,27650644,prof–krzysztof–simon–dostal–policyjna–ochrone–antyszczepionkowcy.html.

Ewick, Patricia; Silbey, Susan. "Subversive Stories and Hegemonic Tales: Toward a Sociology of Narrative". *Law and Society Review* 2 (1995):197–226.

FA, MNIE. "Popiera Konfederację, chce stworzyć Królestwo Podhalańskie. Kim jest Sebastian Pitoń?". *TVP.info*, 12.01.2021. Accessed 30.08.2024. https://web.archive.org/web/20210127042215/https://www.tvp.info/51760945/sebastian–piton–kim–jest–popiera–konfederacje–chce–stworzyc–krolestwo–podhalanskie–inicjator–akcji–goralskie–veto–tvp–info.

Fairclough, Norman. *Critical Discourse Analysis: The Critical Study of Language*. London: Longman, 1995.

Fairclough, Norman. *Discourse and Social Change*. Cambridge: Polity Press, 1992.

Falco, Miriam. "10 infants dead in California whooping cough outbreak". *CNN*, 20.10.2010. Accessed 30.08.2024. https://edition.cnn.com/2010/HEALTH/10/20/california.whooping.cough/index.html.

Falkowska, Beata. "Czy szczepić dzieci?". *Nasz Dziennik*, 10.10.2014. Accessed 06.09.2024. https://naszdziennik.pl/mysl/102489,czy–szczepic–dzieci.html.

Federal Ministry of the Interior, Building and Community (BMI), "Disinformation Related to the Coronavirus", *Federal Ministry of the Interior, Building and Commu-*

nity, no data, accessed 10.10.2025 https://www.bmi.bund.de/SharedDocs/schwerpunkte/EN/disinformation/disinformation-related-to-the-coronavirus.html.

Fenster, Mark. *Conspiracy Theories: Secrecy and Power in American Culture.* 2nd ed. Minneapolis: University of Minnesota Press, 2008.

Ferfecki, Wiktor. "Fałszywa komisja posła Brauna". *Rzeczpospolita*, 28.03.2022. Accessed 06.09.2024. https://www.rp.pl/polityka/art35961971-falszywa-komisja-posla-brauna.

Ferfecki, Wiktor. "Ostrzeżenie przed poglądami posła Roberta Majki". *Rzeczpospolita*, 25.09.2019. Accessed 06.09.2024. https://www.rp.pl/polityka/art1114851-ostrzezenie-przed-pogladami-posla-roberta-majki.

Fijałkowski, Dominik. "Szczepionki? Chcą mieć wybór. Protest rodziców w Inowrocławiu [zdjęcia]". *Express Bydgoski*, 24.08.2017. Accessed 30.08.2024. https://expressbydgoski.pl/szczepionki-chca-miec-wybor-protest-rodzicow-w-inowroclawiu-zdjecia/ar/12415800.

Fiorina, Morris P. and Abrams, Samuel J. "Political Polarization in the American Public". *Annual Review of Political Science* 11, no. 1 (2008):563–588.

Forrester, Georgia. "Kiwi mother and healthy food advocate petitions health star rating". *Stuff*, 14.11.2016. Accessed 06.09.2024. https://www.stuff.co.nz/life-style/wel-good/86300986/kiwi-mother-and-healthy-food-advocate-petitions-health-star-rating.

Foucault, Michael. *Historia szaleństwa w dobie klasycyzmu.* Warsaw: PIW, 1987.

Foucault, Michael. *Nadzorować i karać. Narodziny więzienia.* Warsaw: Aletheia, 2021.

Fraser, Małgorzata. "Eksperci: rosyjska dezinformacja antyszczepionkowa rośnie w siłę wraz z wariantem Delta". *CyberDefence 24*, 09.08.2021. Accessed 30.08.2024. https://www.cyberdefence24.pl/eksperci-rosyjska-dezinformacja-antyszczepionkowa-rosnie-w-sile-wraz-z-wariantem-delta.

Fuchs, Christian. *Communicating Covid-19: Everyday Life, Digital Capitalism, and Conspiracy Theories in Pandemic Times.* Leeds, Emerald Publishing Limited, 2021.

Fundacja Życie i Rodzina. "O liniach płodowych z aborcji – dr n. med. Dorota Jarczewska". *YouTube*, 21.10.2022. Accessed 06.09.2024. https://www.youtube.com/watch?v=e4gOX4vSAPs.

Furmaniuk, Anna. "O mnie". *Furmaniuk.com*, no data. Accessed 06.09.2024. https://www.furmaniuk.com.pl/id/6/o-mnie.

Furtak, Ewa. "Lekarka niezłomna. Rodzice polecają ją sobie, gdy chcą zaświadczenie o przeciwwskazaniach do szczepienia". *Wysokie obcasy*, 20.11.2021. Accessed 06.09.2024. https://www.wysokieobcasy.pl/wysokie-obcasy/7,173910,27825864,katarzyna-bross-walderdorff-lekarka-niezlomna-rodzice.html.

Furtak, Ewa. "Protestowała w Poznaniu, wspiera ruch antyszczepionkowy. Kim jest lekarka Katarzyna Bross-Walderdorff?". *Wyborcza.pl*, 02.08.2021. Accessed 06.09.2024. https://www.bielskobiala.wyborcza.pl/bielskobiala/7,88025,27402965,leczy-homeopatia-i-wspiera-ruch-antyszczepionkowy-kim-jest.html.

Fursai, Oleksandra and Danylenko, Serhiy. "Vaccinodemic as a Component of the Global Hybrid Conflict Between Democracy and Autocracy: The Case of Ukraine." *Rocznik Instytutu Europy Środkowo-Wschodniej* 20, no. 2 (2022): 19–45.

Gadamer, Hans-Georg. *Truth and Method.* London: Bloomsbury Academic, 2004.

Gądek, Jacek. "Ewa Stankiewicz i Grzegorz Braun. Kim są". *Onet,* 21.11.2014, https://wiadomosci.onet.pl/tylko–w–onecie/ewa–stankiewicz–i–grzegorz–braun–kim–sa/ykkfd.

Gelzok, Zbigniew. "Zbigniew Gelzok Geniusza Historia Prawdziwa". *Prawo do Życia,* no data. Accessed 06.09.2024. https://web.archive.org/web/20161103183404/http://prawodozycia.pl/services/zbigniew–gelzok–historia–prawdziwa/.

Georgiou, Aristos. "Anti–vax Movement Listed by World Health Organization as One of the Top 10 Health Threats for 2019". *Newsweek,* 15.01.2019. Ac. https://www.newsweek.com/world-health-organization-who-un-global-health-air-pollution-anti-vaxxers-1292493.

Gill, Ann. *Rhetoric and Human Consciousness: A History.* Long Grove, IL: Waveland Press, 2006.

Global Affairs Canada, *Foreign Interference Threats to Canada's Democratic Process* (Ottawa: Government of Canada, 2021) accessed 13.04.2025 https://www.canada.ca/en/security-intelligence-service/corporate/publications/foreign-interference-threat-to-canadas-democratic-process.html.

Głuchowski, Piotr. "»Tworzymy listy śmierci!«. Wojciech Olszański i Kamractwo Rodaków na drodze do Sejmu". *Wyborcza.pl,* 17.10.2022., www.wyborcza.pl/duzyformat/7,127290,29022949,kamractwo–rodakow–na–drodze–do–sejmu–pis–to–dla–nich–zdrajcy.html.

Głuchowski, Piotr. "W PiS antyszczepionkowcy górą: Kiedyś, nie daj Boże, czip będzie konieczny, żeby wyjść z domu". *Wyborcza.pl,* 20.01.2022. Accessed 5.09.2024. https://wyborcza.pl/duzyformat/7,127290,28022089,w–pis–antyszczepionkowcy–gora–kiedys–nie–daj–boze–czip–bedzie.html.

Głuchowski, Piotr; Bielecka, Katarzyna. "Putin's Agents of Chaos: How Pro-Russian Trolls Poison Poland's Infosphere". *Wyborcza.pl,* 18.08.2023. Accessed 01.09.2024. https://wyborcza.pl/7,173236,30092711,putin-s-agents-of-chaos-how-pro-russian-trolls-poison-poland-s.html.

Godlee, Fiona; Smith, Jane; Marcovitch, Harvey. "Artykuł Wakefielda wiążący szczepionkę MMR z autyzmem był oszustwem". *Pacjent.pl,* 19.09.2018. Accessed 30.08.2024. https://www.mp.pl/pacjent/szczepienia/sytuacje-szczegolne/59132,artykul-wakefielda-wiazacy-szczepionke-mmr-zautyzmem-byl-oszustwem

Godlewska, Ilona. "Kontrowersyjne tezy lekarza. »Mimo 300 lat istnienia szczepień nie udowodnili, że one cokolwiek działają«". *Wyborcza.pl,* 29.11.2016. Accessed 06.09.2024. https://trojmiasto.wyborcza.pl/trojmiasto/7,35612,21031892,kontrowersyjne-tezy-lekarza-mimo-300-lat-istnienia-szczepien.html.

Goldenberg, Maya J. *Vaccine hesitancy public trust, expertise, and the war on science.* Pittsburgh, Pittsburgh University Press, 2021.

Górniak, Katarzyna. "Lekarka rozpowszechniała teorie spiskowe. Została ukarana, ale zapewnia, że nie spocznie", *TVN24*. 05.11.2022. Accessed 06.09.2024. https://www.fakty.tvn24.pl/zobacz-fakty/lekarka-anna-furmaniuk-rozpowszechniala-teorie-spiskowe-zostala-ukarana-ale-zapewnia-ze-nie-spocznie-ra1123698-6324178.

Górski, Rafał. "Dlaczego Rosja zaatakowała Ukrainę? Jak bronić Polski?". *Tygodnik Spraw Obywatelskich*, 24.05.2022. Accessed 06.09.2024. https://instytutsprawobywatelskich.pl/dlaczego-rosja-zaatakowala-ukraine-jak-bronic-polski.

Górski, Rafał. "Elektrosmog – wyrok sądu, rak mózgu i telefon komórkowy". *Tygodnik Spraw Obywatelskich*, 21.02.2023. Accessed 06.09.2024. https://instytutsprawobywatelskich.pl/elektrosmog-wyrok-sadu-rak-mozgu-i-telefon-komorkowy.

Górski, Rafał. "O Klausie Schwabie, Światowym Forum Ekonomicznym i globalistach z Davos". *Tygodnik Spraw Obywatelskich*, 21.06.2022. Accessed 06.09.2024. https://instytutsprawobywatelskich.pl/o-klausie-schwabie-swiatowym-forum-ekonomicznym-i-globalistach-z-davos.

Górski, Rafał. "O nauce i Nauce". *Tygodnik Spraw Obywatelskich*, 12.05.2020. Accessed 06.09.2024. https://instytutsprawobywatelskich.pl/o-nauce-i-nauce.

Górski, Rafał. "Pandemia udawana, pieniądze prawdziwe". *Tygodnik Spraw Obywatelskich*, 05.06.2020. Accessed 06.09.2024. https://instytutsprawobywatelskich.pl/gorski-prowokuje-pandemia-udawana-pieniadze-prawdziwe.

Górski, Rafał. "Polowania na lekarzy. Dr Martyka, COVID i utajnione rozprawy sądowe". *Tygodnik Spraw Obywatelskich*, 11.04.2023. Accessed 06.09.2024. https://instytutsprawobywatelskich.pl/polowania-na-lekarzy-dr-martyka-covid-i-utajnione-rozprawy-sadowe.

Górski, Rafał. "Szczepienia. Trzy pytania do Ministra Zdrowia". *Tygodnik Spraw Obywatelskich*, 18.05.2021. Accessed 06.09.2024. https://instytutsprawobywatelskich.pl/szczepienia-trzy-pytania-do-ministra-zdrowia.

Górski, Rafał. "Piotr Witczak: System szuka na mnie »haka«". *Tygodnik Spraw Obywatelskich*, 05.12.2023. Accessed 06.09.2024. https://instytutsprawobywatelskich.pl/piotr-witczak-system-szuka-na-mnie-haka.

Górski, Rafał (editor). "Ratuj Dzieci! – Edukacja technologiczna: e-uzależnienia i elektrosmog". *Instytut Spraw Obywatelskich*, 2020. Accessed 06.09.2024. https://instytutsprawobywatelskich.pl/publikacje/ratuj-dzieci-edukacja-technologiczna-e-uzaleznienia-i-elektrosmog.

Górski, Rafał; Swolkień, Olaf; Majchrzak, Anna; Doliwa-Górska, Magdalena. "Do Członków Stowarzyszenia Obywatele Obywatelom". *Silo Tips*, 23.11.2017. Accessed 06.09.2024. https://silo.tips/download/od-do-czonkow-stowarzyszenia-obywatele-obywatelom.

Grabiec, Piotr. "Nadajnik Play podpalony. Przeciwnicy 5G świętują, a razem z nimi asystent jednego z posłów". *Spider's Web*, 26.05.2020. Accessed 06.09.2024. https://spidersweb.pl/2020/05/play–stacja–bazowa–pozar.html.

Greenfield, Adam. *Radical Technologies. The Design of Everyday Life*. London, Verso Books, 2017. Kindle.

Guziak–Nowak, Magdalena. "Antykoncepcja – druga strona medalu", *Przewodnik Katolicki*, no data. Accessed 06.09.2024. https://www.przewodnik-katolicki.pl/Archiwum/2016/Przewodnik-Katolicki-16-2016/Spoleczenstwo/Antykoncepcja-druga-strona-medalu.

Hałas, Teresa. "Zapytanie nr 1659 do prezesa Rady Ministrów w sprawie instalacji sieci 5G". *Sejm Rzeczpospolitej Polskiej*, 12.10.2020, Accessed 5.09.2024. http://www.sejm.gov.pl/sejm9.nsf/InterpelacjaTresc.xsp?key=BUJB4K.

Hake, Witold. "Generał KGB i agent Mosadu o mind control – Witold Hake (25)". *YouTube*, 11.12.2020. Accessed: 30.08.2024. https://www.youtube.com/watch?v=IX5YjIjuyWw.

Hansson, Sten; Madisson, Mari-Liis; and Ventsel, Andreas. "Discourses of Blame in Strategic Narratives: The Case of Russia's 5G Stories." *European Security* 32, no. 1 (2022): 62–84. https://doi.org/10.1080/09662839.2022.2057188.

Hansson, Sten; Madisson, Mari-Liis; and Ventsel, Andreas. "Russia's Strategic Blame Narratives: Comparative Analysis of Domestic and International Media Coverage About 5G." In *Disinformation and Digital Media as a Challenge for Democracy*, edited by Kramer, Michael; Kobylińska, Anna; and Ženko, Maja. Cham: Springer, 2021. https://doi.org/10.1007/978-3-030-73955-3_14.

Harłukowicz, Jacek. "Macierewicz u Rydzyka o eksperymentach z bronią elektromagnetyczną na Polakach". *Wyborcza.pl*. 15.03.2016, Accessed 5.09.2024. https://wroclaw.wyborcza.pl/wroclaw/7,35771,19770604,macierewicz–u–rydzyka–o–eksperymentach–z–bronia–elektromagnetyczna.html.

Hausman, Bernice L. *Antivax. Reframing the vaccine controversy*. Ithaca, ILR Press, 2019.

Häussling, Roger. "A Network Analytical Four-Level Concept for an Interpretation of Social Interaction in Terms of Structure and Agency." In *Mixed Methods Social Networks Research: Design and Applications*, edited by Silvia Domínguez and Betina Hollstein, 90–118. Cambridge: Cambridge University Press, 2014.

Heiss, Raphael; Nanz, Andreas; Matthes, Jörg. "Social media infaormation literacy: Conceptualization and associations with information overload, news avoidance and conspiracy mentality". *Computers in Human Behavior* 148 (2023):107908.

Heilweil, Rebecca. "How the 5G coronavirus conspiracy theory went from fringe to mainstream". *Vox*, 24.04.2020. Accessed 30.08.2024. https://www.vox.com/recode/2020/4/24/21231085/coronavirus–5g–conspiracy–theory–covid–facebook–youtube.

Hermanova, Marie. "From Fringe to Mainstream: QAnon Narratives in Central European Digital Spaces." *AoIR Selected Papers of Internet Research* (March 2022). https://doi.org/10.5210/spir.v2022i0.13020.

Hilts, Donna. "TV Report On Vaccine Stirs Bitter Controversy". *The Washington Post*, 27.04.1982. Accessed 30.08.2024. https://www.washingtonpost.com/archive/local/1982/04/28/tv–report–on–vaccine–stirs–bitter–controversy/80d1fc8a–1012–4732–a517–7976c86ab52d.

Hoc, Czesław. "Interpelacja nr 1736 do ministra cyfryzacji w sprawie lokowania stacji bazowych telefonii komórkowej w strefie uzdrowiskowej A w oparciu o przepisy tzw. megaustawy". *Sejm Rzeczpospolitej Polskiej.* 27.01.2020. Accessed 5.09.2024. http://www.sejm.gov.pl/sejm9.nsf/InterpelacjaTresc.xsp?key=BLFJWP.

Holland, Max. "How Moscow Undermined the Warren Commission". *Washington Decoded*, 30.03.2007. Accessed 30.08.2024. https://www.washingtondecoded.com/site/2007/03/how_moscow_unde_1.html.

Howard, Philip N. and Schliebs, Marcel. "How Russia's Internet Research Agency Built Its Disinformation Campaign." *The Political Quarterly* 92, no. 3 (2021): 532–544.

Hugo Bader, Jacek. *Strażnicy Wolności.* Warsaw, Agora, 2022.

Hybrid CoE, (European Centre of Excellence for Countering Hybrid Threats), *European Centre of Excellence for Countering Hybrid Threats*, Working Paper 29, March 2024, accessed 13.04.2025, https://www.hybridcoe.fi/wp-content/uploads/2024/03/20240306-Hybrid-CoE-Working-Paper-29-The-impact-of-Kremlin-disinformation-WEB.pdf.

Hyra, Michał. "Antyamerykańska i prorosyjska kampania rozpoczęta. Sykulski z Pitoniem powołali w Częstochowie Polski Ruch Antywojenny". *Wyborcza.pl*, 03.02.2023. Accessed 30.08.2024. https://czestochowa.wyborcza.pl/czestochowa/7,48725,29431413,prorosyjska-i-antyamerykanska-kampania-rozpoczeta-sykulski.html.

Indi. "Kulkami w chorobę". *OX.pl*, 13.05.2014. Accessed 06.09.2024. https://www.wiadomosci.ox.pl/kulkami–w–chorobe,26578.

Ingram, David. "Facebook says 126 million Americans may have seen Russia–linked political posts". *Reuters*, 30.10.2017. Accessed 30.08.2024. http://www.reuters.com/article/us-usa-trump-russia-socialmedia-idUSKBN1CZ2OI.

Institute for Strategic Dialogue, *Behind the Curtain: Unveiling Pro-Kremlin Ecosystems in Europe* (London: ISD, January 2025), accessed 13.04.2025 https://www.isdglobal.org/wp-content/uploads/2025/01/Behind-the-curtain-Unveiling-pro-Kremlin-ecosystems-in-Europe.pdf.

Institut Montaigne, *Information Manipulations Around Covid-19: France Under Attack*, Institut Montaigne, Policy Paper, June 2021, accesed 13.04.2025 https://www.institutmontaigne.org/ressources/pdfs/publications/information-manipulations-around-covid-19-france-under-attack-policy-paper_0.pdf.

Istel, Michał. "Krakowskie rodziny »mieszkań nie dostaną«, bo zasiedlą w nich Ukraińców? To fałszywy przekaz". *Konkret24*, 11.03.2022. Accessed 06.09.2024. https://konkret24.tvn24.pl/polska/krakowskie-rodziny-mieszkan-nie-dostana-bo-zasiedla-w-nich-ukraincow-to-falszywy-przekaz-ra1098931-ls5791257.

Iwaniuk, Edyta. "Zdejmij maseczkę, pogoń PiS. Antycovidowcy, antyszczepionkowcy i anty-5G wzywają do wyjścia na ulice". *Wyborcza.pl*, 07.10.2020. Accessed 06.09.2024. https://wyborcza.pl/7,162657,26375707,zdejmij-maseczke-i-pogon-pis-antycovidowcy-antyszczepionkowcy.html.

Jabłonowski, Krzysztof. "Wywiad o nieskuteczności testów na COVID-19: wyjaśniamy manipulacyjne tezy". *TVN24*, 03.09.2020. Accessed 06.09.2024. https://konkret24.tvn24.pl/zdrowie/wywiad-o-nieskutecznosci-testow-na-covid-19-wyjasniamy-manipulacyjne-tezy-ra1028628-ls5789901.

Jackowska, Anna. "Ilu lekarzy-antyszczepionkowców ukarał samorząd lekarski?". *Co w zdrowiu*, 15.12.21. Accessed 06.09.2024. https://www.cowzdrowiu.pl/aktualnosci/post/ilu-lekarzy-antyszczepionkowcow-ukaral-samorzad-lekarski-dane.

Jackowska, Anna. "Lekarka straciła PWZ na rok. Jak często zapadają takie wyroki?". *Cowzdrowiu.pl*, 03.11.2022. Accessed 06.09.2024. https://cowzdrowiu.pl/aktualnosci/post/lekarka-stracila-pwz-na-rok-jak-czesto-zapadaja-takie-wyroki.

Jakubiak, Marek. "Technologia 5G", *Facebook*, 07.07.2019. Accessed 06.09.2024. https://www.facebook.com/watch/?ref=external&v=539788803222400.

Jakubowski Łukasz and Tatar, Anna. *Jak policzyć nienawiść? Hejterzy o Ukraińcach.* Warszawa, Stowarzyszenie "Nigdy Więcej", February 2023. Accessed 06.09.2024. https://www.nigdywiecej.org/docstation/com_docstation/172/jak_policzyc_nienawisc._hejterzy_o_ukraincach.pdf. Kindle.

Jamieson, Kathleen Hall. *Cyberwar: How Russian Hackers and Trolls Helped Elect a President; What We Don't, Can't, and Do Know.* New York: Oxford University Press, 2018.

Janiszewska, Anna. "Protest antyszczepionkowców na ulicy Piotrkowskiej. Zebrali się przy dawnej »Magdzie« ZDJĘCIA". *Express Ilustrowany*, 16.01.2022. Accessed 30.08.2024. https://expressilustrowany.pl/protest-antyszczepionkowcow-na-ulicy-piotrkowskiej-zebrali-sie-przy-dawnej-magdzie-zdjecia/ar/c1-15999603.

Jankowska, Małgorzata. "Blaski i cienie 5G, czyli jak ucywilizować cyfrową cywilizację (1)". *Instytut Spraw Obywatelskich*, 20.12.2019. Accessed 06.09.2024. https://instytutsprawobywatelskich.pl/blaski-i-cienie-5g-czyli-jak-ucywilizowac-cyfrowa-cywilizacje-1.

Janusz, Alina. "Protest przeciwników szczepień przeciwko COVID-19 w centrum Kielc. Mieli flagi, krzyczeli: »Leczyć, a nie szczepić«". *Wyborcza.pl*, 15.01.2022. Accessed 30.08.2024. https://kielce.wyborcza.pl/kielce/7,47262,28005876,protest-p`rzeciwnikow-szczepien-w-centrum-kielc-mieli-flagi.html.

Januszek, Paweł. "[KORONA LUBLIN / WYKŁAD] Dr n. rol. Emilia Sykut-Domańska: »Żyć zdrowo w niezdrowych czasach«". *Postaw na wolność*, 15.12.2021. Accessed 06.09.2024. https://www.postawnawolnosc.pl/2021/12/15/korona-lublin-wyklad-dr-n-rol-emilia-sykut-domanska-zyc-zdrowo-w-niezdrowych-czasach.

Jarynowski, Andrzej. "Pro-Kremlin German Twitter Users Are More Likely to Be Involved in Both Anti-Lockdown and Anti-Vaccine Discourse Than Anti-Kremlin Users." Preprint, *Social Science Research Network*, April 1, 2022. Accessed 13.04.2025. https://papers.ssrn.com/sol3/papers.cfm?abstract_id=4079045.

Jarynowski, Andrzej; Wójta-Kempa, Monika; Belik, Vitaly. "Percepcja "koronawirusa" w polskim Internecie do czasu potwierdzenia pierwszego przypadku zakażenia SARS-CoV-2 w Polsce". *Pielęgniarstwo i Zdrowie Publiczne* 10, no. 2 (2020):89–106.

Jarczewska, Dorota Łucja. "Linie płodowe i ich związek ze szczepieniami przeciwko Covid-19 – aspekty etyczne", *in: Ograniczenia praw i wolności wprowadzone w Polsce w związku z epidemią Covid-19. Część druga – aspekty zdrowotne*. Edited by Bernaciński Łukasz and Kwaśniewski, Jerzy. *Ograniczenia praw i wolności wprowadzone w Polsce w zw. z epidemią Covid-19. Aspekty prawne, medyczne i ekonomiczne*. Warszawa, Wydawnictwo Naukowe Instytutu na rzecz Kultury Prawnej Ordo Iuris, 2022.

Jaskółka, Janusz. "Góralskie Veto i Sebastian Pitoń wspierają Grzegorza Brauna.". *YouTube*, 08.06.2021. Accessed 30.08.2024. https://www.youtube.com/watch?v=vfXlRNDRqzo.

Jelonek, Michał. "Niedawny gość wRealu24 – Rosjanin Ivan Komarenko". *Facebook*, 14.06.2022. Accessed 06.09.2024. https://www.facebook.com/watch/?v=3316023765345327.

Jeżak, Łukasz. "Doradczyni Jacka Majchrowskiego: Sieć 5G posłuży przede wszystkim rozrywce kosztem zdrowia innych osób [Rozmowa]". *Love Kraków*, 09.06.2020. Accessed 06.09.2024. https://lovekrakow.pl/aktualnosci/siec-5g-posluzy-przede-wszystkim-rozrywce-i-zabawie-kosztem-zdrowia-innych-osob-rozmo157.html.

Jędrysik, Miłada. "Dr Basiukiewicz, lockdowny i odporność zbiorowiskowa, czyli dlaczego Polska nie jest Szwecją". *OKO.Press*, 13.04.2021, za: https://www.oko.press/basiukiewicz-lockdown-nie-dzial.

Jędrysik, Miłada. "Ofensywa »niezależnych lekarzy i naukowców«: o pandemii postanowiono w Davos, amantadyna leczy". *OKO.Press*, 07.07.2021., za https://www.oko.press/niezalezni-lekarze-i-naukowcy-o-pandemii.

Johnston, Kirsty. "»It's like a cult«: How anti-vaccine »mumfluencers« are fuelling the Parliament occupation". *Stuff*, 22.02.2022. Accessed 06.09.2024. https://www.stuff.co.nz/national/health/coronavirus/300520631/its-like-a-cult-how-antivaccine-mumfluencers-are-fuelling-the-parliament-occupation.

Johnson, Neil F.; Velasquez, Nicolas; Johnson Restrepo, Nicholas; Leahy, Rhys; Sear, Richard; Gabriel, Nicholas; Larson, H.; Lupu, Yonatan. "Mainstreaming of Conspiracy Theories and Misinformation." *arXiv* (2021). https://arxiv.org/abs/2102.02382.

Jolley, Daniel; Paterson, Jenny L. "Pylons ablaze: Examining the role of 5G COVID–19 conspiracy beliefs and support for violence". *British Journal of Social Psychology* 59, no. 3, (June 2020):628–640;

Józefczyk, Andrzej. "Starosta mówi NIE dla nadajnika 5G w Korczynie". *TerazKrosno.pl.* 02.07.2020. Accessed 30.08.2024. https://www.terazkrosno.pl/wiadomosci/10260–starosta–mowi–nie–dla–nadajnika–5g–w–korczynie.

Kaczmarek, Aneta. "Prezydent Siemianowic na czele ruchu. Chce walczyć z segregacją sanitarną", *PortalSamorządowy.pl,* 6.10.2021. Accessed 5.09.2024. https://www.portalsamorzadowy.pl/ochrona-zdrowia/prezydent-siemianowic-na-czele-ruchu-chce-walczyc-z-segregacja-sanitarna,317519.html

Kaczmarek, Anna. "Czego naucza Jerzy Zięba w Polskim Radiu? Przesłuchałam audycję i boję się, że ktoś uwierzy w to, co on tam wygaduje". *naTemat.pl,* 31.10.2016. Accessed 06.09.2024. https://natemat.pl/zdrowie/193489,czego-naucza-jerzy-zieba-w-polskich-radiu-tym-razem-wzial-sie-za-psychiatrie.

Kaczmarek, Marcin. "Krem anty–5G. Miliony dolarów zysku ze sprzedaży". *ITBiznes,* 29.05.2020. Accessed 06.09.2024. https://itbiznes.pl/technologie/krem–anty–5g–miliony–dolarow–zysku–ze–sprzedazy.

Kaczmarek, Michał. "Protest antyszczepionkowców przed gabinetem prof. Parczewskiego. »Stop segregacji sanitarnej«". *wSzczecinie.pl,* 18.05.2021. Accessed 30.08.2024. https://wszczecinie.pl/index.php/protest-antyszczepionkowcow-przed-gabinetem-prof-parczewskiego-stop-segregacji-sanitarnej/37812.

Kaki. "Demonstracja antyszczepionkowców pod domem prezydenta Wałbrzycha. Nazwali go doktorem Mengele". *Wyborcza.pl,* 03.05.2021. Accessed 30.08.2024. https://wroclaw.wyborcza.pl/wroclaw/7,35771,27045236,antyszczepionkowcy–demonstruja–w–walbrzychu–poszli–pod–dom.html.

Kak. "Lekarka głośno powiedziała, co myśli o pandemii. Zakazano jej wykonywania zawodu". *Fronda.pl,* 07.09.2020. Accessed 06.09.2024. https://www.fronda.pl/a/lekarka–glosno–powiedziala–co–mysli–o–pandemii–zakazano–jej–wykonywania–zawodu,149823.html.

Kaługa, Mateusz. "Marsz »Stop ukrainizacji Polski« w Warszawie z udziałem radnego z Szydłowca. Arkadiusz Sokołowski pokazany ... w rosyjskiej telewizji". *Echodnia.eu,* 03.10.2022. Accessed 06.09.2024. https://echodnia.eu/radomskie/marsz–stop–ukrainizacji–polski–w–warszawie–z–udzialem–radnego–z–szydlowca–arkadiusz–sokolowski–pokazany–w–rosyjskiej–telewizji/ar/c1–16928313

Kamińska, Agnieszka. "Marsz antyszczepionkowców ulicami Gdańska. Na transparentach: »My się nie szczepimy«, »Stop segregacji sanitarnej«". *Nowiny24,* 23.10.2021. Accessed 30.08.2024,. https://nowiny24.pl/marsz-antyszczepionkowcow-ulicami-gdanska-na-transparentach-my-sie-nie-szczepimy-stop-segregacji-sanitarnej/ar/c1-15867695.

Karami, Amir; Lundy, Morgan; Webb, Frank; Turner-McGrievy, Gabrielle; McKeever, Brooke W.; McKeever, Robert. "Identifying and Analyzing Health–Related Themes in Disinformation Shared by Conservative and Liberal Russian Trolls on Twitter". *International Journal of Environmental Research and Public Health* 18, no. 4 (2021):2159.

Karpieszuk, Wojciech. "Konwencja Konfederacji przed wyborami do europarlamentu. »Zatrzymamy zalew imigrantów«. *Wyborcza.pl*, 06.04.2019. Accessed 06.09.2024. http://warszawa.wyborcza.pl/warszawa/7,54420,24622911,konwencja-konfederacji-przed-wyborami-do-europarlamentu-zatrzymamy.html.

Karpiuk, Dawid. "»Kamraci! To jest kij na poselski ryj!« – krzyczy, wymachując pałką. Tłum zaraz chwyta". *Newsweek*, 03.03.2022. Accessed 30.08.2024. https://www.newsweek.pl/polska/spoleczenstwo/antyszczepionkowcy-i-radykalna-prawica-ida-reka-w-reke-dlaczego/23l9v2z.

Kartografia Ekstremalna. "Kiedy miesiąc temu opublikowaliśmy mapę nadmiarowych zgonów ... ". *Facebook*, 18.12.2021. Accessed: 30.08.2024. https://www.facebook.com/153690641483461/posts/1808514839334358.

Kasprak, Alex. "Did a 5G Cellular Network Test Cause Hundreds of Birds to Die?". *Snopes*, 13.11.2018. Accessed 30.08.2024. https://www.snopes.com/fact-check/5g-cellular-test-birds.

Kaszuba-Zwoińska, Jolanta; Gremba, Jerzy; Gałdzińska-Calik, Barbara; Wójcik-Piotrowicz, Karolina; Thor, Piotr J. "Electromagnetic field induced biological effects in humans". *Przegląd Lekarski* 72, no 11 (2015):636–641.

Katolickie Stowarzyszenie Lekarzy Polskich. "Listy otwarte do Prezydenta i Premiera RP Katolickiego Stowarzyszenia Lekarzy Polskich ws. publikacji orzeczenia TK". *Radio WNET*, 10.12.2020. Accessed 06.09.2024. https://www.wnet.fm/2020/12/10/listy-otwarte-do-prezydenta-i-premiera-rp-katolickiego-stowarzyszenia-lekarzy-polskich-ws-publikacji-orzeczenia-tk.

Keegan, Katrina. "Clarity for Friends, Confusion for Foes: Russian Vaccine Propaganda in Ukraine and Serbia." *Harvard Kennedy School Misinformation Review* 3, no. 3 (25.04.2022). Accessed 13.04.2025. https://misinforeview.hks.harvard.edu/article/clarity-for-friends-confusion-for-foes-russian-vaccine-propaganda-in-ukraine-and-serbia.

Keller, Lech. "Nowa Chazaria, czyli tajny plan Putina–Szarona?". *Wolna Palestyna*, 07.06.2014. Accessed 06.09.2024. https://palestyna.wordpress.com/2014/06/07/nowa-chazaria-czyli-tajny-plan-putina-szarona.

Kendall, Gavin; Wickham, Gary. *Using Foucault's methods*. London, Thousend Oaks, New Dehli: SAGE, 1999.

Kennedy Jr, Robert F. "What are we doing to our children?". *Twitter*, 21.03.2019. Accessed 06.09.2024. http://www.twitter.com/RobertKennedyJr/status/1108734709041938432.

Key, Jonathan. *Among the Truthers: A Journey Through America's Growing Conspiracist Underground* (New York: Harper Collins, 2011. Kindle.

Khalil Lydia and Roose, Joshua. "Anti-Government Extremism in Australia: Understanding the Australian Anti-Lockdown Freedom Movement as a Complex Anti-Government Social Movement". *Perspectives on Terrorism* 17, no. 1 (2024).

Kirk, Katherine. "How Russia Sows Confusion in the U.S. Vaccine Debate". *Foreign Policy*, no data. Accessed: 30.08.2024. https://foreignpolicy.com/2019/04/09/in-the-united-states-russian-trolls-are-peddling-measles-disinformation-on-twitter.

Kitarovic, Katarina. "Polio in a Post-Vaccine World: The Consequences of Anti-Vaccination Sentiments". *Harvard International Review*, 08.03.2023. Accessed 30.08.2024. https://hir.harvard.edu/polio-in-a-post-vaccine-world-the-consequences-of-anti-vaccination-sentiments.

Kisielewska, Agnieszka. "PROF. ZW. DR HAB. ROMAN ZIELIŃSKI "O SZCZEPIONCE GENETYCZNEJ PFIZERA I TESTACH PCR". *Biuletyn Informacji Publicznej Starostwa powiatowego w Siedlcach*, no data. Accessed 06.09.2024. https://www.bip.powiatsiedlecki.pl/plik,8953,zalacznik-nr-10-prof-zw-dr-hab-roman-zielinski-szczepienia-inzynieria-genetyczna-sars-cov-2.pdf.

KJ. "Protest w Elblągu. »Wyłącz telewizor, włącz myślenie«, »Przestańcie szczepić, zacznijcie leczyć« (film i zdjęcia)". *Info.Elblag.pl*, 24.01.2022. Accessed 30.08.2024. https://www.info.elblag.pl/37,68460,Protest-w-Elblagu-Wylacz-telewizor-wlacz-myslenie-Przestancie-szczepic-zacznijcie-leczyc-film-i-zdjecia.html.

Klinsky, Chris. *Szczepionka na koronawirusa – zagrożenie dla ludzkości?* Warszawa: Wydawnictwo św. Tomasza z Akwinu, 2021.

KMJ Archiwum Polityczne. "Co z tym 5G? »Zostaliśmy tyle razy oszukani, że dmuchamy na zimne«. Jarosław Sachajko". *YouTube*, 11.03.2022. Accessed 06.09.2024. https://www.youtube.com/watch?v=uNsbJfiu7wg.

KMJ Archiwum Polityczne. "Kto w polskim parlamencie ukrywa prawdę o 5G Poseł Robert Majka wRealu24 cz 1". *YouTube*, 25.08.2019. Accessed 06.09.2024. https://www.youtube.com/watch?v=19CaR6LXm80.

Koalicja Polska wolna od 5G. "5G – Stop 5G – Dlaczego protestujemy – Janusz Zagórski w rozmowie z prof. Jacek Nowak". *YouTube*, 20.01.2020. Accessed 06.09.2024. https://www.youtube.com/watch?v=tOAhMLRcU5Y.

Kobeszko, Łukasz. "Metropolie i prowincja. Polityczna wojna światów". *Rzeczpospolita*, 07.08.2020. Accessed 30.08.202. https://www.rp.pl/plus-minus/art585231-metropolie-i-prowincja-polityczna-wojna-swiatow.

Koehler, Daniel "Siren calls of anti-government extremism: Far-right influences on the German anti-vax (»Querdenken«) protest milieu through music". *Behavioral Sciences of Terrorism and Political Aggression* 16, no. 2 (2023):1–22.

Kokoszkiewicz, Mateusz. "Policja przyznała ochronę prezydentowi Wałbrzycha. Po hejcie i groźbach od antyszczepionkowców". *Wyborcza.pl*, 06.05.2021. Accessed

30.08.2024. https://walbrzych.wyborcza.pl/walbrzych/7,178336,27056710,policja-przyznala-ochrone-prezydentowi-walbrzycha-po-hejcie.html.

Konarska, Iwona. "Pierwszy zastępca przewodniczącego Okręgowego Sądu Lekarskiego w Tarnowie i jego poglądy". *Termedia*, 14.12.21. Accessed 06.09.2024. https://www.termedia.pl/koronawirus/Pierwszy-zastepca-przewodniczacego-Okregowego-Sadu-Lekarskiego-w-Tarnowie-i-jego-poglady,44843.html.

Konfederacja. "Badania nad szczepionką Pfizer-BioNTech trwały tylko kilka miesięcy". *Twitter*, 07.12.2020. Accessed 06.09.2024. https://www.twitter.com/KONFEDERACJA_/status/1335968878355226626.

Konfederacja. "Musimy natychmiast zatrzymać ten chory lockdown, całkowicie otworzyć gospodarkę". *Twitter*, 13.02.2021. Accessed 06.09.2024. https://www.twitter.com/KONFEDERACJA_/status/1360609171050270725.

Konfederacja. "Tego potrzebuje teraz PolskaPL! Jeśli się zgadzasz, podaj dalej!!". *Twitter*, 13.09.2021. Accessed 06.09.2024. https://twitter.com/KONFEDERACJA_/status/1437334131550167040.

Konfederacja. "UWAGA, koniec z leczeniem #COVID-19 amantadyną!". *Twitter*, 03.12.2020. Accessed 06.09.2024. https://twitter.com/KONFEDERACJA_/status/1334424821720109056.

Konopka, Krzysztof. "Protest antyszczepionkowców w Katowicach. Część osób znów w mundurach. »My jesteśmy czwartą falą«". *Gazeta.pl*, 07.08.2021. Accessed 30.08.2024. https://wiadomosci.gazeta.pl/wiadomosci/7,114883,27426268,protest-antyszczepionkowcow-w-katowicach-stop-segregacji.html.

Kontrrewolucja.net. "Dr Anna Martynowska oficjalnie wraca do wykonywania zawodu. Wygrała w sądzie". *Kontrrewolucja.net*, 26.08.2021.. Accessed 06.09.2024. https://www.kontrrewolucja.net/wiadomosci/dr-anna-martynowska-oficjalnie-wraca-do-wykonywania-zawodu-wygrala-w-sadzie.

Kontrrewolucja.net. "Dr Witczak przypomina, że na temat walki z tzw. pandemią »nie było żadnej debaty eksperckiej«. »Walka z wirusem zabiła więcej ludzi niż on sam«". *Kontrrewolucja.net*, 01.10.2021. Accessed 06.09.2024. https://www.kontrrewolucja.net/wiadomosci/dr-witczak-przypomina-ze-na-temat-walki-z-tzw-pandemia-nie-bylo-zadnej-debaty-eksperckiej-walka-z-wirusem-zabila-wiecej-ludzi-niz-on-sam.

Korab-Karpowicz, Włodzimierz Julian; Rutkowski, Ryszard; Sulkowski, Stanisław; Sobaniec-Łotowska, Maria; Kasacka, Irena; Lisiecka-Opalko, Krystyna; Bartyzel, Jacek et al. "Apel naukowców i lekarzy w sprawie szczepień na koronawirusa SARS-CoV-2". *Instytut Spraw Obywatelskich*, 30.11.20. Accessed 06.09.2024. https://instytutsprawobywatelskich.pl/apel-naukowcow-i-lekarzy-w-sprawie-szczepien-na-koronawirusa-sars-cov-2.

Korab-Karpowicz, Włodzimierz Julian; Hałat, Zbigniew; Rutkowski, Ryszard; Borawska, Maria; Śliwiński, Artur; Zajączkowski, Ryszard; Błochowiak, Mariusz et al.

"Trzy pytania do Ministra Zdrowia w sprawie bezpieczeństwa i skuteczności szczepień na SARS–CoV–2". *Tygodnik Spraw Obywatelskich*, 29.04.21. Accessed 06.09.2024. https://instytutsprawobywatelskich.pl/trzy–pytania–do–ministra–zdrowia–w–sprawie–bezpieczenstwa–i–skutecznosci–szczepien–na–sars–cov–2

Korczarowski, Piotr. "Ponownie publikuję mój wczorajszy komentarz, w którym ośmieliłem się postawić znak równości wobec śmierci żyda i goja". *Facebook*, 23.02.2022. Accessed 06.09.2024. https://www.facebook.com/piotr.korczarowski/posts/4489834687787128.

Korolkiewicz, Leszek. "Elektrosmog – puszka Pandory?". *Lepsze Zdrowie* (blog), no data. Accessed 06.09.2024. http://www.lepszezdrowie.info/elektrosmog.

Korzeniewska, Ewa; Krawczyk, Andrzej; Jacek Stańdo. "Pola torsyjne – przykład pseudonaukowej koncepcji w fizyce". *Przegląd Elektrotechniczny* 1 (2021):196–199.

Kosiniak–Kamysz, Władysław. "Czy to prawda, że @AndrzejDuda dogadał się z Amerykanami o 5G". *Twitter*, 25.06.2020. Accessed 06.09.2024. https://twitter.com/KosiniakKamysz/status/1276054582515453953.

Kościn, Żmicier. "Mieszkańcy wsi Puchły nie chcą wieży 5G w swojej miejscowości". *Polskie Radio Białystok*, 15.07.2020. Accessed 30.08.2024. https://www.radio.bialystok.pl/wiadomosci/index/id/186742.

Kowalczyk, Klaudia. "Pacyfizm czy rosyjska racja stanu? Profesor KUL zbiera na antywojenne bilbordy". *Jawny Lublin*, 17.04.23. Accessed 06.09.2024. https://jawnylublin.pl/pacyfizm-czy-rosyjska-racja-stanu-profesor-kul-zbiera-na-antywojenne-bilbordy/.

Kowalewicz, Tomasz. "Profesor KUL przeciw powszechnej akcji szczepień. »Bóg przystosował człowieka do życia w wirusach«". *Wyborcza.pl*, 15.12.20. Accessed 06.09.2024. https://lublin.wyborcza.pl/lublin/7,48724,26610185,profesor-kul-w-szczepionce-na-koronawirusa-widzi-wieksze-zagrozenie.html.

Kowalewicz, Tomasz. "»Szczepienia obniżają IQ i powodują autyzm«. Skandaliczny wykład na KUL–u". *Wyborcza.pl*, 16.06.2017. Accessed 06.09.2024. https://lublin.wyborcza.pl/lublin/7,48724,21969310,szczepienia-obnizaja-iq-i-powoduja-autyzm-skandaliczny-wyklad.html .

Kowalewski, Zbigniew Marcin; Zgliczyński, Stefan; Pankowski, Rafał. "LIST DO KIEROWNICTW ATTAC–POLSKA I ATTAC–FRANCJA". *Archiwum LBC*, 06.12.01. Accessed 06.09.2024. http://www.geocities.ws/lbc_1917_2/128attaclist.htm.

Koz. "Antyszczepionkowcy mają swój Trybunał Narodowy i rozporządzenie: »Nie ma obowiązku wstrzykiwania do ciała obcego ciała«". *Wyborcza.pl*, 01.02.2022. Accessed 30.08.2024. https://wroclaw.wyborcza.pl/wroclaw/7,35771,28058951,antyszczepionkowcy–maja–trybunal–narodowy–i–rozporzadzenie.html.

Koz, UJ. "Protest antyszczepionkowców we Wrocławiu. Z pl. Solnego musiał odjechać szczepciobus [ZDJĘCIA]". *Wyborcza.pl*, 15.01.2022. Accessed 30.08.2024. https://

wroclaw.wyborcza.pl/wroclaw/7,35771,28004636,atyszczepionkowcy-we-wroclawiu.html.

Kozakiewicz, Karolina. "Putinowska propaganda na marszu narodowców we Wrocławiu. Kim są ludzie, którzy zastąpili Międlara?". *Wyborcza.pl*, 12.11.2022. Accessed 06.09.2024. https://www.wroclaw.wyborcza.pl/wroclaw/7,35771,29133285,rosyjska-propaganda-na-marszu-narodowcow-we-wroclawiu-kim-sa.html.

Kozioł, Magdalena. "Antycovidowcy mają własną telewizję. Wiele programów negujących koronawirusa bije rekordy popularności". *Wyborcza.pl*, 14.10.2020. Accessed 06.09.2024. https://wroclaw.wyborcza.pl/wroclaw/7,35771,26393182,wroclawski-trzon-antycovidowego-protestu-ufolog-sygnalistka.html.

Kp. "Protesty w Sieradzu i Zduńskiej Woli". *Nasze.fm*, 15.01.2022. Accessed 30.08.2024. https://www.nasze.fm/news,40015.

Kraushar, Karolina. "Sanok. Konfederacja zorganizowała protest »Stop ukrainizacji Polski«. I nikt na niego nie przyszedł". *Gazeta.pl*, 26.09.2022. Accessed 06.09.2024. https://wiadomosci.gazeta.pl/wiadomosci/7,114883,28955056,sanok-stop-ukrainizacji-polski-na-protescie-organizowanym.html

Kromer, Oktawia. "Rodacy Kamraci grożą śmiercią Igorowi Isajewowi. »Lepszy byłby dla ciebie sznur i gałąź pod ciężarem zgięta«". *Wyborcza.pl*, 04.08.2022. Accessed 30.08.2024. https://warszawa.wyborcza.pl/warszawa/7,54420,28754692,olszanski-i-osadowski-namawiaja-do-bicia-ukraincow-i-groza-smiercia.html.

Kuczyńska, Magda. "SPRAWOZDANIE Z BERLIŃSKIEGO PROTESTU PRZECIW PLANDEMII". *Konfederacja Korony Polskiej*, no data. Accessed 06.09.2024. https://konfederacjakoronypolskiej.pl/sprawozdanie-z-berlinskiego-protestu-przeciw-plandemii.

Kunert, Jan. "Konta antyszczepionkowe pracują dla Kremla. »Potwierdziło się, że paliwo było ze Wschodu«". *Konkret24*, 05.04.2022. Accessed 06.09.2024. https://konkret24.tvn24.pl/polityka/konta-antyszczepionkowe-pracuja-dla-kremla-potwierdzilo-sie-ze-paliwo-bylo-ze-wschodu-ra1100795-ls5791231.

Kursa, Weronika. "Biała księga pandemii koronawirusa – podsumowanie chorych czasów". *Tygodnik Spraw Obywatelskich*, 27.12.2023. Accessed: 06.09.2024. https://instytutsprawobywatelskich.pl/biala-ksiega-pandemii-koronawirusa-podsumowanie-chorych-czasow.

Kurzak, Tomasz. "Gadżety chroniące przed 5G. Oszuści wykorzystują obawy przed nową techniką". *Komputer Świat*, 30.05.2020. Accessed 06.09.2024. https://www.komputerswiat.pl/artykuly/redakcyjne/gadzety-chroniace-przed-5g-oszusci-wykorzystuja-obawy-przed-nowa-technika/l4d7wwn#slajd-2.

Kwiatkowska, Barbara. "Dolny Śląsk. »Rzekoma epidemia«. Lekarka nie wierzy w COVID-19. Odsunięta od zawodu". *Wp.pl*, 04.08.2020. Accessed 06.09.2024. https://wiadomosci.wp.pl/wroclaw/dolny-slask-rzekoma-epidemia-lekarka-nie-wierzy-w-covid-19-odsunieta-od-zawodu-6539494913538880a.

Laf, Ml. "Konfederacja chce uznania Mosbacher za osobę niepożądaną w Polsce". *Polsat News*, 15.07.2020. Accessed 06.09.2024. https://www.polsatnews.pl/wiadomosc/2020-07-15/mosbacher-persona-non-grata-postulat-konfederacji.

Langguth, Johannes; Filkuková, Petra; Brenner, Stefan; Schroeder, Daniel Thilo; Pogorelov, Konstantin. "COVID-19 and 5G conspiracy theories: long term observation of a digital wildfire". *International Journal of Data Science and Analytics* 15 (2023):329–346.

Lasecki, Ronald. "Światopogląd biocentryczny". *Tygodnik Spraw Obywatelskich*, 06.09.2022. Accessed 06.09.2024. https://instytutsprawobywatelskich.pl/swiatopoglad-biocentryczny.

Lee, Sun Kyong; Sun, Juhyung; Jang, Seulki; Connelly, Shane. "Misinformation of COVID-19 vaccines and vaccine hesitancy". *Scientific Reports* 12 (2022). Accessed 30.08.2024. https://www.nature.com/articles/s41598-022-17430-6

LeszekBubelTV. "GRZEGORZ BRAUN – PSEUDOCHRZEŚCIJANIN I UZURPATOR, CZĘŚĆ 1 – TOBIASZ ŻUCHEWICZ". *BitChute*, 08.12.2019. Accessed 06.09.2024. https://www.bitchute.com/video/Tpb15zEL1gRc.

Lewandowski, Arkadiusz. "Antysystemowość ruchu Kukiz'15 w okresie kampanii wyborczej 2015 roku". In *Nowe partie i ruchy polityczne. Praktyka i idee współczesnej demokracji*. Edited by Walecka, Katarzyna and Wojtas, Kinga. Warszawa, Wydawnictwo Naukowe UKSW, 2017.

Leśniewski, Bartłomiej. "Marsz antyszczepionkowców w Warszawie". *Termedia*, 04.06.2017. Accessed 30.08.2024. https://www.termedia.pl/mz/Marsz-antyszczepionkowcow-w-Warszawie,26348.html.

Lindberg, John C.H.; Archer, Denali. "Radiophobia: Useful concept, or ostracising term?". *Progress in Nuclear Energy*, 149 (2022) 1,11.

Linvill, Darren, and Warren, Patrick. "Engaging with Others: How the IRA Coordinated Information Operation Made Friends." *Harvard Kennedy School (HKS) Misinformation Review*, February 14, 2020. Accessed March 30, 2025. https://misinforeview.hks.harvard.edu/article/engaging-ira-coordinated-information-operation/.

Linvill, Darren; Patrick Warren; Steven Sheffield; Leland Fecher; Kylie Gilbert; Jonathon Greco; Alexa Gubanich, et al. *Writers of the Storm: Who's Behind the Ongoing Production of Pro-Russian False Narratives*. Clemson University Media Forensics Hub, October 2024. Accessed March 30, 2025. https://open.clemson.edu/mfh_ci_reports/10.

Lipka, Katarzyna. "Nie, podejrzanymi o zabójstwo na Nowym Świecie nie są obcokrajowcy". *FakeNews.pl*, 20.05.2022. Accessed 06.09.2024. https://fakenews.pl/spoleczenstwo/nie-podejrzanymi-o-zabojstwo-na-nowym-swiecie-nie-sa-obcokrajowcy.

Lukow24. "Prywatny inwestor chce budować maszt telefonii 5G". *Lukow24*.pl, 28.02.2020. Accessed 30.08.2024. https://www.lukow24.pl/wiadomosci/prywatny-inwestor-chce-budowac-maszt-telefonii-5g/XjaoyKLZibKAZL3fq3JB.

Luq. "»Jestem listonoszem, który przyniósł niektórym lekarzom złą wiadomość«. Zięba w *Skandalistach*". *Polsat News*, 29.04.2017. Accessed 06.09.2024. https://www.polsatnews.pl/wiadomosc/2017-04-29/jerzy-zieba-w-skandalistach.

Lurka, Krystian. "Sienkiewicz przestrzega przed szczepieniami. UMB i białostocki OIL wyciągną konsekwencje". *Termedia*, 15.11.2018. Accessed 06.09.2024. https://www.termedia.pl/mz/Sienkiewicz-przestrzega-przed-szczepieniami-UMB-i-bialostocki-OIL-wyciagna-konsekwencje,32090.html.

Lurka, Krystian. "Skutecki: Wielu Polaków wie więcej o szczepieniach niż lekarze". *TerMedia*, 11.06.2018. Accessed 06.09.2024. https://www.termedia.pl/mz/Skutecki-Wielu-Polakow-wie-wiecej-o-szczepieniach-niz-lekarze,30337.html.

Lyotard, Jean-François. "Kondycja ponowoczesna". *Sztuka i Filozofia* 13 (1997):25–35.

Lynas, Mark. "Anti-vaxxers and Russia behind viral 5G COVID conspiracy theory". *Alliance for Science*, 08.04.2020. Accessed 30.08.2024. https://allianceforscience.org/blog/2020/04/anti-vaxxers-and-russia-behind-viral-5g-covid-conspiracy-theory.

Łopatniuk, Paulina. "Wyleczy raka i powiększy biust, czyli Zięba na uniwersytecie", *Patolodzy na klatce* (blog), no data. Accessed 06.09.2024. https://www.web.archive.org/web/20170607024459/https://patolodzynaklatce.wordpress.com/2015/11/24/wyleczy-raka-i-powiekszy-biust-czyli-zieba-na-uniwersytecie.

Łopatyński, Sławomir. "Pikieta antyszczepionkowców w Toruniu. Lekarz: – Pod respiratorami umiera 99 proc. niezaszczepionych". *Wyborcza.pl*, 15.01.2022. Accessed 30.08.2024. https://torun.wyborcza.pl/torun/7,48723,28004722,rozaniec-pod-pomnikiem-kopernika-w-toruniu-intencja-wstrzymanie.html.

Łuniewski, Marcin. "Polscy przyjaciele Aleksandra Łukaszenki". *Rzeczpospolita*, 19.09.2020. Accessed 30.08.2024. https://www.rp.pl/Plus-Minus/309119989-Polscy-przyjaciele-Aleksandra-Lukaszenki.html.

Maciuszczak, Tomasz. "Profesor KUL gwiazdą antycovidowego marszu w Lublinie. Co na to uczelnia?". *Dziennik Wschodni*, 12.10.20. Accessed 06.09.2024. https://www.dziennikwschodni.pl/lublin/profesor-kul-gwiazda-antycovidowego-marszu-w-lublinie,n,1000276120.html.

MacFarquhar, Neil. "A Powerful Russian Weapon: The Spread of False Stories". *The New York Times*, 29.08.2016. Accessed: 30.08.2024. http://www.nytimes.com/2016/08/29/world/europe/russia-sweden-disinformation.html.

Maciej Poręba Kanał Oficjalny. "Maciej Poręba w CEP – co dalej z Ukrainą, plany Putina, czy będzie wojna w Donbasie, Новороссия". *YouTube*, 25.02.2022. Accessed 06.09.2024. https://youtu.be/LpwMH3tkonU.

Maciejasz, Dominika. "Przedsiębiorcy i Strajk Kobiet razem wychodzą na ulice. Lempart: Południe Polski zaczęło obywatelskie nieposłuszeństwo", *Wyborcza.pl*,

19.01.2021. Accessed 30.08.2024. https://wyborcza.biz/biznes/7,177151,26700795, przedsiebiorcy-i-pioruny-wyjda-na-ulice-marta-lempart-poludnie.html.

Maciejewski, Jakub. "Guru antyszczepionkowców fotografuje młode "dziewice", promuje rosyjskie treści, twierdzi, że szczepionka przyciąga magnes". *wPolityce.pl*, 28.12.2021. Accessed 06.09.2024. https://wpolityce.pl/polityka/579752-antyszczepionkowe-fakenewsy-zbigniewa-halata.

Majka, Robert. "Interpelacja nr 28857 do prezesa Rady Ministrów w sprawie realizacji planu wcielenia w życie w Polsce w 2020 r. technologii nowej generacji 5G w kontekście realizacji ustawy z druku nr 1066 i jej wpływu na populację Polek i Polaków". *Sejm Rzeczypospolitej Polskiej*, 13.01.2019. Accessed 06.09.2024. http://www.sejm.gov.pl/sejm8.nsf/InterpelacjaTresc.xsp?key=B8PJWS.

MAK, pp, gg. "Antyszczepionkowcy otoczyli mobilny punkt szczepień. Medycy wezwali policję, będzie 15 wniosków do sądu". *TVN24*, 01.08.2021. Accessed 30.08.2024. https://tvn24.pl/pomorze/koronawirus-w-polsce-gdynia-antyszczepionkowcy-otoczyli-szczepibus-5163021.

Makowska, Beata. "Protest antyszczepionkowców we Wrocławiu". *Radio Wrocław*, 04.12.2021. Accessed 30.08.2024. https://www.radiowroclaw.pl/articles/view/114727/Protest-antyszczepionkowcow-we-Wroclawiu.

Małecki, Mikołaj. "Sejm niemal jednogłośnie wezwał UE do niewydawania wiz Rosjanom. Dwóch posłów wstrzymało się od głosu". *Dziennik. Gazeta Prawna*, 29.09.2022. Accessed 06.09.2024. https://www.gazetaprawna.pl/artykuly/8558510, wizy-dla-rosjan-ue-sejm-uchwala-braun-berkowicz.html.

Manning, Martin J.; Romerstein, Herbert. "Disinformation", in: *Historical Dictionary of American Propaganda*, edited by Manning, Martin J.; Romerstein, Herbert. Greenwood: Bloomsbury Academic, 2004.

Marcinkiewicz, Paweł. "Nalot ABW na mieszkania Kamratów i antyszczepionkowców. W tle współpraca z obcym wywiadem i podżeganie do zabójstwa". *Wyborcza.pl*, 13.01.2024. Accessed 30.08.2024. https://warszawa.wyborcza.pl/warszawa/7,54420,30588856,koniec-bezkarnosci-kamratow-interwencja-abw-w-mieszkaniach.html.

Marczyk, Krzysztof. "Protest »Stop segregacji sanitarnej« w Koszalinie. Sprzeciwiają się przymusowi szczepień [ZDJĘCIA]". *GK24*.pl, 15.01.2022. Accessed 30.08.2024. https://gk24.pl/protest-stop-segregacji-sanitarnej-w-koszalinie-sprzeciwiaja-sie-przymusowi-szczepien-zdjecia/ar/c1-15998779.

Markiewicz, Kamil. "Protestują przeciw budowie masztu 5G. Konstrukcja ma stanąć w okolicach ulicy Prostej". *TVP3 Kielce*, 13.07.2020. Accessed 30.08.2024. https://kielce.tvp.pl/48955032/protestuja-przeciw-budowie-masztu-5g-konstrukcja-ma-stanac-w-okolicach-ulicy-prostej.

Martin, Rachel. "Former Anti-Vaccine Mom Explains How Movement Pulled Her In, And How She Left". *NPR*, 18.01.2021. Accessed 06.09.2024. https://www.npr.org/

2021/01/18/957981974/how-anti-vaccine-movement-could-hurt-efforts-to-end-pandemic.

Martyka, Zbigniew. "Efekty wiary w pandemię". *Agencja Informacyjna*, 01.02.21. Accessed 06.09.2024. https://www.agencja-informacyjna.com/dr-zbigniew-martyka-efekty-wiary-w-pandemie.

Marwick, Alice E. "Why Do People Share Fake News? A Sociotechnical Model of Media Effects." *Georgetown Law Technology Review* 2, no. 2 (2018): 474–512.

Marwick, Alice E., and Partin, William Clyde. "Constructing Alternative Facts: Populist Expertise and the QAnon Conspiracy." *New Media & Society* 26, no. 5 (2022): 2535–2555. https://doi.org/10.1177/14614448221090201.

Mayring, Philipp. *Qualitative Content Analysis: Theoretical Foundation, Basic Procedures and Software Solution*. Klagenfurt: SSOAR, 2014. Accessed 20.04.2025. https://www.ssoar.info/ssoar/bitstream/handle/document/39517/ssoar-2014-mayring-Qualitative_content_analysis_theoretical_foundation.pdf.

Mb. "Gorzów Wlk. Groźby wobec członków Rady Medycznej na wiecu antyszczepionkowców. Sprawą zajmuje się policja". *Gazeta.pl*, 16.01.2022. Accessed 30.08.2024. https://wiadomosci.gazeta.pl/wiadomosci/7,173952,28007676,gorzow-wlk-grozby-pod-adresem-rady-medycznej-na-wiecu-antyszczepionkowcow.html.

Meese, James; Firth, Jordan and Wilken, Rowan. "COVID-19, 5G Conspiracies and Infrastructural Futures." *Media International Australia* 177, no. 1 (2020): 30–46.

Medexpress. "Kukiz'15 nie ma nic wspólnego z projektem ustawy znoszącej obowiązek szczepień". *Medexpress.pl*, 16.07.2018. Accessed 06.09.2024. https://www.medexpress.pl/ochrona-zdrowia/kukiz-15-nie-ma-nic-wspolnego-z-projektem-ustawy-znoszacej-obowiazek-szczepien-71051.

Media. "Lekarze sceptyczni wobec szczepień. Naczelna Izba Lekarska wzywa na przesłuchania". *Onet*, 09.04.2021. Accessed 06.09.2024. https://www.wiadomosci.onet.pl/bialystok/szczepienia-naczelna-izba-lekarska-wzywa-na-przesluchania-sceptycznych-lekarzy/mzdx51s.

Media Narodowe. "5G – realne czy kreowane ZAGROŻENIE? Dostawca 5G z zarzutami korupcyjnymi | dr T. Rzymkowski". *CDA*, no data. Accessed 06.09.2024. https://www.cda.pl/video/48391401c.

Media Narodowe. "Dr Zbigniew Hałat gościem Tematu Dnia | MEDIA NARODOWE". *CDA*, no data. Accessed 06.09.2024. https://www.cda.pl/video/7769911e6.

Media Narodowe. "Media Narodowe – Wojsko WYMUSZA szczepienia? Piotrowski OSTRO do Błaszczaka: TAK NIE MOŻE BYĆ! – 27. 07. 2021". *CDA*, no data. Accessed 06.09.2024. https://www.cda.pl/video/80448802d.

Media Narodowe. "SZOKUJĄCE, ALE CZY PRAWDZIWE? Jak wprowadza się 5G w Polsce? | M. Jarosiewicz". *CDA*, no data. Accessed 06.09.2024. https://www.cda.pl/video/3673192d0.

Medyczne Grono. "Harmonia Kosmosu 2022: prof. Andrzej Frydrychowski o leczeniu chorób przewlekłych". *Medyczne Grono*, 28.06.2022. Accessed 06.09.2024.

https://www.medycznegrono.pl/wydarzenia/harmonia-kosmosu-2022-prof-andrzej-frydrychowski-o-leczeniu-chorob-przewleklych.

Meltzer, Marisa. "QAnon's Unexpected Roots in New Age Spirituality". *The Washington Post Magazine*, 29.03.2021. Accessed 07.09.2024. http://washingtonpost.com/magazine/2021/03/29/qanon-new-age-spirituality.

MG. "Mimo protestów, starostwo wydaje pozytywne opinię dla nowych stacji bazowych". *WadowiceOnline.pl*, 23.07.2020. Accessed 30.08.2024. https://wadowiceonline.pl/biznes/16612-mimo-protestow-starostwo-wydaje-pozytywne-opinie-dla-nowych-stacji-bazowych.

Miastokolobrzeg. "Protest przeciwko sieci 5G". *YouTube*, 26.04.2020. Accessed 30.08.2024. https://www.youtube.com/watch?v=73a8483itqE.

Mider, Daniel. "Open Source Intelligence on the Internet – Categorisation and Evaluation of Search Tools." *Przegląd Bezpieczeństwa Wewnętrznego* 31 (2024): 383–412.

Mierzyńska, Anna. "»Czwarta noga Konfederacji«. Antyszczepionkowcy budują ugrupowanie Zjednoczeni". *OKO.Press*, 05.06.2020. Accessed 06.09.2024. https://www.oko.press/antyszczepionkowcy-maja-byc-czwarta-noga-konfederacji.

Mierzyńska, Anna. "Epidemia dezinformacji w Polsce. Część fałszywek wskazuje na aktywność Rosji". *OKO.Press*, 14.03.2020. Accessed: 30.08.2024. https://oko.press/koronawirus-wywolal-dezinformacyjna-powodz.

Mierzyńska, Anna. "O co walczy Grzegorz Braun? Zbudował radykalny polityczny ekosystem i ma coraz większe wpływy". *OKO.Press*, 29.04.2023. Accessed 06.09.2024. https://www.oko.press/co-robi-grzegorz-braun.

Mierzyńska, Anna. "Poseł Braun wykorzystuje Sejm, by nakręcać antyukraińskie nastroje. To spodoba się w Rosji". *OKO.Press*, 19.07.2022. Accessed 06.09.2024. https://www.oko.press/posel-braun-wykorzystuje-sejm-by-nakrecac-antyukrainskie-nastroje-to-spodoba-sie-w-rosji.

Mierzyńska, Anna. "Poseł polskiego Sejmu promuje w Warszawie prorosyjskie oświadczenie obwiniające Ukrainę i nato". *OKO.Press*, 26.06.2022. Accessed 06.09.2024. https://oko.press/posel-polskiego-sejmu-promuje-prorosyjskie-oswiadczenie.

Mierzyńska, Anna. "Przeciwnicy szczepień grożą, straszą, oskarżają. Są zdeterminowani i eskalują działania". *OKO.Press*, 26.01.2022. Accessed 30.08.2024. https://oko.press/przeciwnicy-szczepien-groza-strasza-oskarzaja-sa-zdeterminowani-i-eskaluja-dzialania.

Mierzyńska, Anna. "Przedsiębiorcy, bezmaskowcy i antyszczepionkowcy nawołują do antyrządowej rewolucji. Zaskakujący miks". *OKO.Press*, 11.05.2020. Accessed 30.08.2024. https://oko.press/przedsiebiorcy-bezmaskowcy-i-antyszczepionkowcy-nawoluja-do-antyrzadowej-rewolucji-zaskakujacy-miks.

Mierzyńska, Anna. "Rosyjski ślad. Prokremlowscy działacze wśród uczestników napaści z udziałem antyszczepionkowców". *OKO.Press*, 06.08.2021. Accessed

06.09.2024. https://www.oko.press/prokremlowscy-dzialacze-wsrod-antyszczepionkowcow.

Mierzyńska, Anna. "Rosyjski troll sieje chaos na polskim Facebooku. Jest wśród antyszczepionkowców". *OKO.Press*, 14.10.2018. Accessed: 30.08.2024. https://www.oko.press/rosyjski-troll-sieje-chaos-na-polskim-facebooku-jest-wsrod-antyszczepionkowcow.

Mierzyńska, Anna. "Tak skrajna prawica wykorzystała fake newsa o morderstwie, by szczuć Polaków przeciwko Ukraińcom". *OKO.Press*, 14.05.2022. Accessed 06.09.2024. https://oko.press/tak-skrajna-prawica-wykorzystala-fake-newsa-o-morderstwie-by-szczuc-polakow-przeciwko-ukraincom.

Mierzyńska, Anna. "Ujawniamy prawdziwą »piątkę Konfederacji«. Czego nienawidzą kandydaci skrajnej prawicy?". *OKO.Press*, 26.07.2023. Accessed 06.09.2024. https://oko.press/prawdziwa-piatka-konfederacji.

Mierzyńska, Anna. "Uwaga na niby-pacyfistyczną narrację! Korzysta na niej Kreml, w Polsce szerzy ją Sykulski". *OKO.Press*, 11.02.2023. Accessed 30.08.2024. https://oko.press/uwaga-na-niby-pacyfistyczna-narracje-sykulski.

Mierzyńska, Anna. "Zmienny jak poseł Chruszcz. Kolejny narodowiec dołącza do PiS. Czy na pewno nie lubi już Putina?". *OKO.Press*, 03.04.2019. Accessed 5.09.2024. https://oko.press/zmienny-jak-posel-chruszcz-kolejny-narodowiec-dolacza-do-pis-czy-na-pewno-nie-lubi-juz-putina.

Miesnik, Magda. "Wybory do PE 2019. »WiceZięba« dr Hubert Czerniak startuje do europarlamentu". *WP*, 16.04.2019. Accessed 06.09.2024. https://wiadomosci.wp.pl/wybory-do-pe-2019-wicezieba-dr-hubert-czerniak-startuje-do-europarlamentu-6370935095813761a.

Miller, Wiesław; Łagocki, Marek; Macal, Dawid. "W Koszalinie powstanie 5G? Mieszkańcy protestują! Mamy stanowisko ratusza". *KoszalinInfo.pl*, 20.04.2020. Accessed 30.08.2024. https://koszalininfo.pl/w-koszalinie-powstanie-5g-mieszkancy-protestuja-mamy-stanowisko-ratusza.

Mishler, Elliott G. "Models of Narrative Analysis: A Typology". *Journal of Narrative and Life History* 5, no. 2 (1995):87–123.

MJM, MW. "Lekarka głosiła, że pandemia to »ściema«. Odebrano jej prawo wykonywania zawodu". *Puls Medycyny*, 02.11.2022,, za: https://www.pulsmedycyny.pl/lekarka-glosila-ze-pandemia-to-sciema-odebrano-jej-prawo-wykonywania-zawodu-1168346.

Mk. "Konfederacja: PiS po cichu wprowadza państwo policyjne". *Tygodnik Bydgoski*, 12.01.2021. Accessed 06.09.2024. https://www.tygodnikbydgoski.pl/aktualnosci/konfederacja-pis-po-cichu-wprowadza-panstwo-policyjne.

Mk. "Na zdrowie ... brzoza i seks? Czyli na co skarżą się pacjenci ośrodka w Gomunicach". *Radomsko. Nasze Miasto*, 09.11.2015. Accessed 06.09.2024. https://www.radomsko.naszemiasto.pl/na-zdrowie-brzoza-i-seks-czyli-na-co-skarza-sie-pacjenci/ar/c1-3565602.

Ml. "Kandydat na prezydenta Wrocławia, przeciwnik obowiązkowych szczepień gościem »Skandalistów« o 20:00". *Polsat News*, 06.10.2018. Accessed 06.09.2024. https://www.polsatnews.pl/wiadomosc/2018-10-06/kandydat-na-prezydenta-wroclawia-przeciwnik-obowiazkowych-szczepien-gosciem-skandalistow-o-2000.

Młodziejewska, Nicole. "Pseudonaukowa konferencja medyczna tuż przy szpitalu tymczasowym i punkcie szczepień? Stowarzyszenie »Otwarte Ramiona« jest przeciwne". *Głos Wielkopolski*, 25.06.2021. Accessed 06.09.2024. https://www.gloswielkopolski.pl/pseudonaukowa-konferencja-medyczna-tuz-przy-szpitalu-tymczasowym-i-punkcie-szczepien-stowarzyszenie-otwarte-ramiona-jest/ar/c1-15682027.

Molga, Tomasz. "Powstał miesięcznik szarlatanów. Lekarze to »skończeni, sadystyczni bandyci«". *WP*, 17.12.2018. Accessed 06.09.2024. https://wiadomosci.wp.pl/powstal-miesiecznik-szarlatanow-lekarze-to-skonczeni-sadystyczni-bandyci-6328518195103361a.

Moment w Atmosferze. "Jerzy Zięba – Pola torsyjne (14 grudnia 2023)". *YouTube*, 15.12.2023., https://youtu.be/yukyZ4tpRbU.

Mon. "Łódzcy ekolodzy: dostali pieniądze, zakończyli protest". *WP*, 06.09.2001. Accessed 06.09.2024. https://wiadomosci.wp.pl/lodzcy-ekolodzy-dostali-pieniadze-zakonczyli-protest-6109071767045249a.

Mossakowski, Rafał. "Ewa Pawela: Kontrola umysłu trafiła pod strzechy – aktualne ciekawe książki i ważne teksty artykułów". *YouTube*, 12.09.2023. Accessed: 30.08.2024. https://www.youtube.com/watch?v=HDybkiodF-4.

Mosse, George L. *Kryzys ideologii niemieckiej*. Warsaw: Czytelnik, 1972.

Motyka, Piotr. "Anna Maria Siarkowska wyraźnie o segregacji sanitarnej: Stop sanitaryzmowi! [NASZ WYWIAD]". *Media Narodowe*, 26.11.21. Accessed 06.09.2024. https://medianarodowe.com/2021/11/26/anna-maria-siarkowska-wyraznie-o-segregacji-sanitarnej-stop-sanitaryzmowi-nasz-wywiad.

Mouzo, Jessica. "Parents of diphtheria-stricken boy feel »tricked« by anti-vaccination groups". *El Pais*, 05.06.2015. Accessed 30.08.2024. www.elpais.com/elpais/2015/06/05/inenglish/1433512717_575817.html.

Moy, Wesley R.; Gradon, Kacper. "COVID-19 Effects and Russian Disinformation Campaigns". *Homeland Security Affairs Journal* Special COVID-19 Issue (2020):2–27.

Mnookin, Seth. *The panic virus. The true story behind the vaccine-autism controversy*. New York: Simon & Schuster, 2011.

MP. "»Plandemia to ściema«. Naczelny Sąd Lekarski surowo ukarał lekarkę". *Salon24*, 02.11.2022. Accessed 06.09.2024. https://www.salon24.pl/newsroom/1261719,plandemia-to-sciema-naczelny-sad-lekarski-surowo-ukarl-lekarke.

Mroszczak, Stanisław. "Manipulacje o »turystyce emerytalnej« ukraińskich seniorów". *Wojownicy Klawiatury*, 04.05.2023. Accessed 06.09.2024. https://wojownicyklawiatury.pl/manipulacje-o-turystyce-emerytalnej-ukrainskich-seniorow.

MS. "Antyszczepionkowcy chcą usunięcia obowiązku szczepień. Zebrali 120 tys. podp isów pod projektem ustawy". *Newsweek*, 12.07.2018. Accessed 06.09.2024. https://www.newsweek.pl/polska/polityka/antyszczepionkowcy-zlozyli-w-sejmie-pod pisy-pod-projektem-ustawy/7ze4pw2.

Ms/PAP."Warszawa. Protest antyszczepionkowców. »Zdejmijcie w końcu te obostrzenia«". *Polsat News*, 22.05.2021. Accessed: 30.08.2024. https://www.polsatnews.pl/wiadomosc/2021-05-22/protest-antyszczepionkowcow-w-warszawie-zdejmijc ie-w-koncu-te-obostrzenia.

Mucha, Joanna. "Interpelacja nr 9556 do ministra cyfryzacji, ministra spraw wewnętrznych i administracji, ministra – członka Rady Ministrów, koordynatora służb specjalnych w sprawie sieci telekomunikacyjnej i instalacji urządzeń firmy Huawei". *Sejm Rzeczpospolitej Polskiej*. 10.08.2020. Accessed 5.09.2024. http://www.sejm.gov.pl/sejm9.nsf/InterpelacjaTresc.xsp?key=BSN9Y5.

Nadler, Anthony; Crain, Matthew and Donovan, Joan . *Weaponizing the Digital Influence Machine: The Political Perils of Online Ad Tech*. New York: Data & Society Research Institute, 2018. Accessed 14.04.2025. https://datasociety.net/wp-content /uploads/2018/10/DS_Digital_Influence_Machine.pdf.

NAI Warszawa. "Jaka Polska w wieku Azji? – konferencja". *Evenea.pl*, no data. Accessed 30.08.2024. https://app.evenea.pl/event/nai?source=youtube.Wsp.

Nasz Dziennik. "lek. med. Dorota Łucja Jarczewska – Wykład – 14 maja 2016 r.". *YouTube*, 31.10.2016. Accessed 06.09.2024. https://www.youtube.com/watch?v=lkSJ u8wMqSI.

National Security and Intelligence Committee of Parliamentarians (NSICOP), *Annual Report 2020* (Ottawa: Government of Canada, April 2021) accessed 30.03.2025 https://www.nsicop-cpsnr.ca/announcements/ann-2021-04-09/ann-2021-04-09-en .html.

NATULI dzieci są ważne. "Antybiotyki są nadużywane! Rozmowa z Katarzyną Bross-Walderdorff". *NATULI. Dzieci są ważne*, no data. Accessed 06.09.2024. https://www.dziecisawazne.pl/antybiotyki-sa-naduzywane-rozmowa-katarzyn a-bross-walderdorff.

Näsi, Matti; Tanskanen, Maiju; Kivivuori, Janne; Haara, Paula; Reunanen, Essa. "Crime News Consumption and Fear of Violence: The Role of Traditional Media, Social Media, and Alternative Information Sources". *Crime & Delinquency* 67, no. 4, (2021): 574–600.

Network Contagion Research Institute (NCRI), *The Future of Disinformation Operations: Emerging Narratives and Trends in Online Manipulation*, NCRI February 2024, accessed 13.04.2025 https://networkcontagion.us/reports/the-future-of-disin formation-operations/.

Neuman, Scott. "Russia Using Disinformation To »Sow Discord In West«, Britain's Prime Minister Says". *National Public Radio*, 14.11.2017. Accessed 30.08.2024, http://www.npr.org/sections/thetwo-way/2017/11/14/564013066/russia-using-disinformation-to-sow-discord-in-west-britains-prime-minister-says.

Nigdy Więcej Szurów. "Sebastian Pitoń: Wybudowałbym mur na granicy z Ukrainą i pomógł Putinowi". *YouTube*, 25.09.2022. Accessed 30.08.2024. https://www.youtube.com/watch?v=yGyMTVA_G98.

Niezależna Telewizja. "5G powstrzymać zbrodniczą falę _ Janusz Zagórski i Ewa Pawela". *BitChute*, 03.06.2020. Accessed 06.09.2024. https://www.bitchute.com/video/yzEMZyrvug2v.

Nijakowski, Lech M. *Świat po apokalipsie: Społeczeństwo w świetle postapokaliptycznych tekstów kultury popularnej*. Warszawa: Wydawnictwo Naukowe Scholar, 2019.

Nimmo, Ben. "Question That: RT's Military Mission". *Medium*, 08.01.2018. Accessed 30.08.2024. https://medium.com/dfrlab/question–that–rts–military–mission–4c4bd9f72c88.

NISS TV (Narodowy Instytut Studiów Strategicznych). "»Nie planuję samobójstwa« – P. Szlachtowicz, o czym mówi A. Wolska, to się samo komentuje – G. Płaczek". *YouTube*, 17.01.2022. Accessed 06.09.2024. https://www.youtube.com/watch?v=7UY5WeHIscQ.

NISS TV (Narodowy Instytut Studiów Strategicznych). "»Wirus« w służbie ponadnarodowych i ponadpaństwowych korporacji. Protesty w całej Polsce 10.10.2020!". *YouTube*, 10.10.2020. Accessed 06.09.2024. https://www.youtube.com/watch?v=jdbKcMwcJgg.

NISS TV (Narodowy Instytut Studiów Strategicznych). "Bitwa o ludzką duszę–Jadwiga Łopata. Konferencja –"Od Fałszu i Faszyzmu do Prawdy i Wolności, Razem!". *YouTube*, 01.03.2023. Accessed 06.09.2024. https://www.youtube.com/watch?v=XamEzKa49Ao.

NISS TV (Narodowy Instytut Studiów Strategicznych). "Dorota Gudaniec – Przestańcie się bać. Medycyna naturalna – największym zagrożeniem dla big farmy!". *YouTube*, 14.09.2020. Accessed 06.09.2024. https://www.youtube.com/watch?v=oCOiWI6edlc.

NISS TV (Narodowy Instytut Studiów Strategicznych). "Dr Anna Martynowska oskarża posła RP. Również o: Ludobójcach, złodzieju, psychopatach i TW SANEPID–u". *YouTube*, 08.10.2022. Accessed 06.09.2024. https://www.youtube.com/watch?v=mH2A7fDtmVU.

NISS TV (Narodowy Instytut Studiów Strategicznych). "Dwie drogi do suwerenności Polski – synergiczne wizje. H. Kazahari, P. Jankowski, W. Dobrzyński". *YouTube*, 14.07.2020. Accessed 06.09.2024. https://www.youtube.com/watch?v=PLws5FrplJM.

NISS TV (Narodowy Instytut Studiów Strategicznych). "Grzyby cz.1. Mózg jest plastyczny! Soplówka jeżowata regeneruje i oczyszczać mózg. Synestezja.". *YouTube*, 02.06.2023. Accessed 06.09.2024. https://www.youtube.com/watch?v=Dpfhj3kpIzI.

niss tv (Narodowy Instytut Studiów Strategicznych). "Kandydat na Prezydenta o NWO, o pieniądzu, o skradzionych podpisach, o bojkocie wyborów.". *YouTube*, 12.06.2020. Accessed 06.09.2024. https://www.youtube.com/watch?v=52y__On cHWA.

NISS TV (Narodowy Instytut Studiów Strategicznych). "Jedwabne – 81 lat po ... Nadal czekamy na prawdę. Domagamy się ekshumacji!". *YouTube*, 12.07.2022. Accessed 06.09.2024. https://www.youtube.com/watch?v=C--4wIDsM4I.

NISS TV (Narodowy Instytut Studiów Strategicznych). "Kongres Pokoju. Różne poglądy jeden cel ... POKÓJ! cz.1". *YouTube*, 13.03.2023. Accessed 06.09.2024. https://www.youtube.com/watch?v=DIurXny7vhU.

NISS TV (Narodowy Instytut Studiów Strategicznych). "Kongres Pokoju. Pokój – różne poglądy jeden cel. Liberalizm – pokojowy kolonializm, pokojowy podbój.". *YouTube*, 11.03.2023. Accessed 06.09.2024. https://www.youtube.com/watch?v=rhiez HyeyDw.

NISS TV (Narodowy Instytut Studiów Strategicznych). "Koronawirus III Wojna Światowa – wojna biologiczna". *YouTube*, 03.03.2020. Accessed 06.09.2024. https://www.youtube.com/watch?v=tCQz4hKGRPA.

NISS TV (Narodowy Instytut Studiów Strategicznych). "Precz z maseczkami, żadnej drugiej fali »pandemii«!". *YouTube*, 28.07.2020. Accessed 06.09.2024. https://www.youtube.com/watch?v=UgWyyHB_Efo.

NISS TV (Narodowy Instytut Studiów Strategicznych). "Nawigatorzy Państwa Polskiego obrali kurs na wojnę"– dr Leszek Sykulski specjalista ds. geopolityki". *YouTube*, 24.10.2022. Accessed 06.09.2024. https://www.youtube.com/watch?v= MmuNNaMmyWQ.

NISS TV (Narodowy Instytut Studiów Strategicznych). "Piotr Szlachtowicz o zniewolonym umyśle. Marcin Flach o pieniądzu, który zniewala.". *YouTube*, 20.10.2022. Accessed 06.09.2024. https://www.youtube.com/watch?v=Kx1yw5id9Sc.

NISS TV (Narodowy Instytut Studiów Strategicznych). "S. Pitoń – Globaliści a obóz azjatycki. Kryzysy nas straszą. Człowiek umarł ze strachu przed ZUS–em!". *YouTube*, 09.10.2022. Accessed 06.09.2024. https://www.youtube.com/watch?v=XE1o tsvRGDE.

NISS TV (Narodowy Instytut Studiów Strategicznych). "Stop NOP – demonstracja w Warszawie (Justyna Socha, Grzegorz Braun, Paweł Nogal, Piotr Jankowski)". *YouTube*, 08.06.2020. Accessed 06.09.2024. https://www.youtube.com/watch?v= A1NQwFjK5Sk.

NISS TV (Narodowy Instytut Studiów Strategicznych). "TI DAY. Cyfryzacja człowieka. Transhumanizm. Cywilizacja Człowieka nad przepaścią.". *YouTube*, 04.09.2022, accesttps://www.youtube.com/watch?v=16USMtJ995Qsed 06.09.2024, h.

NISS TV (Narodowy Instytut Studiów Strategicznych). "Uwolnienie zawodu lekarza i naukowca od zniewolenia. Dr Agata Osiniak. Kraków 21 sierpnia 2022 roku.". *You-*

Tube, 09.09.22. Accessed 06.09.2024. https://www.youtube.com/watch?v=i5Qbrr–5Oyo.

NISS TV (Narodowy Instytut Studiów Strategicznych). "W. Dobrzyński – Świadomość dwuwalutowości. J. Zagórski – NOWA ZIEMIA, początek kolejnego etapu drogi". *YouTube*, 27.09.2022. Accessed 06.09.2024. https://www.youtube.com/watch?v=vqhRDTuwUcA.

NISS TV (Narodowy Instytut Studiów Strategicznych). "Ważny apel Justyny Sochy. Sobota – demonstracje w Polsce przeciw segregacji sanitarnej. Godzina W19.". *YouTube*, 12.01.2022. Accessed 06.09.2024. https://www.youtube.com/watch?v=SIyT7RhBZgw.

NISS TV (Narodowy Instytut Studiów Strategicznych). "Wszystkim Polakom należy się Bezwarunkowa Dywidenda Narodowa! NISS w Centrum Edukacyjnym Polska.". *YouTube*, 30.04.2020. Accessed 06.09.2024. https://www.youtube.com/watch?v=vSTAAIUPSsI.

NISS TV (Narodowy Instytut Studiów Strategicznych). "Zatrzymujemy NWO – tworzymy alternatywne społeczeństwo! Zjazd Wolnych Ludzi w Warszawie. 14.09.2020". *YouTube*, 10.09.2020. Accessed 06.09.2024. https://www.youtube.com/watch?v=SuHi2DkkQSg.

NISS TV (Narodowy Instytut Studiów Strategicznych). "Zbigniew Gelzok. 5G a Covid–19 na Śląsku. Czy telekomy zarabiają na pornografii i pedofilii?". *YouTube*, 12.05.2020. Accessed 06.09.2024. https://www.youtube.com/watch?v=JoXs2yDb3–A.

Nocny Pingwin. "Danuta Grzyb – numer 12 na liście Konfederacji". *Twitter*, 15.09.2023. Accessed 06.09.2024. https://twitter.com/Nocny_Pingwin/status/170261847076287030.

Nocny Pingwin. "Danuta Leheńka–Das – numer 12 na liście Konfederacji w okręgu 2". *Twitter*, 14.09.2023. Accessed 06.09.2024. https://twitter.com/Nocny_Pingwin/status/1702323945763164293.

Nocny Pingwin. "Diana Ruchniewicz – czwarta na liście Konfederacji". *Twitter*, 11.09.2023. Accessed 06.09.2024. https://twitter.com/Nocny_Pingwin/status/1701229831160045646.

Nocny Pingwin. "Dorota Rycak – numer 20 na liście Konfederacji w okręgu 7". *Twitter*, 19.09.2023. Accessed 06.09.2024. https://twitter.com/Nocny_Pingwin/status/1704059308583841795.

Nocny Pingwin. "Kandydat Konfederacji przyznał, że wolałby". *Twitter*, 16.02.2024. Accessed 06.09.2024. https://twitter.com/Nocny_Pingwin/status/1758496973936595272.

Nocny Pingwin. "Kandydat Konfederacji udostępnia dużo postów". *Twitter*, 01.03.2024. Accessed 06.09.2024. https://twitter.com/Nocny_Pingwin/status/1763575689884639491.

Nocny Pingwin. "Karolina Pikuła – numer 30 na liście Konfederacji". *Twitter*, 10.10.2023. Accessed 06.09.2024. https://twitter.com/Nocny_Pingwin/status/1711631396803141796/photo/2.

Nocny Pingwin. "Krystyna Jarocka – kandydatka Konfederacji do senatu z okręgu 9". *Twitter*, 12.09.2023. Accessed 06.09.2024. https://twitter.com/Nocny_Pingwin/status/1701607205152370748.

Nocny Pingwin. "Krzysztof Olędzki – numer 21 na liście Konfederacji w okręgu 19". *Twitter*, 09.10.2023. Accessed 06.09.2024. https://twitter.com/Nocny_Pingwin/status/1711252650979664234.

Nocny Pingwin. "Marek Banyś – numer 11 na liście Konfederacji w okręgu 8". *Twitter*, 20.09.2023. Accessed 06.09.2024. https://twitter.com/Nocny_Pingwin/status/1704417778814382386.

Nocny Pingwin. "Mariusz Bożyk – numer 16 na liście Konfederacji w okręgu 14". *Twitter*, 29.09.2023. Accessed 06.09.2024. https://twitter.com/Nocny_Pingwin/status/1707584784396935423.

Nocny Pingwin. "Na profilu Budnioka znajdziemy też sporo antynaukowych bzdur". *Twitter*, 01.03.2024. Accessed 06.09.2024. https://twitter.com/Nocny_Pingwin/status/1763575693810565291.

Nocny Pingwin. "Oto przykładowe fragmenty filmów, jakie udostępnia kandydat". *Twitter*, 01.03.24. Accessed 06.09.2024. https://twitter.com/Nocny_Pingwin/status/1763575697560174674.

Nocny Pingwin. "Po akcji Grzegorza Brauna z gaśnicą". *Twitter*, 16.02.2024. Accessed 06.09.2024. https://twitter.com/Nocny_Pingwin/status/1758496979833720916.

Nocny Pingwin. "Rafał Czubak – numer 5 na liście Konfederacji w okręgu 17". *Twitter*, 08.10.2023. Accessed 06.09.2024. https://twitter.com/Nocny_Pingwin/status/1710851681775161403.

Nocny Pingwin. "Rafał Foryś – kandydat Konfederacji na prezydenta Radomia". *Twitter*, 16.02.2024. Accessed 06.09.2024. https://twitter.com/Nocny_Pingwin/status/1758496975870197875.

Nocny Pingwin. "Standardem na profilach Konfederatów są fałszywe informacje". *Twitter*, 01.03.2024. Accessed 06.09.2024. https://twitter.com/Nocny_Pingwin/status/1763575685828739414.

Nocny Pingwin. "Stanisław Bąk – numer 8 na liście Konfederacji w okręgu 7". *Twitter*, 21.09.2023. Accessed 06.09.2024. https://twitter.com/Nocny_Pingwin/status/1704781279680450629.

Nocny Pingwin. "Sylwia Siłuszyk – numer 15 na liście Konfederacji w okręgu 12". *Twitter*, 02.10.2023. Accessed 06.09.2024. https://twitter.com/Nocny_Pingwin/status/1708723490902200362.

Nocny Pingwin. "Tomasz Walczuk, numer 6 na liście Konfederacji". *Twitter*, 16.09.2023. Accessed 06.09.2024. https://twitter.com/Nocny_Pingwin/status/1703003120152854903.

Nocny Pingwin. Żaneta Nochowicz – płaskoziemka na liście Konfederacji". *Twitter*, 22.09.2023. Accessed 06.09.2024. https://twitter.com/Nocny_Pingwin/status/17050 69820704641406.

Nogaj, Magda. "Antyszczepionkowcy paraliżują pracę szpitala we Wrocławiu. "Nie dają się leczyć, boją się nawet termometru". *Wyborcza.pl*, 23.12.2021. Accessed 30.08.2024. https://wroclaw.wyborcza.pl/wroclaw/7,35771,27939908,antyszczepio nkowcy-paralizuja-prace-szpitala-we-wroclawiu-nie.html

Nowak, Mateusz. "Bojówka "Anty 5G" złapana ze szlifierką kątową w ręce. Próbowano uszkodzić krakowską stację bazową". *Spider's Web*, 04.10.2019. Accessed 30.08.2024. https://spidersweb.pl/2019/10/anty-5g-niszczenie-nadajnikow.html.

NTV. "Debata – Punkt zwrotny dla Polski i Świata – prof. M. Matyja, J. Zięba, W. Sumliński, J. Zagórski". *YouTube*, 22.06.2022. Accessed 06.09.2024. https://www.youtube.com/watch?v=L9EM47tkY44.

NTV. "Obrona suwerenności zdrowotnej – Justyna Socha". *YouTube*, 24.06.2022. Accessed 06.09.2024. https://www.youtube.com/watch?v=K7sFiTdupvA.

NTV. "Wystąpienie dr Huberta Czerniaka". *YouTube*, 25.06.2022. Accessed 06.09.2024. https://www.youtube.com/watch?v=acvfCTSHSEI.

Nyczka, Tomasz. "Zamieszanie w Konfederacji w Poznaniu. Socha ostatecznie skreślona, uderza w Mentzena i Bosaka". *Wyborcza.pl*, 29.07.2023. Accessed 06.09.2024. https://poznan.wyborcza.pl/poznan/7,36001,30018814,zamieszanie-w-konfeder acji-w-poznaniu-socha-ostatecznie-skreslona.html.

Offit, Paul A. *Deadly Choices. How the Anti-Vaccine Movement Threatens Us All.* New York: Basic Books, 2010.

Offit, Paul A. *The Cutter Incident. How America's First Polio Vaccine Led to the Growing Vaccine Crisis.* New Haven and London: Yale University Press, 2005.

Ognyanova, Katherina; Lazer, David; Robertson, Ronald E.; Wilson, Christo. "Misinformation in action: Fake news exposure is linked to lower trust in media, higher trust in government when your side is in power". *Harvard Kennedy School Misinformation Review* 4, no. 1 (2020), n. pag.

Olo. "Marta Patena odchodzi z Platformy Obywatelskiej, kruszeje koalicja w magistracie". *Wyborcza.pl*, 08.06.2016. Accessed 06.09.2024. https://krakow.wyborcza.pl/krakow/7,44425,20207354,marta-patena-odchodzi-z-platformy-obywatelskiej -kruszeje-koalicja.html.

Opas, Joanna. "Protest Przeciw Przymusowi Szczepień". *Polskie Radio Katowice*, 02.06.2018. Accessed 30.08.2024. https://www.radio.katowice.pl/zobacz,36901,Pro test-Przeciw-Przymusowi-Szczepien.html#.WxedZzSFOUk.

Opas, Radosław. "Protest antyszczepionkowców w Warszawie. »Nie pozwolimy na kolejny lockdown«". *WP*, 11.09.2020. Accessed 30.08.2024. https://wiadomosci.wp .pl/protest-antyszczepionkowcow-w-warszawie-nie-pozwolimy-na-kolejny-lo ckdown-6553025609075616a.

Opasiak, Anna. "Lublin/Covid-19 – manifestacje przeciwników szczepień". *Radio Plus*, 22.01.2022. Accessed 30.08.2024. https://www.radioplus.pl/lublin/lublin-covid-19-manifestacje-przeciwnikow-szczepien-aa-uVd4-aa7t-1cVG.html.

Ortiz, Jorge L."Anti-vaxxers open door for measles, mumps, other old-time diseases back from near extinction". *USA Today*, no data. Accessed 30.08.2024. https://eu.usatoday.com/story/news/health/2019/03/28/anti-vaxxers-open-door-measles-mumps-old-time-diseases/3295390002.

Ostolski, Adam. "Dyskurs u władzy. Filozofia polityczna u Michaela Foucaulta". *Etyka* 36 (2003):159–172.

Ostropolska, Grażyna. "Prawnik po zawodówce". *Nowości. Dziennik Toruński*, 30.03.2013. Accessed 06.09.2024. https://nowosci.com.pl/prawnik-po-zawodowce/ar/10909193.

Ośrodek Monitorowania Zachowań Rasistowskich i Ksenofobicznych. "OTO DLACZEGO WREALU24 ZOSTAŁO ZABLOKOWANE NA YOUTUBE". *Facebook*, 26.09.2022. Accessed 30.08.2024. https://www.facebook.com/watch/?v=496862065281497.

Ośrodek Monitorowania Zachowań Rasistowskich i Ksenofobicznych. "TYLKO WYSZEDŁ Z ARESZTU I JUŻ WZYWA DO MORDOWANIA KOBIET". *Facebook*, 01.09.2022. Accessed 30.08.2024. https://www.facebook.com/watch/?v=1998688476993339.

P.mal. "Poseł Kukiz'15 Jarosław Sachajko o obowiązku szczepień: Dorośli mają prawo umrzeć". *Rzeczpospolita*, 06.12.2021. Accessed 06.09.2024. https://www.rp.pl/polityka/art19172541-posel-kukiz-15-jaroslaw-sachajko-o-obowiazku-szczepien-dorosli-maja-prawo-umrzec.

Palińska, Ewa. "Szczepionkowy ping-pong doktora z GUMedem". *Trójmiasto.pl*, 23.11.2016. Accessed 06.09.2024. https://www.trojmiasto.pl/zdrowie/Szczepionkowy-ping-pong-doktora-z-GUMedem-n107665.html?&vop=w&strona=4.

PAN. "Wraca temat budowy stacji bazowej. Okoliczni mieszkańcy boją się oddziaływania 5G". *Limanowa.in*, 11.07.2020. Accessed 30.08.2024. https://limanowa.in/aktualnosci/wraca-temat-budowy-stacji-bazowej-okoliczni-mieszkancy-boja-sie-oddzialywania-5g/49749.

Pankowski, Rafał; Tatar, Anna; Dzięgielewski, Jacek. *Wirus nienawiści: "Brunatna Księga" czasu epidemii*. Warszawa: Nigdy Więcej, 2020.

PAP. "Antyszczepionkowcy zaatakowali punkt szczepień w Grodzisku Mazowieckim. Interweniowała policja". *Rynek Zdrowia*, 26.07.2021. Accessed: 30.08.2024. https://www.rynekzdrowia.pl/Prawo/Antyszczepionkowcy-zaatakowali-punkt-szczepien-w-Grodzisku-Mazowieckim-Interweniowala-policja,223509,2.html.

PAP. "Braun uderza w ustawę o współpracy z USA: Kolonizacja plus, a może nawet Okupacja plus". *Dziennik.pl*, 15.09.2020. Accessed 06.09.2024. https://wiadomosci.dziennik.pl/polityka/artykuly/7823966,grzegorz-braun-konfederacja-usa-umowa-wspolpraca-polska-wojsko-armia.html.

PAP. "Byli współpracownicy Kukiza sceptyczni ws. budowy przez niego nowej partii", *Dzieinnik.pl.* 07.03.2020. Accessed 06.09.2024. https://wiadomosci.dziennik.pl/polityka/artykuly/6453215,kukiz-nowa-partia-byli-wspolpracownicy-sceptycyzm.html.

PAP and jd. "Niedzielski o występie Brauna w rosyjskiej telewizji: staje się stronnikiem Putina niszczącego niepodległą Ukrainę". *Wirtualne Media*, 03.03.2022. Accessed 06.09.2024. https://www.wirtualnemedia.pl/artykul/grzegorz-braun-w-rosyjskiej-telewizji-krytyka.

Pastuszka, Luiza. "Warszawa. Protest antyszczepionkowców. »Zakończyć pandemię«". *WP*, 16.08.2020. Accessed 30.08.2024. https://wiadomosci.wp.pl/warszawa/warszawa-protest-antyszczepionkowcow-zakonczyc-pandemie-6543709232253568a.

Patryła, Magdalena. "Protest antyszczepionkowców w Warszawie. Przepychanki z policją". *RMF 24*, 22.05.2021. Accessed 30.08.2024. https://www.rmf24.pl/raporty/raport-koronawirus-z-chin/polska/news-protest-antyszczepionkowcow-w-warszawie-przepychanki-z-polic,nId,5248913.

Pawela, Ewa. "5G narzędzie masowej kontroli umysłu – wykład". *Stowarzyszenie Stop Zet* (home page), 04.03.2019. Accessed: 30.08.2024. https://stopzet.pl/2019/03/04/5g-narzedzie-masowej-kontroli-umyslu-wyklad.

Pawela, Ewa. "Kamienie Milowe NWO (WTC 9/11). *CDA*, 11.09.2023. Accessed: 30.08.2024. https://www.cda.pl/video/17032762f0.

Pawela, Ewa and Hake, Witold. "Międzynarodowy Dzień Targeta w Polsce". *CDA*, 29.08.2022. Accessed 06.09.2024. https://ebd.cda.pl//620x368/1184288056/vfilm.

Pawlik, Paweł. "Antyszczepionkowcy tworzą nowy ruch. Wśród założycieli znane osoby". *Onet*, 27.09.2021. Accessed 5.09.2024. https://wiadomosci.onet.pl/slask/polska-jest-jedna-piech-pospieszalski-i-poslanka-pis-tworza-ruch-antyszczpionkowy/ct26yry.

Pawliszak, Piotr. "Analiza narracyjna dyskursu publicznego. W poszukiwaniu struktur kulturowych i procesów narzucania znaczeń formujących polityką ekologiczną", in: *Analiza dyskursu publicznego. Przegląd metod i perspektyw badawczych.* Edited by Otrocki Michał; Stachowiak Jerzy; Czyżewski Marek and Piekot Tomasz. Warsaw: Wydawnictwo Akademickie SEDNO, 2017.

Patena, Maria. "O mnie", *Patena.pl.* no data. Accessed 06.09.2024. http://www.patena.pl/pl/o-mnie.

Petrilli, Susan. "Multicaultural Societies, Monotheistic Religions, and Globalization: Semioethic Vistas". *The Journal of Dialogic Ethics: Interfaith and Interhuman Perspectives* 1, no. 1 (2022):46–77.

PCh24TV · Polonia Christiana. "Grzegorz Braun: jesteśmy świadkami próby ostatecznego rozwiązania kwestii polskiej!". *YouTube*, 08.02.2018. Accessed 06.09.2024.

https://www.pch24.pl/grzegorz-braun--jestesmy-swiadkami-ostatecznego-roz wiazania-kwestii-polskiej--video,58119,tv.html.

Peltola Heikki and Heinonen, Olli P. "Frequency of true adverse reactions to measles-mumps-rubella vaccine. A double-blind placebo-controlled trial in twins". *Lancet* 1, no. 8487 (1986):939–942.

Piasecki, Maciej. "Rosyjskie trolle nakręcają antyszczepionkową histerię w Polsce? Wysyp anonimowych kont". *OKO.Press*, 31.01.2018. Accessed: 30.08.2024. https://www.oko.press/rosyjskie-trolle-nakrecaja-histerie-antyszczepionkowa -w-polsce-wysyp-anonimowych-kont.

Piątek, Tomasz. "Sojusznik Białorusi i antyszczepionkowców na jednej liście z ministrem Kamińskim". *Wyborcza.pl*, 11.09.2023. Accessed 06.09.2024. https://wyborcza .pl/7,75398,30175157,sojusznik-bialorusi-i-antyszczepionkowcow-na-jednej-lisc ie-z.html.

Piecyk, Marcin. "W Lutoryżu nie chcą masztu 5G. Mieszkańcy protestują, gmina Boguchwała ma związane ręce". *Nowiny24*, 18.07.2020. Accessed 30.08.2024. https://nowiny24.pl/w-lutoryzu-nie-chca-masztu-5g-mieszkancy-protestuja-gmina-boguchwala-ma-zwiazane-rece/ar/c1-15083339.

Pietraszewski, Mikołaj. "»Festiwal propagandy«, »oklaski od Konfederacji«. Znowu burza po słowach Kai Godek". *Radio Zet*, 05.01.2022. Accessed 30.08.2024. https://www.wiadomosci.radiozet.pl/polska/Kaja-Godek-szczepionki-na-COVI D-19-to-korzystanie-ze-zbrodni-aborcji.-Komentarze-ofensywa-antyszczepio nkowcow

Pietraszewski, Mikołaj. "Jako jedyny zagłosował przeciw Finlandii w NATO. Teraz podał powód". *Radio Zet*, 07.07.2022. Accessed 06.09.2024. https://wiadomosci.ra diozet.pl/polska/polityka/Robert-Winnicki-zaglosowal-przeciw-Finlandii-w-NAT O.-Posel-Konfederacji-podal-powod.

Pięta, Martyna. "Braun o Żydach – rabini Chabad-Lubawicz to antychrześcijańska sekta". *Media Narodowe*, 15.01.2019. Accessed 06.09.2024. https://medianarodowe .com/braun-o-zydach-rabini-chabad-lubawicz-to-antychrzescijanska-sekta.

Piłaszewicz, Mira. "Debata dziesięciu uczestników »Nie ma tego złego ... Czyli świat płonie i oczyszcza się w ogniu« u Miry Piłaszewicz [WIDEO]". *Wolność.tv*, 25.04.2022. Accessed 06.09.2024. https://www.wolnosc.tv/debata-dziesieciu-uc zestnikow-nie-ma-tego-zlego-czyli-swiat-plonie-i-oczyszcza-sie-w-ogniu-u -miry-pilaszewicz-wideo.

Pink, Sarah. *Doing Visual Ethnography*. 4th ed. London: SAGE Publications, 2021.

Pitoń, Agnieszka. "Dr Martyka stracił na rok prawo wykonywania zawodu. Awantura z udziałem posłów Konfederacji przed Izbą Lekarską w Krakowie". *Wyborcza.pl*, 19.12.2022. Accessed 06.09.2024. https://www.krakow.wyborcza.pl/krakow/7,44 425,29280014,protest-antyszczepionkowcow-pod-okregowa-izba-lekarska-w-k rakowie.html.

Pitoń, Sebastian. "Pitoń: Prawybory niedemokratyczne". *Myśl Polska*, no data. Accessed 30.08.2024. https://myslpolska.info/2023/03/28/piton-prawybory-niede mokratyczne.

Piwar, Agnieszka. "»Musimy zachować zdrowy rozsądek« – wywiad Agnieszki Piwar z Grzegorzem Braunem". *Myśl Polska*, 21.03.2022. Accessed 06.09.2024. https://my slpolska.info/2022/03/20/grzegorz-braun-musimy-zachowac-zdrowy-rozsadek.

Pk. "Marsz antyszczepionkowców na ulicach Lublina. Krzyczeli: »Cała Polska śpiewa z nami, nie jesteśmy królikami«". *Wyborcza.pl*, 17.09.2021. Accessed 30.08.2024. https://lublin.wyborcza.pl/lublin/7,48724,27583860,marsz-antyszczepionkow cow-na-ulicach-lublina-krzyczeli-cala.html.

Płaszczyca, Marcin. "Czy ludzie w Izdebniku dostaną raka od 5G? Ministerstwo Zdrowia odpowiada: »Nie mamy dowodów«". *Wadowice24*.pl, 03.03.2020. Accessed 30.08.2024. https://www.wadowice24.pl/nowe/pieniadze/13055-czy-ludzie-w-iz debniku-dostana-raka-od-5g.html.

Pmd. "Kim jest Sebastian Pitoń, inicjator »Góralskiego veta«? »Pandemii nie ma, diagnozy lipne«". *Onet*, 12.01.2021. Accessed 30.08.2024. https://wiadomosci.onet .pl/krakow/sebastian-piton-goralskie-veto-kim-jest/t4fre4l.

Pp, ads, katke, rzw, r. "Szarpanina przed punktem szczepień w Grodzisku Mazowieckim. Dwie osoby zatrzymane przez policję". *TVN24*, 25.07.2021. Accessed 30.08.2024. https://tvn24.pl/polska/grodzisk-mazowiecki-antyszczepionkowcy-wywolali-szarpanine-przed-punktem-szczepien-interweniowala-policja-5157 297.

Podjadek, Zbigniew. "Anatomia polskiego ruchu anty5G. Kim są ludzie, którzy walczą z budową sieci nowej generacji?". *Wyborcza.pl*. 16.09.2020. Accessed 06.09.2024. https://wyborcza.biz/biznes/7,177150,26290907,anatomia-polskiego-ruchu-anty 5g-kim-sa-ludzie-ktorzy-walcza.html.

Pog. "Protest przeciwników 5G na Rynku Głównym. »Zabójcza technologia«". *Kraków. Nasze Miasto*, 19.06.2020. Accessed 30.08.2024. https://krakow.naszemiasto .pl/protest-przeciwnikow-5g-na-rynku-glownym-zabojcza/ar/c3-7765204.

Polskie Forum, "Jacek Wilk o szczepieniach, WHO, cenzurze, 5G, Billu Gatesie i Szumowskim! Kto korzysta na pandemii?", *YouTube*, 28.04.2020. Accessed 06.09.2024. https://www.youtube.com/watch?v=DWf1dWGKPJo.

Polskie Sprawy. "Ewa Pawela: Jak się przeciwstawić depopulacji Polaków?". *CDA*, 08.01.2019. Accessed 06.09.2024. https://www.cda.pl/video/1321708011.

PorozmawiajmyTV. "Co powinieneś wiedzieć o szczepieniach – Justyna Socha". *YouTube*, 23.11.2017. Accessed 06.09.2024. https://www.youtube.com/watch?v=bpv 835EHk30.

PorozmawiajmyTV. "Czy szczepić dziecko przeciwko WZW typu B – dr Hubert Czerniak". *YouTube*, 30.04.2018. Accessed 06.09.2024. https://www.youtube.com/ watch?v=1BFrkQiFoCQ.

PorozmawiajmyTV. "Patologie współczesnej medycyny i inne sprawy do naprawy, część 1 – dr Jerzy Jaśkowski". *YouTube*, 06.02.2022. Accessed 06.09.2024. https://www.youtube.com/watch?v=E3HzEDxPFnE.

PorozmawiajmyTV. "Wielokulturowość degraduje i niszczy ciągłość pokoleń – prof. Włodzimierz Julian Korab–Karpowicz". *YouTube*, 31.01.2018. Accessed 06.09.2024. https://www.youtube.com/watch?v=LvkP_QLdW3o&t=57s.

Porozmawiajmy.tv. "Zabójczy wpływ sieci 5G na pola torsyjne organizmów żywych – dr Diana Wojtkowiak". *YouTube*, 10.03.2019. Accessed: 30.08.2024. https://www.youtube.com/watch?v=Qn6JKMrLN5g

Porter, Dorothy; Porter, Roy. "The politics of prevention: Anti–vaccination and public health in 19th century England". *Medical History* 32 (1988):231–252.

Posard, Marek; Smith, Shawn; Ma, Logan and Matthews, Miriam. *Russian Disinformation Efforts on Social Media*. RAND Corporation, 2022. Accessed 13.04.2025. https://www.rand.org/pubs/research_reports/RR4373z2.html.

Posoborowe absurdy. "Wykład »Fałszywa pandemia« – Grzegorz Braun, dr Anna Martynowska, dr Tomasz Wołoszyn i Roman Szostek.". *Gloria.tv*, 17.11.2021. Accessed 06.09.2024. https://www.gloria.tv/post/ypQg84EEbg2b1xMesFgDfzTZA#2080.

Price, Michelle L. "Anti–vaccine activist RFK Jr. challenging Biden in 2024". *Associated Press*, 06.04.2023. Accessed 06.09.2024. https://apnews.com/article/robert-kennedy-jr-presidential-campaign-9fb5ed5c8e1fd31d2a4458d44b086593.

Prier, Jarred. "Commanding the Trend: Social Media as Information Warfare". *Strategic Studies Quarterly* 11, no. 4 (2017):50–85.

PSNLiN. "Dokąd zmierzamy jako ludzkość – prof. Ryszard Zajączkowski Lubartów 29.08.2022". *Rumble*, 29.08.2022. Accessed 06.09.2024. https://rumble.com/v1iyxp1–dokd–zmierzamy–jako–ludzko–prof.–ryszard–zajczkowski–lubartw–29.08.2022.html.

PSNLiN. "Ratujmy nasze dzieci przed eksperymentalną szczepionką inżynierii genetycznej – Warszawa". *Polskie Stowarzyszenie Niezależnych Lekarzy i Naukowców*, 12.07.2021. Accessed 06.09.2024. https://www.psnlin.pl/artykuly,ratujmy–nasze–dzieci–przed–eksperymentalna–szczepionka–inzynierii–genetycznej–warszawa,34,88.html

Ptak–Iglewska, Aleksandra. "Sklep sprzedaje bieliznę, która ochroni przed 5G". *Rzeczpospolita*, 09.07.2019. Accessed 06.09.2024. https://cyfrowa.rp.pl/telekomunikacja/art18197041–sklep–sprzedaje–bielizne–ktora–ochroni–przed–5g.

PZ. "Fizyk w poznańskiej księgarni neguje covid–19. »Maski nieskuteczne, pandemia niegroźna«". *Epoznan.pl*, 18.09.2022. Accessed 06.09.2024. https://www.epoznan.pl/news–news–132558–fizyk_w_poznanskiej_ksiegarni_neguje_covid_19_maski_nieskuteczne_pandemia_niegrozna.

Radio Maryja. "»Oblicza pandemii«: dr Marek Sobolewski". *YouTube*, 29.01.2022. Accessed 06.09.2024. https://www.youtube.com/watch?v=NbxX–yD4oRc.

Raffini, Luca; Penalva-Verdú, Clemente. "The Problematic Relationship Between Science, Politics and Public Opinion in Late Modernity: The Case of the Anti-Vax Movement in Spain and Italy". In: *Populism and Science in Europe*, edited by Hande Eslen-Ziya, Alberta Giorgi, 141–162. Cham: Palgrave Macmillan, 2022.

RAV. "Ofiary broni elektromagnetycznej piszą do Ministerstwa Sprawiedliwości". *Newsweek*, 26.10.2016. Accessed 5.09.2024. https://www.newsweek.pl/polska/ofiary-broni-elektromagnetycznej-pisza-do-ministerstwa-sprawiedliwosci/eq6 5vj9.

Raz. "Mieszkańcy Hałcnowa obawiają się budowy masztu 5G". *Bielsko.info*, 30.06 .2020. Accessed 30.08.2024. https://www.bielsko.info/wiadomosci/23785-miesz kancy-halcnowa-obawiaja-sie-budowy-masztu-5g-bielsko-biala.

Red. "Antyszczepionkowcy grożą śmiercią burmistrzowi Mikołowa. »Rewolucyjna Armia Polaków wydała na ciebie wyrok«". *Gazeta Lubuska*, 14.01.2022. Accessed 30.08.2024. https://gazetalubuska.pl/antyszczepionkowcy-groza-smiercia-bur mistrzowi-mikolowa-rewolucyjna-armia-polakow-wydala-na-ciebie-wyrok/a r/c14-15996143.

Red. "Prof. Fal: szczepienia są największą zdobyczą nauki". *Polsat News*, 29.06.2021. Accessed 06.09.2024. https://www.polsatnews.pl/wiadomosc/2021-06-29/prof -andrzej-fal-i-dr-katarzyna-ratkowska-w-programie-gosc-wydarzen-transmi sja-od-1920.

Red. "Prof. Piotrowski popiera Brauna i mówi o Ferencu: »Na starość zdziwaczał«", *CzytajRzeszów.pl*, 29.04.2021. Accessed 06.09.2024. https://czytajrzeszow.pl/?idd= 26&id=17059&poz=mn1&show=-1&pagebda1_mobile=0.

Redakcja. "»Stop przymusowi szczepień, biedzie i bezprawiu«. Protest w Olsztynie [ZDJĘCIA, WIDEO]". *Olsztyn.com.pl*, 23.01.2022. Accessed 30.08.2024. https://www .olsztyn.com.pl/artykul,stop-przymusowi-szczepien-biedzie-i-bezprawiu-prote st-w-olsztynie-zdjecia-wideo,34158.html.

Redakcja. "Dr Katarzyna Ratkowska: Kto wydał zgodę, żeby Pfizer robił eksperymenty na naszych dzieciach?". *CAI24*.pl, 16.06.2021. Accessed 06.09.2024. https://www .cai24.pl/polska/33200/dr-katarzyna-ratkowska-kto-wydal-zgode-zeby-pfizer-r obil-eksperymenty-na-naszych-dzieciach.

Redakcja. "Kilkanaście osób przed jednostką wojskową na proteście przeciw 5G". *Istotne.pl*, 25.04.2020. Accessed 30.08.2024. https://istotne.pl/boleslawiec/wiado mosc/Xdf/kilkanascie-osob-przed-jednostka-wojskowa-na-protescie-przeciw -5g.

Redakcja. "Protest antyszczepionkowców w Łodzi. Bronili »swojego« lekarza [FILM, ZDJĘCIA]". *Dziennik Łódzki*, 21.12.2017. Accessed 30.08.2024. https://dziennik lodzki.pl/protest-antyszczepionkowcow-w-lodzi-bronili-swojego-lekarza-film -zdjecia/ar/12787207.

Redakcja. "Wczoraj Braun nazwał »zdradą« wojska USA w Polsce. Dziś jego wystąpienie zamieścił Sputnik". *Niezalezna.pl*, 20.11.2019. Accessed 06.09.2024. https://

niezalezna.pl/polska/wczoraj-braun-nazwal-zdrada-wojska-usa-w-polsce-dzis-jego-wystapienie-zamiescil-sputnik/298238.

Redakcja Dentonet. "Wtargnięcie antyszczepionkowców do siedziby Naczelnej Izby Lekarskiej". *Dentonet*, 22.02.2021. Accessed 30.08.2024. https://dentonet.pl/wtargniecie-do-siedziby-nil/#gref.

Redakcja e-grajewo.pl. "Antyszczepionkowcy grożą burmistrzowi Wągrowca". *E-grajewo.pl*, 25.11.2021. Accessed 30.08.2024. https://e-grajewo.pl/wiadomosc,antyszczepionkowcy-groza-burmistrzowi-wagrowca,47621.html.

Redakcja Plonsk24.pl. "»Gdy pójdziemy drogą szczepień, to jaka czeka nas przyszłość?«". *Plonsk24*.pl, 02.01.2021. Accessed 06.09.2024. https://www.plonsk24.pl/aktualnosc-14-3777-gdy_pojdziemy_droga_szczepien_to_jaka.html.

Redakcja Płońsk24. "Stowarzyszenie o pandemii, szczepionkach i szczepieniu dzieci". *Płońsk24*.pl, 10.06.2021. Accessed 06.09.2024. https://www.plonsk24.pl/aktualnosci/news/id/3918.html.

Redakcja Portalu. "»Potrzebujemy spotkań, które dodają odwagi«. Rusza Letnia Akademia Pro-life". *Stacja7.pl*, 31.05.2023. Accessed 06.09.2024. https://www.stacja7.pl/z-kraju/potrzebujemy-spotkan-ktore-dodaja-odwagi-rusza-letnia-akademia-pro-life.

Redakcja Xportal.pl. "Paweł Skutecki dla Xportalu: Hezbollah to najlepsza armia Bliskiego Wschodu". *Xportal.pl*, 23.08.2018. Accessed 06.09.2024. http://xportal.press/?p=33464.

Redmerska, Katarzyna. "Antyszczepionkowcy protestowali przed Sejmem". *Prawo.pl*, 15.09.2021. Accessed 30.08.2024. https://www.prawo.pl/zdrowie/antyszczepionkowcy-protestowali-przed-sejmem,510721.html.

Region Lubelski – media obywatelskie. "Konferencja dla wolności, bezpieczeństwa i zdrowia. Opole Lubelskie, 20.02.2022". *Rumble*, no data. Accessed 06.09.2024. https://www.rumble.com/vvh1d4-konferencja-dla-wolnoci-bezpieczestwa-i-zdrowia.-opole-lubelskie-20.02.2022.html.

Rej. "Wybory parlamentarne 2019. Paweł Skutecki w Konfederacji. Do Sejmu wystartuje w Warszawie". *Wyborcza.pl*, 12.08.2019. Accessed 06.09.2024. https://bydgoszcz.wyborcza.pl/bydgoszcz/7,48722,25081579,wybory-parlamentarne-2019-pawel-skutecki-w-konfederacji-do.html.

Rembiałkowska, Ewa; Metera, Dorota and Janeczko, Łukasz. "Organizmy genetycznie zmodyfikowane. W kierunku ogólnopolskiej debaty o GMO/NGT". *Instytut Spraw Obywatelskich*, sierpień 2023. Accessed 06.09.2024. https://instytutsprawobywatelskich.pl/publikacje/organizmy-genetycznie-zmodyfikowane-w-kierunku-ogolnopolskiej-debaty-o-gmo-ngt.

Reset Obywatelski. "Adam Andruszkiewicz działał w grupie parlamentarnej sympatyków Łukaszenki". *YouTube*, 22.03.2023. Accessed 5.09.2024. https://www.youtube.com/watch?v=TAhxMl5K6jo.

Rewucka, Faustyna. "Dr Ratkowska: Nie można rozwijać nauki kosztem życia ludzi", *Fundacje Życie i Rodzina*. 01.03.2022. Accessed 06.09.2024. https://www.ratujzycie.pl/dr–ratkowska–nie–mozna–rozwijac–nauki–kosztem–zycia–ludzi.

Rewucka, Faustyna. "Covid ujawnił problem wykorzystywania linii płodowych z aborcji", *Fundacja Życie i Rodzina*. 31.05.2022. Accessed 06.09.2024. https://ratujzycie.pl/covid-ujawnil-problem-wykorzystywania-linii-plodowych-z-aborcji.

Reuters Staff. "Fact check: California wildfires were not caused by »powerful lasers«. *Reuters*, 07.08.2020. Accessed: 30.08.2024. http://www.reuters.com/article/uk–fact check–california–wildfireslasers–idUSKBN25R1VH.

Ribberink, Egbert; Achterberg, Peter; Houtman, Dick. "Religious polarization: contesting religion in secularized Western European countries". *Journal of Contemporary Religion* 33, no. 2 (2018):209–227.

Richardson, Richard. *The Microvita Revolution* (Calcutta, Microvita Research Institute, 1990).

Richardson, Richard. *The Microvita Revolution* (Calcutta, Microvita Research Institute, 1990).

RIRM, "W kilku miastach Polski odbędą się protesty przeciwników technologii 5G", *Radio Maryja*, 25.01.2020, accessed 25.11.2025, https://www.radiomaryja.pl/informacje/w-kilku-miastach-polski-odbeda-sie-protesty-przeciwnikow-technologii-5g.

Roberts, Hannah. "Italy and Russia: A love affair that hasn't quite ended". *Politico*, 06.04.2022. Accessed 30.08.2024. https://www.politico.eu/article/italy–russia–love–affair–no–end.

Rodecka, Magdalena. "W Stalowej Woli protest przeciwko restrykcjom związanym z koronawirusem". *Stalowka.net*, 10.10.2020. Accessed 30.08.2024. https://www.stalowka.net/wiadomosci.php?dx=23777.

Rogan, Aaron. "Anti–vaccine group to campaign in Ireland". *The Times*, 20.03.2017. Accessed 06.09.2024. https://www.thetimes.co.uk/article/anti–vaccine–group–to–campaign–in–ireland–jnq5fpfkv.

Rolik, Mateusz and Gabiś, Tomasz. "O psychiatrii. Rozmowa z Robertem Whitakerem". *Tygodnik Spraw Obywatelskich*, 04.04.2023. Accessed 06.09.2024. https://instytutsprawobywatelskich.pl/o–psychiatrii–rozmowa–z–robertem–whitakerem.

Roman, Dorota and Fabjańska, Weronika. "Braun do ministra Niedzielskiego: Będziesz pan wisiał! Elżbieta Witek składa zawiadomienie do prokuratury". *Wyborcza.pl*, 16.09.21. Accessed 06.09.2024. https://www.wyborcza.pl/7,82983,27578646,braun–do–ministra–niedzielskiego–bedziesz–pan–wisial.html.

Rose, Julian. "Cicha broń masowego rażenia", *Zmieniając kurs na życie*, 28.11.2018. Accessed 06.09.2024. https://renesans21.pl/category/artykuly.

Rose, Julian. "Izraelskie zbrodnie przeciwko ludzkości. Podział w sercu konfliktu izraelsko–palestyńskiego". *WNet.pl*, 14.07.2018. Accessed 06.09.2024. https://wnet

.fm/kurier/elita–z–krwi–zawsze–u–steru–bezposrednie–przedluzenie–iii–rzeszy–wraz–z–jej–zamiarami–ludobojstwa–i–eugeniki.

Rose, Julian. "NATO – prowokator nieustającej wojny". *Zmieniając kurs na życie*, 03.08.2016. Accessed 06.09.2024. https://renesans21.pl/category/artykuly.

Rose, Julian. "Polska do wzięcia?". *Nowy Obywatel*, 07.05.2012. Accessed 06.09.2024. https://nowyobywatel.pl/2012/07/02/polska–do–wziecia.

Rose, Julian. "Polska do wzięcia?". *Nacjonalista.pl*, 23.05.2012. Accessed 06.09.2024. https://www.nacjonalista.pl/2012/05/23/julian–rose–polska–do–wziecia.

Rotkiewicz, Marcin. *W królestwie monszatana. GMO, gluten i szczepionki*. Czarne: Wołowiec, 2022.

Rozbicka, Sylwia. "Schizofrenia – przyczyny, rodzaje, terapia". *Mind Health*, no data. Accessed 06.09.2024. https://mindhealth.pl/co–leczymy/schizofrenia.

Rozwadowska, Adriana. "Półwiejska: Protest przeciw obowiązkowym szczepieniom [AKTUALIZACJA]". *Głos Wielkopolski*, 21.04.2012. Accessed 30.08.2024. https://gloswielkopolski.pl/polwiejska–protest–przeciw–obowiazkowym–szczepieniom–aktualizacja/ar/558845.

Rudy, Mieczysław. "Konfederacja w sprawie projektu dot. weryfikacji covidowej: Potencjalnie przestępczy". *Dziennik. Gazeta Prawna*, 05.01.2022, za: https://www.gazetaprawna.pl/wiadomosci/kraj/artykuly/8327759,konfederacja–w–sprawie–projektu–dot–weryfikacji–covidowej–potencjalnie–przestepczy.html.

Rusiniak, Krystian. "Znany dziennikarz Marcin Rola opublikował książkę o manipulacjach przy epidemii koronawirusa. Jeden z redaktorów TV Media Narodowe stał się bohaterem. Uchwycił on ciekawą wpadkę autora [FOTO]". *Media Narodowe*, 09.06.23. Accessed 06.09.2024. https://tvmn.pl/Znany-dziennikarz-Marcin-Rola-opublikowal-ksiazke-o-manipulacjach-przy-epidemii-koronawirusa-Jeden-z-redaktorow-TV-Media-Narodowe-stal-sie-bohaterem-Uchwycil-on-ciekawa-wpadke-autora-2112a.

Russell, Francis. "Pox populi: Anti–vaxx, anti–politics". *Journal of Sociology* 59, no. 3 (2023):699–715.

Rutka, Marek; Gill–Piatek, Hanna;Kopiec, Maciej; Kotula Katarzyna. "Interpelacja nr 7110 do ministra cyfryzacji w sprawie wykorzystania sieci 5G do zmniejszenia wykluczenia cyfrowego w Polsce". *Sejm Rzeczpospolitej Polskiej*, 30.05.2020. Accessed 5.09.2024. http://www.sejm.gov.pl/sejm9.nsf/InterpelacjaTresc.xsp?key=BQ9DPW.

Rutkowski, Ryszard. "COVID–19. List prof. Rutkowskiego do ministra zdrowia". *Tygodnik Spraw Obywatelskich*, 03.11.2020. Accessed 06.09.2024. https://instytutsprawobywatelskich.pl/covid–19–list–prof–rutkowskiego–do–ministra–zdrowia.

Rzeczkowski, Grzegorz. *Katastrofa posmoleńska. Kto rozbił Polskę*. Warszawa: Ta rcza, 2019.

Rzeczkowski, Grzegorz. "Służba Putina do zadań specjalnych zatruwa Polskę. Tak działa RISI". *Wyborcza.pl*, 20.04.2022. Accessed: 30.08.2024. https://www.wy

borcza.pl/7,75398,28351668,sluzba-putina-do-zadan-specjalnych-zatruwa-polske-tak-dziala.html.

Sachajko, Jarosław. "Interpelacja nr 9600 do ministra zdrowia w sprawie badań nad COVID-19, sekcji zwłok osób zmarłych, opracowania krajowych standardów, leków i badania szczepionek oraz zagadnień poruszanych w liście otwartym prof. dr hab. Ryszarda Rutkowskiego". *Sejm Rzeczpospolitej Polskiej*, 12.08.2020. Accessed 06.09.2024. https://www.sejm.gov.pl/Sejm9.nsf/InterpelacjaTresc.xsp?key=BSN9ZH.

Sachajko, Jarosław. "Interpelacja nr 16657 do ministra zdrowia w sprawie szczepień obowiązkowych oraz Programu Elektronicznego Nadzoru nad Dystrybucją Szczepionek". *Sejm Rzeczpospolitej Polskiej*, 23.10.2017. Accessed 06.09.2024. https://www.sejm.gov.pl/Sejm8.nsf/InterpelacjaTresc.xsp?key=3F6DFCFD.

Sachajko, Jarosław. "Interpelacja nr 32367 w sprawie bezpieczeństwa wprowadzenia technologii 5G do Ministra Zdrowia". *Sejm Rzeczpospolitej Polskiej*, 01.07.2019. Accessed 06.09.2024. https://www.sejm.gov.pl/Sejm8.nsf/InterpelacjaTresc.xsp?key=BE6C88.

Sachajko, Jarosław. "Interpelacja nr 32369 w sprawie bezpieczeństwa wprowadzenia technologii 5G do Ministra Obrony Narodowej". *Sejm Rzeczpospolitej Polskiej*, 01.07.2019. Accessed 06.09.2024. https://www.sejm.gov.pl/Sejm8.nsf/InterpelacjaTresc.xsp?key=BE6C8A.

Sachajko, Jarosław. "Interpelacja nr 32370 w sprawie bezpieczeństwa wprowadzenia technologii 5G do Ministra Infrastruktury". *Sejm Rzeczpospolitej Polskiej*, 01.07.2019. Accessed 06.09.2024. https://www.sejm.gov.pl/Sejm8.nsf/InterpelacjaTresc.xsp?key=BE6C8B.

Sachajko, Jarosław. "Interpelacja nr 32371 w sprawie bezpieczeństwa wprowadzenia technologii 5G do Ministra Sprawiedliwości i Ministra Cyfryzacji", *Sejm Rzeczpospolitej Polskiej*, 01.07.2019. Accessed 06.09.2024. https://www.sejm.gov.pl/Sejm8.nsf/InterpelacjaTresc.xsp?key=BE6C8C.

Sachajko, Jarosław. "Interpelacja nr 32373 w sprawie bezpieczeństwa wprowadzenia technologii 5G do Ministra Środowiska". *Sejm Rzeczpospolitej Polskiej*, 01.07.2019. Accessed 06.09.2024. https://www.sejm.gov.pl/Sejm8.nsf/InterpelacjaTresc.xsp?key=BE6C8E.

Sachajko, Jarosław and Kozłowski, Jerzy. "Interpelacja nr 15350 do ministra zdrowia w sprawie finansowania szczepionek i obowiązku szczepień w Polsce". *Sejm Rzeczpospolitej Polskiej*, 07.09.2017. Accessed 06.09.2024. https://sejm.gov.pl/Sejm8.nsf/InterpelacjaTresc.xsp?key=399D7B34.

Sachajko, Jarosław and Szramka, Paweł. "Interpelacja nr 970 w sprawie rozwoju sieci 5G i tworzenia spółki celowej "Polskie 5G" do Ministra Cyfryzacji". *Sejm Rzeczpospolitej Polskiej*, 18.12.2019. Accessed 06.09.2024. https://www.sejm.gov.pl/sejm9.nsf/InterpelacjaTresc.xsp?key=BKCE7Q.

Sachajko, Jarosław; Brynkus, Józef; Chrobak, Barbara; Józwiak, Bartosz; Masłowski, Maciej; Romecki, Stefan et al. "Interpelacja nr 28560 do ministra cyfryzacji w sprawie wątpliwości związanych z budową sieci 5G w Polsce". *Sejm Rzeczpospolitej Polskiej*, 02.01.2019. Accessed 06.09.2024. https://www.sejm.gov.pl/Sejm8.nsf/InterpelacjaTresc.xsp?key=B88JVU.

Sadowska-Dubicka, Ewelina. "Marsz i protest w obronie lekarki, która odradza szczepionki". *Białystok Online*, 08.02.2022. Accessed 06.09.2024. https://www.bialystokonline.pl/marsz-i-protest-w-obronie-lekarki-ktora-odradza-szczepionki-zdjecia,artykul,128534,1,1.html.

Sakowski, Łukasz. "Premier Szydło nie wie czy szczepionki są dobre dla dzieci, czy złe. To wsparcie antyszczepionkowców". *OKO.Press*, 24.09.2017. Accessed 5.09.2024. https://oko.press/premier-szydlo-wie-szczepionki-sa-dobre-dla-dzieci-zle-wsparcie-antyszczepionkowcow.

Sałwacka, Sylwia. "36-latka zmarła po pseudoterapii w ośrodku medycyny naturalnej. Odpowie za to pielęgniarka". *Wyborcza.pl*, 04.09.2021. Accessed 06.09.2024. https://www.poznan.wyborcza.pl/poznan/7,36001,27533062,smierc-w-osrodku-medycyny-naturalnej-pielegniarka-odpowie-za.html.

Sałwacka, Sylwia. "114 lekarzy antyszczepionkowców w sądzie. Protest przed izbą lekarską, przyjechali Pospieszalski i Winnicki", *Wyborcza.pl*. 17.11.2023. Accessed 06.09.2024. https://www.poznan.wyborcza.pl/poznan/7,36001,29569297,sprawa-114-lekarzy-antyszczepionkowcow-protest-przed-izba-lekarska.html.

Sałwacka, Maja. "Antyszczepionkowcy pod ratuszem straszyli za zgodą prezydenta: »Szczepionki są robione z płodów ludzkich«". *Wyborcza.pl*, 17.11.2021. Accessed 30.08.2024. https://zielonagora.wyborcza.pl/zielonagora/7,35182,27817685,antyszczepionkowcy-pod-ratuszem-straszyli-szczepionki-sa-robione.html.

Sałwacka, Sylwia. "Co się wydarzyło w Klinice 100? Pacjentka nie żyje, szefowie ośrodka milczą, strona internetowa znika". *Wyborcza.pl*, 23.01.2019. Accessed 06.09.2024. https://www.poznan.wyborcza.pl/poznan/7,36001,24392005,co-sie-wydarzylo-w-klinice-100-pacjentka-nie-zyje-szefowie.html.

Sałwacka, Sylwia. "»Czego Ci lekarz nie powie«. Pseudomedyczny kongres na targach w Poznaniu. Gwiazdami są autorzy fake newsów". *Wyborcza.pl*, 17.06.2021. Accessed 06.09.2024. https://www.poznan.wyborcza.pl/poznan/7,36001,27212801,czego-nie-powie-ci-lekarz-pseudomedyczny-kongres-na-targach.html.

Sałwacka, Sylwia. "Genetyk z Radomia został guru antyszczepionkowców. Naukowcy z PAN: »To ekspert od małży«". *Wyborcza.pl*, 12.01.2020. Accessed 06.09.2024. https://www.poznan.wyborcza.pl/poznan/7,36001,26677114,genetyk-z-radomia-zostal-guru-antyszczepionkowcow-naukowcy.html.

Sałwacka, Sylwia. "»Żadnej skruchy«. Justyna Socha ze Stop NOP mimo wyroku sądu znów oskarża lekarzy". *Wyborcza.pl* 22.01.2019. Accessed 06.09.2024. https://poznan.wyborcza.pl/poznan/7,36001,24387903,justyna-socha-ze-stop-nop-nie-sklada-broni-mimo-wyroku-sadu.html.

Saławacka, Sylwia. "STOP NOP: »Lekarze zabijają na intensywnej terapii«. Wiceminister zdrowia składa zawiadomienie do prokuratury". *Wyborcza.pl*, 18.10.2019. Accessed 06.09.2024. https://poznan.wyborcza.pl/poznan/7,36001,25321009,skandaliczny–wpis–stop–nop–wiceminister–zdrowia–sklada–zawiadomienie.html.

Sanders, Chris and Burnett, Kristin. "The neoliberal roots of modern vaccine hesitancy". *Journal of Health and Social Sciences* 4, no. 2 (2019):149–156.

Saresma, Tuija and Palonen, Emilia. "On the Emergence of Alt–Science Counterhegemony: The Case of the Finns Party". In: *Populism and Science in Europe*, edited by Hande Eslen–Ziya, Alberta Giorgi, 117–140. Cham: Palgrave Macmillan, 2022.

Sayers, Freddie. "Robert Kennedy Jr: America needs a revolution". *UnHerd*, 03 5..2023. Accessed 06.09.2024. https://unherd.com/2023/05/robert-kennedy-jr-america-needs-a-revolution.

Sądej, Jacek. "Eksperymenty na polskich dzieciach". *Nasz Dziennik*, 26.06.2021. Accessed 06.09.2024, https://www.naszdziennik.pl/polska–kraj/239403,eksperymenty–na–polskich–dzieciach.html.

Scarre, Geoffrey and Callow, John. *Witchcraft and Magic in Sixteenth– and Seventeenth–Century Europe*. London: Palgrave Macmillan, 2001.

Scerri, Andy. "Deep Ecology, the Holistic Critique of Enlightenment Dualism, and the Irony of History". *Environmental Values* 25, no. 5 (2016):527–551.

Schafer, Bret. "RT Deutsch Finds a Home with Anti–Vaccination Skeptics in Germany". *Alliance for Securing Democracy*, 05.10.2021. Accessed 30.08.2024. https://securingdemocracy.gmfus.org/rt-deutsch-youtube-antivaccination-germany.

Schmid, Alex P. "Radicalisation, De–Radicalisation, Counter–Radicalisation: A Conceptual Discussion and Literature Review". Haga: The International Centre for Counter–Terrorism, 2013.

Schnirelmann, Victor. "How to become the »Slavic Aryans«: The founders of the Russian Neo–paganism and their ambitions". In: *Religion and Magic in Socialist and Post–Socialist Contexts, I. Historic and Ethnographic Case Studies of Orthodoxy, Heterodoxy, and Alternative Spirituality*, edited by Alexandra Cotofana, James M. Nyce, 75–98. Stuttgart: Ibidem Press, 2017.

Schnirelmann, Victor. "Russian response: archaeology, Russian nationalism and Arctic homeland". In: *Selective remembrance: archaeology in the construction, commemoration, and consecration of national pasts*, edited by Philip L. Kohl, Mara Kozelsky, Nachman Ben–Yehuda, 31–70. Chicago: Chicago University Press, 2007

Schreier, Margrit. *Qualitative Content Analysis in Practice*. London: SAGE Publications, 2012.

Sd. "Na placu Adamowicza: »Stop ukrainizacji Polski« [GALERIA, FILM]". *Kurier Szczeciński*, 07.10.2022. Accessed 06.09.2024. https://24kurier.pl/aktualnosci/wiadomosci/na–placu–adamowicza–stop–ukrainizacji–polski

Sedgwick, Mark. *Against the Modern World. Traditionalism and the Secret Intellectual History of the Twentieth Century*. Oxford: Oxford University Press, 2004.

Senate Standing Committees on Finance and Public Administration, *Foreign Interference through Social Media*, Parliament of Australia, August 2023, accessed 30.03.2025 https://www.aph.gov.au/Parliamentary_Business/Committees/Senate/Foreign_Interference_through_Social_Media/ForeignInterference47/Report.

Sepioło, Mariusz; Flis, Daniel. "Ludwiczek równa krok". *Frontstory*, 13.04.2023. Accesse d 30.08.2024. https://frontstory.pl/osadowski-radykalizmy-prawica-kamraci.

Shultz, Richard H. and Godson, Roy. *Dezinformatsia: Active Measures in Soviet Strategy*. Washington: Pergamon-Brassey's, 1984.

Shafaati, Maryam. "The next pandemic catastrophe: can we avert the inevitable?". *New microbes and new infections* 52 (2023):1–4.

Sikora, Kamil. "Sędzia Trybunału Stanu "ma dowody" na czipowanie ludzi przez Watykan i zdradę państwa przez prezydenta Komorowskiego". *NaTemat*, 24.05.2016. Accessed 06.09.2024. https://natemat.pl/180625,sedzia-trybunalu-stanu-pisze-o-chipowaniu-ludzi-przez-watykan-i-zdradzie-panstwa-przez-komorowskiego.

Simpson, Michael. "Autism and vaccines – 164 science articles say no link (UPDATED)". *Skeptical Raptor* (blog), 10.07.2017. Accessed 06.09.2024. www.skepticalraptor.com/skepticalraptorblog.php/vaccines-and-autism-science-unrelated.

Sitnicka, Dominika. "Transparent »Szczepienie czyni wolnym« to nie wypadek przy pracy. Tak Konfederacja pogrywa epidemią". *OKO.Press*, 18.12.2021. Accessed: 30 .08.2024. https://oko.press/jak-konfederacja-politycznie-pogrywa-epidemia.

Sitnicka, Dominika. "W imię wolności? Jak politycy dla zysku podsycają lęk przed szczepionkami". *OKO.Press*, 30.05.2021. Accessed 5.09.2024. https://oko.press/w-imie-wolnosci-jak-politycy-dla-zysku-podsycaja-lek-przed-szczepionkami.

SK. "Antyszczepionkowcy zaatakowali Olsztyn [ZDJĘCIA]". *Gazeta Olsztyńska*, 31.01.2021. Accessed 30.08.2024. https://gazetaolsztynska.pl/690896,Antyszczepionkowcy-zaatakowali-Olsztyn-ZDJECIA.html.

Skowronek, Maciej. "Marta Patena kandydatką PO na prezydenta Krakowa". *Radio Kraków*, 31.08.2014. Accessed 06.09.2024. https://www.radiokrakow.pl/aktualnosci/krakow/marta-patena-kandydatka-po-na-prezydenta-krakowa.

Skrzydłowska-Kalukin, Katarzyna. "Jak kłamią antyszczepionkowcy". *Wprost*, 16.07.2017. Accessed 06.09.2024. https://www.wprost.pl/tygodnik/10065260/Jak-klamia-antyszczepionkowcy.html .

Skomra, Sławomir. "SLD z sierotami po Lepperze". *Wyborcza.pl*, 20.08.2010. Accessed 06.09.2024. https://wyborcza.pl/7,75398,8273052,sld-z-sierotami-po-lepperze.html.

SL, KOAL. "Prokremlowski propagandysta o posłach Konfederacji: »To moi koledzy«". *TVP.info*, 28.07.2022., za https://web.archive.org/web/20220729142309/https://

www.tvp.info/61539675/bialorus-prokremlowski-propagandysta-chwali-polityk ow-konfederacji.

Sługocki, Waldemar. "Interpelacja nr 8744 do ministra cyfryzacji w sprawie wprowadzenia w Polsce technologii 5G". *Sejm Rzeczpospolitej Polskiej*, 09.07.2020. Accessed 5.09.2024. https://www.sejm.gov.pl/sejm9.nsf/InterpelacjaTresc.xsp?key=BRJA3J&view=null.

Smolińska, Aleksandra. "Hamankiewicz: pacjenci powinni wiedzieć, że homeopatia nie działa". *Polityka Zdrowotna*, 22.11.2016. Accessed 06.09.2024. https://www.politykazdrowotna.com/artykul/hamankiewicz-pacjenci-powinni-wiedziec-ze-homeopatia-nie-dziala/841647.

Smolińska, Izabela. "Dzieci umierają w Polsce po szczepionkach? Sprawa w prokuraturze", *Wprost*, 27.06.2014. Accessed 06.09.2024. https://www.wprost.pl/tylko-u-nas/454092/dzieci-umieraja-w-polsce-po-szczepionkach-sprawa-w-prokuraturze.html.

Smura, Tomasz. "Disinformation and Propaganda in Russia's Information Warfare. Concepts, Resources, Areas of Impact." *Athenaeum. Polish Political Science Studies* 83 (2024): 5–21.

Sobul, Ilona. "Elektrosmog – o co tyle hałasu? Relacja z debaty". *Tygodnik Spraw Obywatelskich*, 20.07.2021. Accessed 06.09.2024. https://instytutsprawobywatelskich.pl/elektrosmog-o-co-tyle-halasu-relacja-z-debaty.

Solecka, Małgorzata. "Kolejny lekarz bez prawa wykonywania zawodu za działania antynaukowe". *MP.pl*, 23.11.2022. Accessed 06.09.2024. https://www.mp.pl/szczepienia/aktualnosci/311437,kolejny-lekarz-bez-prawa-wykonywania-zawodu-za-dzialania-antynaukowe.

Solecka, Małgorzata. "Małopolska kurator jak antyszczepionkowy harcownik". *MP.pl*, 07.02.2022. Accessed 06.09.2024. https://www.mp.pl/szczepienia/aktualnosci/291069,malopolska-kurator-jak-antyszczepionkowy-harcownik.

Sonik, Bogusław. "Interpelacja nr 2356 do ministra cyfryzacji w sprawie budowy sieci 5G". *Sejm Rzeczpospolitej Polskiej*. 14.02.2020. Accessed 5.09.2024. https://www.sejm.gov.pl/sejm9.nsf/InterpelacjaTresc.xsp?key=BLYCLY.

Sonik, Bogusław. "Interpelacja nr 2358 do ministra zdrowia w sprawie wpływu na zdrowie sieci 5G". *Sejm Rzeczpospolitej Polskiej*, 14.02.2020. Accessed 5.09.2024. https://www.sejm.gov.pl/sejm9.nsf/InterpelacjaTresc.xsp?key=BLYCM2.

Sowa, Marek. "Interpelacja nr 2737 do ministra zdrowia w sprawie wpływu stacji bazowych telefonii komórkowej na zdrowie człowieka". *Sejm Rzeczpospolitej Polskiej*, 26.02.2020. Accessed 5.09.2024. https://www.sejm.gov.pl/sejm9.nsf/InterpelacjaTresc.xsp?key=BMDHP6.

Stachura, Ewa, "60-metrowy kolos? Mieszkańcy Myślca nie zgadzają się na budowę przekaźnika GSM". *Sądeczanin*, 16.06.2020. Accessed 30.08.2024, https://sadeczanin.info/wiadomosci/60-metrowy-kolos-mieszkancy-myslca-nie-zgadzaja-sie-na-budowe-przekaznika-gsm.

Stasiuk-Krajewska, Karina and Wenzel, Michał, eds. *Dezinformacja w czasach kryzysu*. Toruń, Wydawnictwo Adam Marszałek, 2024.

Stanuch, Aleksander. "Antyszczepionkowcy pod eskortą policji przeszli przez Szczecin, by głosić, że »pandemia to ściema«". *Wyborcza.pl*, 15.01.2022. Accessed 30.08.2024. https://szczecin.wyborcza.pl/szczecin/7,34939,28004524,protest-antyszczepionkowcow-w-szczecinie-szczepionki-na-covid-19.html.

Streep, Abe. "The Military's Mystery Machine: The High-frequency Active Auroral Research Program, or HAARP, has been called a missile-defense tool and a mind-control device. The truth is a bit less ominous". *Popular Science* 273, no. 1 (2008):60–63.

Ssd. "Polski poseł odpłynął. Mówi o depopulacji Polaków!". *Fakt*, 25.01.2019. Accessed 06.09.2024. https://www.fakt.pl/polityka/posel-robert-majka-wierzy-w-teorie-spiskowe-siec-5g-to-depopulacja/9rnosxt.

SSŃ. "Antyszczepionkowcy zaatakowali. Sędziszowski ratusz zdewastowany". *O2*, 19.01.2022. Accessed 30.08.2024,. https://www.o2.pl/informacje/antyszczepionkowcy-zaatakowali-sedziszowski-ratusz-zdewastowany-6728119703563232a.

Stefanik, Dawid. "Antyszczepionkowcy przeciwko obostrzeniom epidemicznym [FOTO, WIDEO]". *TuLegnica.pl*, 23.01.2022. Accessed 30.08.2024. https://tulegnica.pl/artykul/marsz-foliarzy-przez/1267167.

Suwerenna Myśl. "Wojciech Olszański vel. A. Jabłonowski wyraża wdzięczność ambasadorowi Rosji za … Iwana Sierowa!". *YouTube*, 02.03.2019. Accessed 30.08.2024. https://www.youtube.com/watch?v=SqmFb7Ut5zM&feature=youtu.be.

Suwerenni-TV. "Dr Katarzyna Ratkowska z PSNLiN. EKSPERYMENTY na dzieciach w Polsce? AMANTADYNA pomaga w leczeniu?". *YouTube*, 06.09.21. Accessed 06.09.2024. https://www.youtube.com/watch?v=Bs4YLZpVO8Q.

Swolkień, Olaf. "Brzydka gra Katyniem". *Myśl Polska*, no data. Accessed 06.09.2024. https://myslpolska.info/2023/04/13/brzydka-gra-katyniem.

Swolkień, Olaf. "Rodacy do pracy", *Tygodnik Spraw Obywatelskich*. 13.10.2020. Accessed 06.09.2024. https://instytutsprawobywatelskich.pl/rodacy-do-pracy.

Szagdaj, Nadia. "Lekarka głosiła, że Covid-19 nie istnieje i straciła prawo do wykonywania zawodu. Teraz znów przyjmuje pacjentów". *i.pl*, 27.08.2021.. Accessed 06.09.2024. https://www.i.pl/lekarka-glosila-ze-covid19-nie-istnieje-i-stracila-prawo-do-wykonywania-zawodu-teraz-znow-przyjmuje-pacjentow/ar/c1-15773736.

Szczerbiak, Agata. "Stowarzyszenie wątpiących medyków". *Polityka*, 02.06.2021. Accessed 06.09.2024. https://www.polityka.pl/tygodnikpolityka/kraj/2120787,1,stowarzyszenie-watpiacych-medykow.read .

Szczygieł, Konrad. "ABW ściga, Sejm wpuszcza". *FrontStory*, 23.08.2022. Accessed 06.09.2024. https://frontstory.pl/mateusz-piskorski-szpiegostwo-sejm.

Szejna, Andrzej. "Interpelacja nr 4795 do prezesa Rady Ministrów w sprawie rozporządzenia dotyczącego budowy sieci 5G". *Sejm Rzeczpospolitej Polskiej*, 20.04.2020.

Accessed 5.09.2024. http://www.sejm.gov.pl/sejm9.nsf/InterpelacjaTresc.xsp?key=BP3F2D.

Szejna, Andrzej. "Interpelacja nr 4796 do ministra cyfryzacji w sprawie rozporządzenia dotyczącego budowy sieci 5G". *Sejm Rzeczpospolitej Polskiej*, 20.04.2020. Accessed 5.09.2024. http://www.sejm.gov.pl/sejm9.nsf/InterpelacjaTresc.xsp?key=BP3F2E.

Szejna, Andrzej. "Interpelacja nr 4797 do ministra spraw zagranicznych w sprawie rozporządzenia dotyczącego budowy sieci 5G". *Sejm Rzeczpospolitej Polskiej*, 20.04.2020. Accessed 5.09.2024. http://www.sejm.gov.pl/sejm9.nsf/InterpelacjaTresc.xsp?key=BP3F2F.

Szpunar, Olga. "Antyszczepionkowy film zamówiony przez Barbarę Nowak wciąż na kanale kuratorium. Katarzyna Lubnauer z KO: »Będziemy interweniować«". *Wyborcza.pl*, 20.07.22. Accessed 06.09.2024. https://www.krakow.wyborcza.pl/krakow/7,44425,28705950,antyszczepionkowy-film-zamowiony-przez-barbare-nowak-wciaz-na.html.

Szpyrka, Łukasz and Szczepański, Jakub. "Kulisy antyszczepionkowego buntu w PiS". *Interia*, 23.11.2021. Accessed 5.09.2024. https://wydarzenia.interia.pl/autor/lukasz-szpyrka/news-kulisy-antyszczepionkowego-buntu-w-pis,nId,5663360.

Szymański, Wojciech. "Sankcje na Rosję. Populiści chcą powrotu do normalności". *DW*, 26.09.2022. Accessed 06.09.2024. https://www.dw.com/pl/sankcje-na-rosj%C4%99-populi%C5%9Bci-apeluj%C4%85-o-powr%C3%B3t-do-normalno%C5%9Bci/a-63246522.

Szynkaruk, Klaudia. "»Stop segregacji sanitarnej«. Przeciwnicy przymusu szczepień spotkali się na placu Łuczkowskiego w Chełmie. Zobacz zdjęcia". *Kurier Lubelski*, 29.01.2022. Accessed 30.08.2024. https://kurierlubelski.pl/stop-segregacji-sanitarnej-przeciwnicy-przymusu-szczepien-spotkali-sie-na-placu-luczkowskiego-w-chelmie-zobacz-zdjecia/ar/c14-16021867.

Śmigiel, Martyna. "TAK dla obowiązkowych szczepień! – Lewica z radykalnym pomysłem na walkę z pandemią". *Wyborcza.pl*, 05.12.2021. Accessed 5.09.2024. https://warszawa.wyborcza.pl/warszawa/7,54420,27878096,tak-dla-obowiazkowych-szczepien-lewica-z-radykalnym-pomyslem.html.

Świech, Michał. "Dumnie walczył z 5G. Teraz zostanie posłem". *Telepolis*, 19.07.2023. Accessed 06.09.2024. https://www.telepolis.pl/wiadomosci/wydarzenia/konfederacja-5g-rafal-forys-teorie-spiskowe.

Tajne Komplety. "Pan Grzegorz Braun – Czarny scenariusz judeopolonia cz.4d". *YouTube*, 01.08.2012. Accessed 06.09.2024. https://www.dlapolski.pl/pan-grzegorz-braun-czarny-scenariusz-judeopolonia.

Targalski, Jerzy. "Archipelag szurów". *Niezalezna.pl*, 11.08.2019. Accessed: 30.08.2024. https://niezalezna.pl/index.php/opinie/archipelag-szurow/283347.

Targalski, Jerzy. "Archipelag szurów (2)". *Jerzy Targalski* (journalist home page). Accessed: 30.08.2024. https://jozefdarski.pl/13594-archipelag-szurow-2

Targalski, Jerzy. "Tworki atakują", *Józef Darski*, no data. Accessed 30.08.2024. https://jozefdarski.pl/13612-tworki-atakuja.

Taylor, Adam. "Before »fake news«, there was Soviet »disinformation«". *The Washington Post*, 26.11.2016. Accessed 30.08.2024. https://www.washingtonpost.com/news/worldviews/wp/2016/11/26/before-fake-news-there-was-soviet-disinformation.

Telepolis, "Relacja z protestu przeciwko sieci 5G w Gliwicach TELEPOLIS.PL". *YouTube*, 15.10.2018. Accessed 30.08.2024. https://www.youtube.com/watch?v=y4QLF20HMMM.

Telepolis. "Relacja z protestu przeciwko sieci 5G w Warszawie TELEPOLIS.PL". *YouTube*, 01.05.2019. Accessed 30.08.2024. https://www.youtube.com/watch?v=SxT5ztBnjWs.

The Ferret. "Pro-Russian propaganda spreading in Scots anti-vax group". *The Ferret*, 20.03.2022. Accessed 30.08.2024. https://www.theferret.scot/pro-russia-propaganda-anti-vaccine-white-rose.

The Standing Committee on Public Safety and National Security, *Up to the Task: Strengthening Canada's Security Posture in Relation to Russia* (Ottawa: House of Commons Canada, 2023), accessed 13.04.2025 https://www.ourcommons.ca/DocumentViewer/en/44-1/SECU/report-12.

Theus, Julia. "Pytamy antyszczepionkowców, co by ich przekonało do szczepień przeciw COVID. Odpowiedź jest krótka: nic". *OKO.Press*, 16.04.2021. Accessed 30.08.2024. https://www.oko.press/antyszczepionkowcy-co-by-ich-przekonalo-do-szczepien.

Theus, Julia. "Tak wyglądało najście Kowalskiego i Siarkowskiej na dom dziecka. »Grozili, byli nieprzyjemni«". *OKO.Press*, 15.07.2021. Accessed 5.09.2024. https://oko.press/kowalski-i-siarkowska-nachodza-dom-dziecka.

Thomas, Elise; Zhang, Albert; and Currey, Emilia. *COVID-19 Disinformation and Social Media Manipulation: Pro-Russian Vaccine Politics Drives New Disinformation Narratives*. International Cyber Policy Centre, 2020. Accessed March 30, 2025. https://apo.org.au/node/307763.

Tkwl. "A tyle mówił o zdradzie! Braun: »Fort Xi jest moją odpowiedzią na koncepcję Fort Trump wysuniętą przez polityków warszawskich«". *wPolityce.pl*, 22.11.2019. Accessed 06.09.2024. https://wpolityce.pl/polityka/474174-braun-fort-xi-jest-moja-odpowiedzia-na-fort-trump.

Tokarczyk, Michał. "»W portfelu mam 200 zł, a długu już 300 tysięcy«. Bałtyckie Veto, czyli bunt nadmorskich przedsiębiorców". *Wyborcza.pl*, 19.01.2021. Accessed 30.08.2024,. https://trojmiasto.wyborcza.pl/trojmiasto/7,35612,26698206,rozzaleni-zadluzeni-i-zdesperowani-przedsiebiorcy-znad-morza.html.

Tomás, Juan Pedro. "Vodafone, Ericsson carry out 5G trial in central London". RCR *Wireless News*, 21.12.2017. Accessed: 30.08.2024. https://www.rcrwireless.com/20171221/5g/vodafone-ericsson-carry-out-5g-trial-london-tag23.

Tw. "Polsat News krytykowany po wypowiedzi psychiatry o wyższej śmiertelności zaszczepionych na covid". *Wirtualne Media*, 30.06.2021. Accessed 06.09.2024. www.wirtualnemedia.pl/artykul/polsat-news-krytykowany-po-wypowiedzi-psychiatry-o-wyzszej-smiertelnosci-zaszczepionych-na-covid.

Tw. "W Łodzi podpalono maszt telekomunikacyjny. Play: to przestępstwo, będzie zgłaszane organom ścigania". *Wirtualne Media*, 28.05.2020. Accessed 30.08.2024. https://www.wirtualnemedia.pl/artykul/w-lodzi-podpalono-maszt-telekomunikacyjny-play-to-przestepstwo-bedzie-zglaszane-organom-scigania.

Tuńska, Dorota. "Dr Witczak: Szczepienie dzieci na COVID-19 to skandal". *DoRzeczy*, 16.04.2023. Accessed 06.09.2024. https://www.dorzeczy.pl/zdrowie/428266/szczepienie-dzieci-na-covid-19-to-skandal.html.

Tygodnik. "W Siedlcach będzie pikieta »Zakończyć Plandemię«". *Tygodnik Siedlecki*, 06.10.2020. Accessed 30.08.2024. https://www.tygodniksiedlecki.com/siedlce/t58287-w-siedlcach-bedzie-pikieta-zakonczyc-plandemie-htm.

U.S. Department of Homeland Security, *Homeland Threat Assessment*, Washington, DC: Department of Homeland Security (October 2020), accessed 30.03.2025 https://www.dhs.gov/sites/default/files/publications/2020_10_06_homeland-threat-assessment.pdf.

U.S. Department of State, *GEC Special Report: Pillars of Russia's Disinformation and Propaganda Ecosystem*, U.S. Department of State, 04.08.2020, archived March 31, 2021, accessed 13.04.2025, https://web.archive.org/web/20210331233359/https://www.state.gov/wp-content/uploads/2020/08/Pillars-of-Russia%E2%80%99s-Disinformation-and-Propaganda-Ecosystem_08-04-20.pdf.

University of Edinburgh. "Virus ancestry could aid bid to predict next pandemic, study finds". *ScienceDaily*, 05.02.2024. Accessed 07.09.2024. http//www.sciencedaily.com/releases/2024/02/240205165913.htm

Urazińska, Agnieszka. "Antyszczepionkowcy nie będą bezkarni? Prokuratura wszczęła śledztwo, atakowany lekarz dostał ochronę". *Wyborcza.pl*, 17.08.2021. Accessed 30.08.2024, . https://lodz.wyborcza.pl/lodz/7,35136,27460523,antyszczepionkowcy-nie-beda-bezkarni-atakowany-lekarz-ma-ochrone.html.

Usidus, Mirosław. "Jak przetrwać tsunami z 120 bitami w głowie? Ludzki mózg i ogrom informacji". *Młody Technik*, no data. Accessed 07.09.2024. https://mlodytechnik.pl/technika/31258-jak-przetrwac-tsunami-z-120-bitami-w-glowie-ludzki-mozg-i-ogrom-informacji.

Vanian, Jonathan. "Russian disinformation campaigns are trying to sow distrust of COVID vaccines, study finds". *Fortune*, 23.07.2021. Accessed 30.08.2024. http://

www.fortune.com/2021/07/23/russian-disinformation-campaigns-are-trying-to-sow-distrust-of-covid-vaccines-study-finds.

Verjovsky, Marina; Porto Barreto, Mariana; Carmo, Isabella; Coutinho, Bruno; Thomer, Lilian; Lifschitz, Sérgio; Jurberg, Claudia. "Political quarrel overshadows vaccination advocacy: How the vaccine debate on Brazilian Twitter was framed by anti–vaxxers during Bolsonaro administration". *Vaccine* 41, no. 39 (2023): 5715–5721.

Vulpe, Simona-Nicoleta; Stoian, Maria-Steluța. "Vaccines: saving lives or depopulating the world? A discourse analysis". *Journal of Comparative Research in Anthropology and Sociology* 9, no. 1 (2018):67–87;

Walczak, Agnieszka. "Kaliski protest antyszczepionkowców z antyrządowymi hasłami. Narodowcy mówili o »bandziorach w rządzie«". *Wyborcza.pl*, 15.01.2022. Accessed 30.08.2024. https://kalisz.wyborcza.pl/kalisz/7,181359,28005858,kaliski–protest–antyszczepionkowcow–z–antyrzadowymi–haslami.html.

Walewski, Paweł. "Zięba i Socha znów w natarciu. Kto zyskuje, a kto traci na zdrowotnej dezinformacji?". *Polityka*, 19.04.2020. Accessed 06.09.2024. https://www.polityka.pl/tygodnikpolityka/nauka/1953210,1,zieba–i–socha–znow–w–natarciu–kto–zyskuje–a–kto–traci–na–zdrowotnej–dezinformacji.read.

Wallerstein, Immanuel. *Decline of American Power: The U.S. in a Chaotic World*. New York, New Press, 2003.

Walter, Dror; Ophir, Yotam; and Jamieson, Kathleen Hall. "Russian Twitter Accounts and the Partisan Polarization of Vaccine Discourse, 2015–2017." *American Journal of Public Health* 110, no. 5 (2020): 718–24.

Waluś, Monika. "Górale otwierają 18 stycznia swoje biznesy. »Panie prezydencie, zapraszamy na narty!«". *Wyborcza.pl*, 14.01.2021. Accessed 30.08.2024. https://krakow.wyborcza.pl/krakow/7,44425,26685711,gorale–otwieraja–18–stycznia–swoje–biznesy–panie–prezydencie.html.

Waluś, Monika. "Pensjonaty, karczmy, ośrodki narciarskie z Podhala na skraju bankructwa. »Traktują nas jak piąte koło u wozu«". *Wyborcza.pl*, 12.01.2021. Accessed 30.08.2024. https://krakow.wyborcza.pl/krakow/7,44425,26678200,pensjonaty–karczmy–osrodki–narciarskie–z–podhala–na–skraju.html.

Wardle, Claire. "Misunderstanding Misinformation." *Issues in Science and Technology* 39, no. 3 (Spring 2023). Accessed 16.04.2025. https://issues.org/misunderstanding-misinformation-wardle.

Warner, Echo L; Barbati, Juliana L; Duncan, Kaylin L; Yan, Kun; Rains, Stephen A. "Vaccine misinformation types and properties in Russian troll tweets". *Vaccine* 40 (2022):953–960.

Wantuch, Dominika. "Rekordowa liczba medyków przed sądem lekarskim: za negowanie pandemii i opowiadanie bzdur o szczepieniach". *Wyborcza.pl*, 01.06.2022. Accessed 06.09.2024. https://www.krakow.wyborcza.pl/krakow/7,44425,28521312,rekordowa–liczba–medykow–przed–sadem–lekarskim–za–negowanie.html.

Wb. "Wybory 2015. Znamy wyniki wyborów w okręgu bydgoskim". *Wyborcza.pl*, 26.10.2015. Accessed 06.09.2024. https://bydgoszcz.wyborcza.pl/bydgoszcz/1,48722,19092194,wybory-2015-znamy-wyniki-wyborow-w-okregu-bydgoskim.html.

Weinberg Dana Beth and Dawson, Jessica. "From Anti-vaxxer Moms to Militia Men: Influence Operations, Narrative Weaponization, and the Fracturing of American Identity". *The Brookings Institution*, 01.11.2020. Accessed 07.09.2024. https://www.brookings.edu/articles/from-anti-vaxxer-moms-to-militia-men-influence-operations-narrative-weaponization-and-the-fracturing-of-american-identity.

Weiss, Sabrina. "5G health risks are the internet's new favorite conspiracy theory". *Wired*, 12.06.2019. Accessed 30.08.2024. https://www.wired.com/story/5g-health-risks-concerns.

Wesołowski, Piotr. "Wybory 2018 Łódź. Rafał Górski kandydatem na prezydenta ruchu Kukiz'15?". *Wyborcza.pl*, 29.03.2018. Accessed 06.09.2024. https://lodz.wyborcza.pl/lodz/7,35136,23200938,rafal-gorski-kandydatem-na-prezydenta-lodzi-ruchu-kukiz-15.html.

Wg. "Grzegorz Braun głosował przeciw uchwałom ws. Nord Stream 2 i Rosji. Solidarna Polska: »Trzeba deputinizacji polskiej polityki«", *Niezalezna.pl*, 25.02.2023. Accessed 06.09.2024. https://niezalezna.pl/476421-grzegorz-braun-glosowa-przeciw-uchwaom-ws-nord-stream-2-i-rosji-solidarna-polska-trzeba-deputinizacji-polskiej-polityki.

Wicher, Patryk; Dorywalski, Bartłomiej; Kurowski, Władysław; Kaczyński, Filip. "Zapytanie nr 254 do ministra zdrowia w sprawie wpływu technologii 5G na zdrowie i życie ludzkie". *Sejm Rzeczpospolitej Polskiej*, 04.02.2020. Accessed 5.09.2024. https://www.sejm.gov.pl/sejm9.nsf/InterpelacjaTresc.xsp?key=BLRC4T.

Wicher, Patryk; Dorywalski, Bartłomiej; Kurowski, Władysław; Kaczyński, Filip. "Zapytanie nr 255 do ministra klimatu w sprawie wpływu technologii 5G na środowisko naturalne". *Sejm Rzeczpospolitej Polskiej*, 04.02.2020. Accessed 5.09.2024. https://www.sejm.gov.pl/Sejm9.nsf/InterpelacjaTresc.xsp?key=BLRC4U.

Wielomski, Adam and Ziętek-Wielomska, Magdalena. "Rozmowy małżeńskie Wielomskich: Bolesław Piasecki". *Konserwatyzm.pl*, 16.01.2021. Accessed 06.09.2024. https://konserwatyzm.pl/rozmowy-malzenskie-wielomskich-boleslaw-piasecki.

Wierciński, Jacek. "Lekarka z Chojnic przed sądem i protest antyszczepionkowców". *Zawsze Pomorze. Niezależny Portal Regionalny*, 22.06.2022. Accessed 06.09.2024. https://www.zawszepomorze.pl/lekarka-z-chojnic-przed-sadem-i-protest-antyszczepionkowcow.

Wilk, Jerzy. "Interpelacja nr 8310 do ministra środowiska w sprawie informacji o skutkach dla zdrowia, jakie powoduje funkcjonowanie anten telefonii komórkowej". *Sejm Rzeczpospolitej Polskiej*, 23.06.2020. Accessed 5.09.2024. http://www.sejm.gov.pl/sejm9.nsf/InterpelacjaTresc.xsp?key=BR3D7M.

Wilk, Jerzy. "Interpelacja nr 8311 do ministra cyfryzacji w sprawie informacji o skutkach dla zdrowia, jakie powoduje funkcjonowanie anten telefonii komórkowej".

Sejm Rzeczpospolitej Polskiej, 23.06.2020. Accessed 5.09.2024. http://www.sejm.gov.pl/sejm9.nsf/InterpelacjaTresc.xsp?key=BR3D7N.

Winter, Hanah; Gerster, Lea;Helmer, Joshua and Baaken, Till. "Disinformation Overdose: A Study of the Crisis of Trust Among Vaccine Sceptics and Anti-Vaxxers." *Institute for Strategic Dialogue*, June 6, 2021. Accessed 13.04.2025. https://www.isdglobal.org/wp-content/uploads/2021/07/Disinformation-Overdose3.pdf.

Witek, Elżbieta. "UCHWAŁA SEJMU RZECZYPOSPOLITEJ POLSKIEJ z dnia 1 grudnia 2022 r. w sprawie upamiętnienia 90. rocznicy Wielkiego Głodu na Ukrainie". *Sejm Rzeczypospolitej Polskiej*, 01.12.2022. Accessed 06.09.2024. https://isap.sejm.gov.pl/isap.nsf/DocDetails.xsp?id=WMP20220001191.

Wilczyńska, Ewa. "Lekarka przekonuje, że żadnego COVID–u nie ma. »Biedaki w kombinezonach bez sensu się trudzili«". *Wyborcza.pl*, 11.07.20. Accessed 06.09.2024. https://wroclaw.wyborcza.pl/wroclaw/7,35771,26109828,lekarka-przekonuje-ze-zadnego-covidu-nie-ma-sa-ludzie-starsi.html.

Witczak, Piotr. "Trybunał Konstytucyjny – oświadczenie dr Piotra Witczaka z PSNLiN nt. obowiązku szczepienia dzieci". *Polskie Stowarzyszenie Niezależnych Lekarzy i Naukowców*, 08.05.2023. Accessed 06.09.2024. https://www.psnlin.pl/video,trybunal–konstytucyjny–oswiadczenie–dr–piotra–witczaka–z–psnlin–nt–obowiazku–szczepienia–dzieci,135.html.

Witkowski, Przemysław. "Faszyzm? My jesteśmy gorsi!". *Krytyka Polityczna*, 02.02.2019. Accessed 06.09.2024. https://krytykapolityczna.pl/kraj/faszyzm–nop.

Witkowski, Przemysław. *Partia rosyjska*. Warsaw: Arbitror, 2023.

Witkowski, Przemysław. "Służby, mafie i loże vs "szczęść boże", czyli magiczny świat Grzegorza Brauna". *Krytyka Polityczna*, 07.10.2019. Accessed 06.09.2024. https://krytykapolityczna.pl/kraj/grzegorz–braun–witkowski.

Witkowski, Przemysław. "Żydzi, masoni i niskie podatki. Braun i Korwin–Mikke: ukryte twarze Konfederacji". *Polityka.pl*, 13.07.2023. Accessed 06.09.2024. https://www.polityka.pl/tygodnikpolityka/kraj/2219070,1,zydzi–masoni–i–niskie–podatki–braun–i–korwin–mikke–ukryte–twarze–konfederacji.read.

Wma. "Dr Paweł Basiukiewicz o Covid–19: "zostaliśmy zaczarowani przez media". Szczepionki nie znosiły transmisji wirusa". *PCh24*.pl, 24.02.2024. Accessed 06.09.2024. https://www.pch24.pl/dr–pawel–basiukiewicz–o–covid–19–zostalismy–zaczarowani–przez–media–szczepionki–nie–znosily–transmisji–wirusa.

Wodak, Ruth. *The Discourse of Politics in Action: Politics as Usual*. Basingstoke: Palgrave Macmillan, 2009.

Wodak, Ruth, and Meyer, Michael, eds. *Methods of Critical Discourse Analysis*. 2nd ed. London: Sage Publications, 2009.

Wojtkowiak, Diana. "The link between the SARS–CoV–2 pandemic and the 5G network". *Gazeta Warszawska*, 29.05.2020. Accessed 06.09.2024. https://mail.gazetawarszawska.com/index.php/scientia–et–arte/5010–diana–wojtkowiak–the–link–between–the–sars–cov–2–pandemic–and–the–5g–network.

Wolfe-Robinson, Maya. "Piers Corbyn arrested on suspicion of calling for MP s' offices to be burned down". *The Guardian*, 19.12.2021. Accessed 30.08.2024. https://www.theguardian.com/uk-news/2021/dec/19/piers-corbyn-arrested-on-suspicion-of-calling-for-mps-offices-to-be-burned-down.

Woliński, Karol. "Poseł Grzegorz Braun prowokuje: »Jesteśmy znów w stanie wojny«". *CzytajRzeszów.pl*, 01.09.2020. Accessed 06.09.2024. https://czytajrzeszow.pl/?idd=10&id=12614.

Wolska, Jolanta. "Czy Dawid może pokonać Goliata?", *Tygodnik Spraw Obywatelskich*, 01.09.20. Accessed 06.09.2024. https://instytutsprawobywatelskich.pl/czy-dawid-moze-pokonac-goliata.

Woroncow, Jakub. "Prawicowy patostreamer fantazjuje o gwałceniu posłanki. Nie jest wybrykiem, jest problemem". *OKO.Press*, 14.12.2022. Accessed 30.08.2024. https://oko.press/prawicowy-patostreamer-fantazjuje-o-gwalceniu-poslanki-nie-jest-wybrykiem-jest-problemem.

Woźniak, Marta. "Prokuratura zajmie się hejterem dr. Bartosza Fiałka. Lekarze chcą, by sprawców ścigano z urzędu". *Wyborcza.pl*, 20.01.2022. Accessed 30.08.2024. https://warszawa.wyborcza.pl/warszawa/7,54420,28016999,prokuratura-zajmie-sie-hejterem-dr-bartosza-fialka-lekarze.html.

Woźnicki, Łukasz. "Antyszczepionkowcy mają swój zespół w Sejmie. Poseł Kukiza: Ja nie szczepię syna". *Wyborcza.pl*, 08.02.2016. Accessed 06.09.2024. https://wyborcza.pl/7,75398,19599598,antyszczepionkowcy-maja-swoj-zespol-w-sejmie-posel-kukiza.html.

Wójtowicz, Paulina. "»Plandemia to ściema«. Lekarka, która zniechęcała do szczepień straciła prawo wykonywania zawodu". *Medonet*, 03.11.2022. Accessed 06.09.2024. https://www.medonet.pl/zdrowie/wiadomosci,plandemia-to-sciema--lekarka--ktora-zniechecala-do-szczepien-stracila-prawo-wykonywania-zawodu,artykul,16113385.html.

Wprost.pl. "Grzegorz Braun zaprezentował skandaliczną inicjatywę, w sieci wrze. »Świadomie działa przeciwko naszemu bezpieczeństwu«". *Wprost*, 15.07.2022. Accessed 06.09.2024. https://www.wprost.pl/polityka/10784413/grzegorz-braun-zaprezentowal-skandaliczna-inicjatywe-w-sieci-wrze-swiadomie-dziala-przeciwko-naszemu-bezpieczenstwu.html.

Wprost.pl. "Skandaliczny wpis Grzegorza Brauna po agresji Rosji na Ukrainę. »NIE otwierać terytorium RP na przestrzał«, »nadchodzcy«". *Wprost*, 24.02.2022. Accessed 06.09.2024. https://www.wprost.pl/polityka/10634365/skandaliczny-wpis-grzegorza-brauna-po-agresji-rosji-na-ukraine-nie-otwierac-terytorium-rp-na-przestrzal-nadchodzcy.html.

WRealu24.pl. "MARSZ PAMIĘCI DRZBIGNIEWA HAŁATA – Wrocław 3 maja 2022 r.". *Polskie Stowarzyszenie Niezależnych Lekarzy i Naukowców*, 04.05.2022. Accessed 06.09.2024. https://www.psnlin.pl/news,marsz-pamieci-dr-zbigniewa-halata-wroclaw-3-maja-2022-r,149.html.

Wroński, Paweł. "Kukiz jednak z PSL? »Rozmowy są na dobrej drodze«", *Wyborcza.pl*, 04.08.2019. Accessed 06.09.2024. https://wyborcza.pl/7,75398,25058036,kukiz-jednak-z-psl-rozmowy-sa-na-dobrej-drodze.html.

Wykowski, Jacek. »Nie każdy doktor nauk medycznych może leczyć, ale każdy może doradzać«. »Skala zjawiska poraża«". *Puls Medycyny*, 18.01.2023. Accessed 06.09.2024. https://www.pulsmedycyny.pl/nie-kazdy-doktor-nauk-medycznych-moze-leczyc-ale-kazdy-moze-doradzac-skala-zjawiska-poraza-1175123.

Wypych, Tomasz. "Lekarka ukarana przez sąd za antyszczepionkowe wpisy. Teraz chce leczyć homeopatią". *Salon24*, 05.11.2022. Accessed 06.09.2024. https://www.salon24.pl/newsroom/1262337,lekarka-ukarana-przez-sad-za-antyszczepionkowe-wpisy-teraz-chce-leczyc-homeopatia.

Wypychowski, Paweł. "Megaustawa 5G – czy ta Księga rzeczywiście jest Biała?", *Instytut Spraw Obywatelskich*, 2019. Accessed 06.09.2024. https://instytutsprawobywatelskich.pl/publikacje/megaustawa-5g-czy-ta-ksiega-rzeczywiscie-jest-biala.

Ww. "»Szczepienie czyni wolnym« na banerze posłów Konfederacji przed Sejmem. Fala komentarzy". *Onet*, 1512.2021, za: https://www.wiadomosci.onet.pl/kraj/szczepienie-czyni-wolnym-na-banerze-poslow-konfederacji-fala-komentarzy/57p4339.

Verjovsky, Marina; Porto Barreto, Mariana; Carmo, Isabella; Coutinho, Bruno; Thomer, Lilian; Lifschitz, Sérgio; Jurberg, Claudia. "Political quarrel overshadows vaccination advocacy: How the vaccine debate on Brazilian Twitter was framed by anti-vaxxers during Bolsonaro administration". *Vaccine* 41, no. 39 (2023): 5715–5721.

Watkinsbooks. "Julian Rose: Organic Farming and Green Wisdom". *YouTube*, 23.02.2014. Accessed 06.09.2024. https://youtu.be/hFaWwQIWd_4.

Wojtkowiak, Diana. "Petycja w sprawie procedowania megaustawy UD192". 12.07.2019. Accessed 06.09.2024. www.docplayer.pl/146010826-W-p-prezydent-rzeczypospolitej-polskiej-andrzej-duda.html.

Wojtkowiak, Diana. "Zawiadomienie o popełnieniu przestępstwa zbrodni ludobójstwa przez Szumowskiego!". *Głos Gminny*, 20.05.2020. Accessed 06.09.2024. https://web.archive.org/web/20201117124610/https://www.glosgminny.pl/zawiadomienie-o-popelnieniu-przestepstwa-zbrodni-ludobojstwa-przez-szumowskiego.

Woods, Roger. *The Conservative Revolution in the Weimar Republic*. London: Palgrave Macmillan, 1996.

Yan, Kun; Barbati, Juliana L; Duncan, Kaylin L; Warner, Echo L; Rains, Stephen A. "Russian Troll Vaccine Misinformation Dissemination on Twitter: The Role of Political Partisanship". *Health Communication* 38 (2023):1591–1600.

Zajkowska, Anna. "W Okręgowym Sądzie Lekarskim w Gdańsku ruszyła sprawa chojnickiej lekarki Eweliny Gierszewskiej". *Weekendfm.pl*, 24.06.2022. Accessed

06.09.2024. https://www.weekendfm.pl/?n=101590&chojnice_-w_okregowym_sadzie_lekarskim_w_gdansku_ruszyla_sprawa_chojnickiej_lekarki_eweliny_gierszewskiej.

Zalewski, Bartosz. "Ustawienia rodzinne Hellingera – szkodliwa szarlataneria?". *Strefa Psyche Uniwersytetu SWPS*, no data. Accessed 06.09.2024. https://web.swps.pl/strefa-psyche/blog/relacje/18620-ustawienia-rodzinne-berta-hellingera-szkodliwa-szarlataneria.

Zannettou, Savvas; Caulfield, Tristan; Setzer, William; Sirivianos, Michael; Stringhini, Gianluca and Blackburn, Jeremy. "Disinformation Warfare: Understanding State-Sponsored Trolls on Twitter and Their Influence on the Web." *arXiv* preprint arXiv:1801.09288 (2018). Accessed 13.04.2025. https://arxiv.org/pdf/1801.09288.

Zapotoczny, Piotr. "Protest w centrum Opola. Straszono technologią 5G, lobby szczepionkowym i politykami na usługach koncernów". *Wyborcza.pl*, 25.01.2020. Accessed 06.09.2024. https://opole.wyborcza.pl/opole/7,35086,25632807,kontrowersyjny–protest–na–pl–wolnosci–narodowiec–traszyl–technologia.html.

Zappone, Chris. "Russian propaganda »very likely« stoking 5G health fears in Australia: expert claims". *The Sunday Morning Herald*, 16.09.2019. Accessed: 30.08.2024. https://www.smh.com.au/world/oceania/russian–propaganda–stoking–5g–health–fears–in–australia–20190916–p52rmc.html.

Zajączkowski, Ryszard. *Globalna zmiana. Co musimy obronić i jak działać?* Warsaw: Fundacja Osuchowa, 2022.

Zajączkowski, Ryszard; Wielomski, Adam; Lichański, Jakub; Piotrowski, Mirosław; Witczak, Piotr; Norkowski, Jacek Maria; Boldireff Strzemiński, Andrzej et al. "MANIFEST LUBELSKI ŚRODOWISK WOLNOŚCIOWYCH", *Ordo Medicus*, 04.10.2021. Accessed 06.09.2024. https://www.ordomedicus.org/manifest–lubelski–srodowisk
–wolnosciowych.

Zespół ekspertów Korony. *Stop ukrainizacji Polski, Konfederacja Korony Polskiej.* July 2022. Accessed 06.09.2024. https://stopukrainizacjipolski.pl/fileadmin/user_upload/SUP_4_podglad.pdf.

Zew. 'Konfederacja: Segregacja sanitarna jest bezprawna i nieskuteczna'. *Rzeczpospolita*, 18.11.2021. Accessed 06.09.2024. https://www.rp.pl/polityka/art19114141–konfederacja–segregacja–sanitarna–jest–bezprawna–i–nieskuteczna.

Zdziarski, Paweł. "Psychiatra: Szczepienia dzieci i nastolatków są niedopuszczalne". *Do Rzeczy*, 11.06.2021. Accessed 06.09.2024. https://www.dorzeczy.pl/opinie/187785/psychiatra–szczepienia–dzieci–i–nastolatkow–sa–niedopuszczalne.html.

Zdr, ml. "Inicjator Góralskiego Veta Sebastian Pitoń: chorowałem na koronawirusa. To lekka, przyjemna choroba". *Polsat News*, 12.01.2021. Accessed 30.08.2024. https://www.polsatnews.pl/wiadomosc/2021–01–12/inicjator–goralskiego–veta–sebastian–piton–chorowalem–na–koronawirusa–to–lekka–przyjemna–choroba.

Zespół Pravdy. "Cała prawda o antyszczepionkowcach ze STOP NOP. Czy można z nimi wygrać?". *Pravda*, 12.06.2023. Accessed 06.09.2024. https://pravda.org.pl/artykuly/antyszczepionkowcy-stop-nop.

Zespół wGospodarce.pl. "W Gniechowicach lęk przed 5G. Mieszkańcy nie chcą masztu". *wGospodarce.pl*, 27.06.2020. Accessed 30.08.2024. https://wgospodarce.pl/informacje/81577-w-gniechowicach-lek-przed-5g-mieszkancy-nie-chca-masztu.

Zhang, Yafei; Wang, Lin; Zhu, Jonathan J.H.;Wang, Xiaofan. "Conspiracy vs. Science: A Large-Scale Analysis of Online Discussion Cascades." *World Wide Web* 24, no. 2 (2021): 585–606. https://doi.org/10.1007/s11280-021-00862-x.

Zieliński, Rafał. "Lekarka z Dolnego Śląska o pandemii: Psychopatyczne plany, to było przygotowywane latami. Na szczęście mamy swoich księży". *Wyborcza.pl*, 17.05.2021. Accessed 06.09.2024. https://wroclaw.wyborcza.pl/wroclaw/7,35771,27095329,lekarka-z-dolnego-slaska-o-pandemii-psychopatyczne-plany.html.

Zieliński Roman and Polok, Kornelia. *Kowidowy zawrót głowy*. Wiry: Wydawnictwo Indygobook, 2023.

Zięba, Jerzy. "PRZERAŻAJĄCE, NIGDZIE NIENAGŁAŚNIANE FAKTY O SZCZEPIONCE MRNA". *Jerzy Zięba TV*, 03.01.2021. Accessed 06.09.2024. https://jerzyzieba.com/szczepienia.

ZKM, WIL. "Zdewastowana elewacja Beskidzkiej Izby Lekarskiej. Prezesi Okręgowych Izb Lekarskich wystosowali wspólne stanowisko". *Wielkopolska Izba Lekarska*, 08.01.2022. Accessed 30.08.2024,. https://wil.org.pl/zdewastowana-elewacja-beskidzkiej-izby-lekarskiej-prezesi-okregowych-izb-lekarskich-wystosowali-wspolne-stanowisko.

Zma, bas. "Bydgoszcz. Groźby śmierci na wiecu antyszczepionkowców. Poseł PO składa zawiadomienie do prokuratury". *Polsat News*, 31.01.2022. Accessed 30.08.2024. https://www.polsatnews.pl/wiadomosc/2022-01-31/bydgoszcz-grozby-smierci-na-wiecu-antyszczepionkowcow-posel-po-sklada-zawiadomienie-do-prokuratury.

Zollo, Fabiana; Quattrociocchi, Walter; Caldarelli, Guido; Scala, Antonio and Del Vicario, Michela. "Emotional Dynamics in the Age of Misinformation." *PLOS ONE* 10, no. 9 (2015): e0138740. https://doi.org/10.1371/journal.pone.0138740.

Zwaap, René. "Lachen om samenzweringen is er een onderdeel van". *De Groene Amsterdammer*, 30.11.1994. Accessed 30.08.2024. https://www.groene.nl/artikel/lachen-om-samenzweringen-is-er-een-onderdeel-van--2.

Zygmuntowski, Jan. "Prof. Andrzej Zybertowicz: smartfon to nie narzędzie, to władca". *Polskie Radio*, 28.10.2021. Accessed 5.09.2024. https://trojka.polskieradio.pl/artykul/2836312,prof-andrzej-zybertowicz-smartfon-to-nie-narzedzie-to-wladca.

Żabska, Joanna. "Protest przeciwko segregacji sanitarnej w Kłodzku". *24klodzko.pl*, 15.01.2022. Accessed 30.08.2024. http://24klodzko.pl/protest-przeciwko-segregacji-sanitarnej-w-klodzku.

Żuk, Piotr and Żuk, Paweł. "Right-wing populism in Poland and anti-vaccine myths on YouTube: Political and cultural threats to public health". *Global Public Health* 15, no. 6 (2020):790–804

Żuk, Piotr; Żuk, Paweł; Lisiewicz-Jakubaszko, Justyna. "The anti-vaccine movement in Poland: The socio-cultural conditions of the opposition to vaccination and threats to public health". *Vaccine* 37, no. 11 (2019):1491–1494.

Żytnicki, Piotr. "Policja zatrzymała trzy osoby po antysemickim skandalu w Kaliszu. Wśród nich Piotr Rybak". *Wyborcza.pl*, 15.11.2021. Accessed 30.08.2024. https://kalisz.wyborcza.pl/kalisz/7,181359,27806550,antysemicki-skandal-w-kaliszu-policja-zatrzymala-trzy-osoby.html.

Корчмарек, Наталья. "С нас хватит! Даже в Польше задумались об отмене санкций против России". *Life*, 18.08.2016. Accessed 06.09.2024. https://life.ru/p/892033#S.embed_link-K.C-B.1-L.2.zw.

Кузина, Светлана. "Психотронное оружие все-таки существует?". *KP*, 09.08.2007. Accessed: 30.08.2024. https://www.kp.ru/daily/23948.4/71317.

Index

www.ingramcontent.com/pod-product-compliance
Lightning Source LLC
Chambersburg PA
CBHW051401240920
51852CB00002B/81
* 9 7 8 9 0 0 4 7 4 2 2 2 2 *